T0347402

Housing, Race and Law

Housing, Race and Law
The British Experience

Martin MacEwen

Routledge
London and New York

First published 1991
by Routledge
2 Park Square, Milton Park, Abingdon, Oxon, OX14 4RN

Simultaneously published in the USA and Canada
by Routledge
a division of Routledge, Taylor & Francis
270 Madison Ave, New York NY 10016

Transferred to Digital Printing 2005

© 1991 Martin MacEwen

All rights reserved. No part of this book may be reprinted or
reproduced or utilized in any form or by any electronic, mechanical, or
other means, now known or hereafter invented, including photocopying
and recording, or in any information storage or retrieval system, without
permission in writing from the publishers.

British Library Cataloguing in Publication Data

MacEwen, Martin
    Housing, Race and Law : The British Experience.
    1. Great Britain. Ethnic minorities. Housing. Law
    I. Title
    344.1046363599

    ISBN 0-415-00063-7

Library of Congress Cataloging in Publication Data

MacEwen, Martin, 1943–
    Housing, Race, and Law : The British Experience/Martin MacEwen.
        p. cm.
    Includes bibliographical references.
    ISBN 0-415-00063-7
    1. Discrimination in housing – Law and legislation – Great Britain.
    I. Title.
    KD1184.M33   1990
    344.41'0636351–dc20
    [344.104636351]   90-8377
                                                            CIP

TO JESSICA,

AND FIONA AND ZOE

CONTENTS

Contents

Contents

Contents

TABLES AND FIGURES

Tables

Figures

ACKNOWLEDGMENTS

This book is avowedly eclectic and has drawn on the work of a large number of authors. But some require specific mention. First, Geoffrey Bindman and Antony Lester (<u>Race and Law</u>, 1972) have provided a wealth of background information to which liberal reference has been made in Chapters Three and Four. Second, Handy and Alder (1987) and Niner (NFHA, various) in their writing on housing associations have proved invaluable in respect of Chapter Nine. Third, Parmar (1988) on gender (Chapter One), Thomas (1986) on urban renewal (Chapter Six), London Against Racism in Housing (1988) on the private rented sector (Chapter Ten), Goulbourne (1985) on solicitors (Chapter Eleven), and Edwards (1987) on positive discrimination (Chapter Thirteen) have all been plundered. Fourth, although increasingly dated, Brown (<u>Black and White Britain</u>, 1984) remains the most comprehensive and reliable source of information on the housing characteristics of ethnic minorities and is the major reference point for the second chapter. Unfortunately it does not extend to Scotland and the returns from a survey being conducted on behalf of the Scottish Office (1988-1989) were not available at the time of publication.

In addition to published works this book has depended upon a large number of unpaid but not undervalued assistance, notably from Sally Gribb, Francis Deutch, Colin Hann, Dev Sharma (all of the CRE), Arun Mizra (NFHA), and Cliff Hague, Anne Yanetta and Alan Tibbett (Edinburgh College of Art/Heriot-Watt University). Acknowledgments are also due to my colleagues in the Equal Opportunity Law and Policy Group, namely Robin White (Dundee University), Kenny Miller (Strathclyde University) and Ian McKenna (McGill University) who each contributed to a draft synopsis of the Race Relations Act 1976 which formed the basis of Chapter Seven. Given that that draft, albeit in different

Acknowledgments

form, is published by the Commission for Racial Equality as part of their Race Discrimination Law Reports, my thanks are also due them for permission to publish, as with other copyright material owned by them.

My employers, the Edinburgh College of Art and the Department of Town and Country Planning, in addition to encouraging the establishment of the Scottish Ethnic Minorities Research Unit as a joint venture with Glasgow College, also facilitated a term's leave, a substantial part of which went to this venture. This, I trust, will not precipitate any change in our contractual relationship, nor indeed that with my publisher, Routledge, who had the vision to reject this book, as an outline proposal, but the misfortune to inherit responsibility along with the ownership of Croom Helm who accepted it: for their demise I can be but partly to blame.

I should record the valiant attempts of Mark Johnson of the Centre for Research in Ethnic Relations at Warwick (whose crate of whisky is in the post) and Routledge, through Alan Jarvis, to inject order, brevity and cohesion: any failing in this respect, as with all others, the reviewers will readily acknowledge, must, without prejudice of course, be mine.

Penultimately, daunting as the length of the text may be to the reader and writer, it has surely been more so to Lexi McDonald, my typist, whose intelligence, keyboard skills and equanimity of temperament proved invaluable. My wife, Jessica, whose recent attendance at assertiveness training courses enabled her to decline that post, may, once my own training is complete, rue her recalcitrance but (and don't tell anyone) I still love her: her tolerance, as with her partnership, proved marvellous to one and inexplicable to all others. My hope is that the reader is posited between such extremes and lasts the pace, exhausted, no doubt, but some distance from the starting point.

ABBREVIATIONS

ADA	Association of District Councils
AMA	Association of Metropolitan Authorities
BSA	Building Studies Association
CAB	Citizens' Advice Bureau
CARD	Campaign Against Racial Discrimination
CBI	Confederation of British Industry
CHAC	Central Housing Advisory Committee
COSLA	Convention of Scottish Local Authorities
CRC	Community Relations Council
CRE	Commission for Racial Equality
DAFS	Department of Agriculture and Fisheries for Scotland
DoE	Department of the Environment
EAT	Employment Appeal Tribunal
EC	European Community
EOC	Equal Opportunities Commission
FBHO	Federation of Black Housing Organisations
GLC	Greater London Council
HAC	Home Affairs Committee; housing advice centre
HAG	Housing association grant
HAS	Housing Advice Switchboard
HMSO	Her Majesty's Stationery Office
IAP	Inner area programme
ICD	Inner Cities Directorate
IDS	Industry Department for Scotland
IEHO	Institution of Environmental Health Officers
ILPA	Immigration Law Practitioners Association
IoH	Institute of Housing
IRR	Institute of Race Relations
LAG	Legal Action Group
LARH	London Against Racism in Housing
LCC	London County Council
LHU	London Housing Unit

Abbreviations

LPAC	London Planning Advisory Committee
LRC	London Research Centre
LRHRU	London Race and Housing Research Unit
LSS	Law Society of Scotland
MSC	Manpower Services Commission
NACRC	National Association of Community Relations Councils
NACRO	National Association for the Care and Resettlement of Offenders
NCCI	National Committee for Commonwealth Immigrants
NCCL	National Council for Civil Liberties
NCWP	New Commonwealth and Pakistan
NDCF	National Development Control Forum
NFHA	National Federation of Housing Associations
NHPRA	North Hyde Park Residents Association
OMCS	Office of the Minister for the Civil Service
PAG	Planning Advisory Group
PAN	Planning Advice Note
PAS	Public Attitude Surveys
PEP	Political and Economic Planning (later PSI)
PSI	Policy Studies Institute (formerly PEP)
RRB	Race Relations Board
RSG	Rate Support Grant
RSRG	Radical Statistics Race Group
RTPI	Royal Town Planning Institute
SCOLAG	Scottish Legal Action Group
SCPR	Social and Community Planning and Research
SCRE	Scottish Council for Racial Equality
SDA	Sex Discrimination Act
SDD	Scottish Development Department
SED	Scottish Education Department
SEMRU	Scottish Ethnic Minorities Research Unit
SFHA	Scottish Federation of Housing Associations
SHHD	Scottish Home and Health Department
SI	Statutory instrument
TCPA	Town and Country Planning Association
TRO	Tenancy Relations Officer
UDP	Unitary development plan
UKIAS	United Kingdom Immigration Advisory Service
YOP	Youth Opportunity Programme
YTS	Youth Training Scheme
WLDF	Women's Legal Defence Fund

THE READER'S CAVEAT

A writer of limited vision
embarked on a book with a mission.
As he fumbled and faltered
the facts he had altered
to match arguments which had arisen.

The product was limped and lame:
just a gloss on a skeletal frame.
It didn't add much
but a thought or two, such
that its length gave weight a bad name.

A nose for the truth, well he knew it
would sniff out the lean from the suet.
But the fiction he wrote
gave facts a footnote.
And his nose for the truth? Well, he blew it.

INTRODUCTION

According to a 1989 survey, a majority of the British population believe that we live in an unjust society. A prime facet of injustice is racial discrimination. Together with employment and education, housing is one of the most important aspects of an individual's life: moreover, to the extent that it facilitates or inhibits family cohesion and support, housing frequently constitutes a precondition of an individual's educational, social and economic development. The success or failure of the legal provisions designed to promote equality of opportunity in housing, and the policies and practices which further or retard their implementation, must be viewed as a litmus test of social justice.

A plethora of public, private and personal considerations affect the tenure, location, type, condition and quality of our housing. The isolation of race or colour is clearly problematic. This fact often legitimises or gives plausibility to non-discriminatory explanations for racial disadvantage in housing.

Clearly, too, law is an imperfect instrument for effecting social change. Nevertheless, to be given any meaning, law must arrogate a claim to its social impact, a proposition emphasised by the Conservative Government's constant endorsement of the primacy of the rule of law. Consequently the question 'Has the law achieved its purpose?' demands a response, even if no Government has been willing to provide one. This book attempts to examine the question and to provide a tentative answer.

In its first Annual Report, the Race Relations Board (RRB, 1967) summarised the role of legislation as follows:

1. A law is an unequivocal declaration of public policy.
2. A law gives support to those who do not wish to discriminate, but who feel compelled to do so by social reasons.

3. A law gives protection and redress to minority groups.
4. A law thus provides for the peaceful and orderly adjustment of grievances and the release of tension.
5. A law reduces prejudice by discouraging the behaviour in which prejudice finds expression (para. 65).

While this statement of purpose largely reflects the tenor of Parliamentary debates for those advocating such legislation it makes a number of assumptions which require to be tested.

First, the extent to which law represents an unequivocal declaration of public policy will be a reflection, in part, of the specific provisions within the legislation. Thus the 1965 Race Relations Act, in restricting its ambit to places of public resort, remained implicitly equivocal concerning all other aspects of racial discrimination including direct discrimination in both public and private housing. Similarly the Race Relations Act 1968, in ignoring practices which, while not overtly discriminatory, had the effect of racial disadvantage, equivocated on the scope of unlawful practice, including allocation policies in public housing.

Second, while in a formal sense the law may be unequivocal in its declaration of public policy, in reality the declaration cannot be meaningfully severed from how public policy is to be interpreted and given effect. Thus the debates in Parliament on Section 71 of the Race Relations Act 1976 demonstrate that the 'duty' imposed by that section on local authorities was acknowledged to be so vacuous that enforcement was not practicable. Indeed, Lord Hailsham's description of Section 71 as 'cosmetic' has been borne out by the fact that local authorities may choose to ignore that section and by the fact that the Home Secretary has never sought to prosecute its observance.

Third, while legislation may provide protection and redress in respect of unlawful acts, the extent and quality of such measures may reflect not only the draftsman's skills but also the political conviction of the legislature itself.

Certainly the multiplicity of aims listed recognises that different views and priorities may conjoin to enact Race Relations legislation but, and this underscores a fourth assumption, there is no recognition that such motives may conflict and the legislation in its final form represents compromises obscuring unresolved conflict. Thus, while the law may equally meet all the objectives outlined, the fourth, that of providing for a peaceful adjustment of grievances and release of tension, has no necessary correlation with the other objectives. If the primary purpose is conflict management and the orderly release of tension, the other objectives may be viewed as a necessary psychological payment in the bartering process. The down-payment is the legislation but the rental, in the form of its implementation, will be subject to period review and varied in accordance with prevailing circumstances and, more importantly, the prevailing view as to its relevance. The Brixton disturbances in 1981, and the Scarman Report which followed, pointed to racial discrimination and disadvantage as a structural component in urban unrest. The report did not suggest a strengthening of the Race Relations Act but it did indicate a need for complementary measures of support, some of which followed in the form of improved community policing and a sharpening of urban programme assistance. In contrast, following the disturbances in Handsworth in 1985, Government called merely for a police inquiry: the Home Secretary, Douglas Hurd, declared that the riots were not 'a cry for help but a cry for loot'. As Gaffney (1987) has shown, Government clearly denied other structural causes of the riots - including racism and institutional discrimination - and sought 'a clear view' in terms of social control and policing, thereby distancing itself from any association between unemployment and social problems and the riots. In this light the rental payment for social control, while clearly demanded in respect of discrimination, was not to be effected but was to be paid in terms of policing. The perceived need for a peaceful adjustment of grievances and release of tension had been superseded by a perception that adjustments in policing would solve the problem: in doing so it substituted a narrow functionalist view of policing society for a broader structural view of underlying causes to the disturbances.

Such a change of course, without rational explanation, constituted not merely a refutation that racial discrimination and racial disadvantage were a principal cause of social unrest but, in its failure to identify any other rational explanation, it appeared to add weight to the view that the black community was pathologically predisposed to riot. Hurd did not say blacks cried for loot and not for help but the inference was unequivocal.

This conclusion may question the parameters of the last stated objective of legislation, that it reduces prejudice by discouraging the behaviour in which it finds expression. Given that Hurd's view was not supported by evidence and preceded the report of the limited 'authorised' inquiry (the Silverman inquiry and the Black review supplied major 'unauthorised' versions), it is both prejudicial and racial in its connotation (Gaffney, 1987). Given that it was made by the Secretary of State responsible for race relations in terms of the Race Relations Act 1976, it was significant as a pointer to Government's response to urban unrest and to racial discrimination. With equal clarity, however, his utterance (and behaviour in terms of policing responses) was irrelevant in terms of legislative control over racial discrimination. In short, while the law relating to discrimination is likely to influence behavioural norms and the prejudices which underpin them, the countervailing expression of prejudice, not covered by legislation, may well sustain prejudice in other areas, such as employment, education or housing, where its expression in behavioural norms would be unlawful. The result will be either a severance of beliefs from behavioural norms - thinking one thing and doing another - or a surreptitious confluence of behaviour with beliefs. In the former instance, prejudice remains severed from but unaltered by changed compliant behaviour and in the latter, a lack of enforcement negates the legislative intention of changing behavioural norms even when they are in the form of unlawful behaviour.

Some critics of the law have been accused of separating the law from social life, which has led them to an expectation that it can direct social change. As Hepple (1987) has observed, the ineffectiveness of the Race Relations Act 1976 is clearly demonstrated by the PSI Third Survey (Brown, 1984) showing a continued gap not only in the unemployment rates, job levels, earnings and household income but also in the quality of housing between ethnic minority and white people; an outcome confirmed by the 1986 Labour Force Survey (Department of Employment, 1988). The Act has had some effect in breaking down barriers in access to jobs, housing and services in that overt expressions of discrimination which were familiar twenty years ago are no longer so explicit and equal opportunity expressed as a norm of social behaviour is rarely challenged. Nonetheless frustration is expressed that discrimination has been driven underground and entrenched attitudes and behaviour emerge in the resultant disadvantage experienced by Britain's ethnic minority communities. However, to

4

criticise the legislation in this area by reference to some obvious procedural and substantive weaknesses and to suggest that substantial change would be effected by sympathetic revision would appear to ignore the insights provided by sociologists into law as an instrument of social change.

Given the relative weakness of ethnic minority groups in the United Kingdom in relation to economic and social power, there are significant structural handicaps. First, this handicap is evidenced in the difficulty of negotiating effective race relations legislation without significant concessions either within the legislation itself or in related legislation, as illustrated, in relation to immigration, by the 1962, 1968, 1971 and 1988 Acts and in relation to citizenship by the British Nationality Act 1981. Second, this handicap is evidenced in the difficulty of securing not merely adequate instruments of enforcement but also the social ambience which would act as a lever effectively to match objectives with the reality of practice whether in terms of administration or in terms of judicial attitudes and approaches.

One does not have to link the relatively weak economic and social position of ethnic minority groups with the approach of the pluralists, who would interpret legislation as a compromise between conflicting social groups, or with Marxist or other theories of power which interpret the legislation as a mechanism for maintaining social control, to appreciate that the law is an expression of power both in its formulation and in its implementation. Thus Hepple (1987) has argued that the more powerful social position of women along with the influence of EEC law has resulted in the Sex Discrimination Act 1975, which shares almost identical wording[1] with the Race Relations Act 1976, being given much more liberal and sympathetic interpretation by the judiciary. Furthermore although such power relationships will be reflected in social structures and informal policies and practices which emanate from such structures in housing as elsewhere, as Henderson and Karn (1987) have demonstrated, it is not possible to understand the extent and forms of racially discriminatory outcomes purely by reference to such policies. They have argued that discrimination in the distribution of council housing is not predominantly a result of the application of particular allocation policies but, contrary to the prevailing view of racial discrimination, primarily a consequence of individual and collective attitudes.

Introduction

In addressing solely one aspect of unlawful racial discrimination, in this instance housing, there is an inherent danger that its interdependence with other aspects of racial disadvantage, principally employment and education, is obscured. Moreover the extent to which racial disadvantage in housing has broader social and economic causes than one encompassed by a concept of racial discrimination, however defined, begs fundamental questions regarding the meaning of equality of opportunity and affirmative action. Following the Housing Act 1949, reference to the 'working class' in legislation virtually disappeared but class or a perception of class remains an influential determinant of opportunity. To discriminate on the basis of class is not unlawful: so far as ethnic minority groups are located within broader social structures of British society, as opposed to a distinct housing class, racial disadvantage will result irrespective of racial discrimination. Housing authorities, estate agents, building societies and landlords frequently base decisions on the economic circumstances of their applicants, clients and members. Indeed it would be imprudent and on occasion unlawful (see for example the Building Societies Act 1962) to do otherwise. Again, where such decisions are shown to be justifiable in economic terms and untinged by racial discrimination, racial disadvantage is likely to result because of the disproportionate representation of ethnic minority groups in lower income brackets. The Race Relations Act 1976, inadequate as it has proved to be in tackling direct and indirect unlawful racial discrimination, does not constitute an assault on discrimination based on socio-economic class. Its objective of achieving equality of opportunity is consequently purely relative: the conferring of benefits and subjection to disbenefits in access to decent housing, where based on class or status, recognise no legal master. Accordingly the boundaries of racial justice according to law define a limited area for intervention. It is acknowledged that the focus of this book on housing, race and law has irrational limits reflecting the artificiality of legislation itself. Nonetheless an appreciation of the capabilities of law as an instrument of social policy, it is hoped, will inform the debate beyond such boundaries and in turn enable the law to become more effective.

But even within such confines the structure and approach to legislation require explanation. This is attempted in Part I, 'The Framework'. The first chapter attempts to highlight central arguments concerning legislation and race.

Rather than constructing one theoretical hypothesis as an analytic tool in dissecting the discussion of policy and practice, it attempts to outline a multiplicity of perspectives, which not infrequently conflict and which anticipate the contradictions in law and practice exposed, both explicitly and implicitly, in the discussion of various facets of public and private housing affected by race legislation. In attempting to describe a context of contemporary legislation, the second chapter refers to the pattern of ethnic minority settlement and outlines some demographic and social characteristics of minorities relevant to housing. The third chapter attempts a legal backdrop to UK anti-discrimination legislation. It outlines some issues and approaches, examines the earlier responses of the courts to anti-discrimination and public policy in the interpretation of common law and puts the case for anti-discrimination legislation.

The fourth chapter describes the provisions and effect of the first (1965) and second (1968) Race Relations Acts, attempts a brief assessment of the latter in respect of its impact on housing and outlines proposals for change which led to the 1976 Act.

The fifth chapter attempts a concise summary of the Race Relations Act 1976 in respect of its application to housing, incorporating references to subsequent judicial decisions and legislative amendments. The aim is to provide a reasonably comprehensive but concise outline of contemporary housing rights and duties prescribed by the 1976 Act and, in so doing, to enable an assessment of its potential and actual relevance in its application to the subject areas that follow.

Part II, 'The Law in Practice', deals with the experience of the legal provision by reference to subject areas: urban planning (Chapter Six), homelessness (Chapter Seven), public housing allocation (Chapter Eight), housing associations (Chapter Nine), and private housing (Chapter Ten). The attempt here is to provide some integrity within each chapter, enabling the reader to acquire an appreciation of law and practice on race relating to the subject under discussion without, of necessity, having to refer constantly to the various theoretical perspectives, or to the general description of the 1976 Act provided in Part I. The subject chapters describe what the law is and its evident failings. Inevitably in each subject area the treatment lacks uniformity, reflecting not only the author's own predilections and knowledge, or lack of it, but also the selective nature of information available (and accessible).

Part III, 'Evaluation and Reconstruction', attempts to regroup, evaluate and progress. The purpose of addressing 'Institutional Responses' (Chapter Eleven) is twofold. First, in recognising that the law in its formulation and implementation reflects mongrel interests, it attempts to outline the expression of such interests affecting race and housing in the key institutions in government and administration and in the housing and legal professions. Second, while the subject areas dealt with in earlier chapters may refer to such interests, their segregation provides an opportunity both to avoid repetition and to attempt rationalisation. The purpose here is not to analyse government and the professions nor to advocate institutional reforms to such bodies, desirable as they may be, but to outline in what way their current practices on race and housing restrict or facilitate the implementation of the race relations legislation.

The penultimate chapter (Twelve) attempts an assessment of the Race Relations Act 1976 in respect of housing and the concluding chapter (Thirteen) posits proposals for reconstruction in a broader social context.

This structure is designed to serve a number of related purposes. First, for the legal adviser it attempts to state, in a reasonably comprehensive fashion, what the law is and how it has been interpreted. Almost inevitably in attempting to provide a useful legal tool, by referring to the substance and interpretation of the major legal provisions of the Race Relations Act 1976 in Chapter Five, while keeping the text to a manageable length, what is sacrificed there is an assessment of the relevance of these provisions to contemporary issues in housing. To counter-balance such legalistic description and to emphasise the relevance of the legislation to the housing profession, a broader contextual approach is given to the subsequent chapters which focus on distinct areas of housing concern. The purpose here is not merely to provide a broader based approach for the lawyer, but to provide a useful reference for housing officials, local government officers, race relations advisers and other service providers. Whether or not the discussion of the policy implications of the legislation and the opportunities and strategies for change manages to transcend such 'professional' interests, suffice it to say that some attempt has been made to place the discussion of race, housing and law in a broader context. Earlier drafts included a chapter on racial harassment, a key concern to potential victims and a growing number of housing authorities. Its omission is due to three factors; first, it does not fall within the enforcement provisions of the race relations legislation; second, there is a growing literature on the subject area; and, third, space demanded excisions. Its omission is neither a denial of its importance nor of its relevance particularly in respect of public housing allocation policies.

The approach to the subject of race, housing and law adopted in this book is essentially eclectic. Evidently, however, the author's attitude will influence the selection of issues and how they are debated. The text is not an encyclopaedic compendium of theory and practice but a partial and subjective statement based on a number of assumptions. These include the following: first, that racial discrimination is a debilitating practice which requires to be challenged both at a personal and societal level; second, that while law may not be the sole, or even the most effective, sanction against racial discrimination, it provides, at minimum, a potential for its control; third, that its relative influence as a mechanism of control is largely conditioned by policies and practices of central and local government; fourth, that such policies and practices are a reflection of ideology and of the power relationships from which ethnic minorities are largely excluded; and lastly such exclusion is likely to ensure the persistence of racial discrimination as legal controls clash with countervailing ideology.

To question the efficacy of existing legislation, even within such a general framework of assumptions, requires some disaggregation of issues to be addressed. The following questions are implicit recurrent themes throughout the text:

1. What is the nature of racial disadvantage in housing?
2. What is the relationship between such racial disadvantage and the law, principally as expressed as unlawful discrimination in terms of the Race Relations Act 1976?
3. How has the law been interpreted and applied?
4. In the context of the stated objectives of the legislation, how successful has it been in respect of housing provision?
5. To what extent has such success been promoted or counteracted by attitudes, policies and practices of

9

central and local government and others involved in the housing process?

6.   In the light of this analysis, are the objectives for legislation practicable? If not, in what way should they be revised?

7.   What recommendations for change in legislation and policy result from this discussion?

Martin MacEwen
Edinburgh

Part I

THE FRAMEWORK

Chapter one

THEORETICAL PERSPECTIVES

RACE AND RACISM

In dealing with the issue of race it is essential to accommodate a parallel and apparently mutually exclusive dualism. In the first place as Miles (1982) has demonstrated there is a fundamental challenge to the legitimacy of the use of the term 'race'. The term has been defined (Bulloch and Stallybrass, 1977) as a classificactory term broadly equivalent to sub-species. Applied most frequency to human beings, it indicates a group characterised by closeness of common descent and usually also by some shared physical distinctiveness such as colour of skin. Biologically, the concept has only limited value. Most scientists today recognise that all humans derive from a common stock and that groups within the species have migrated and intermarried constantly. Human populations, therefore, constitute a genetic continuum where racial distinctions are relative, not absolute. Any remaining categorisation of races, then, relates only to gradients of frequency delineating the varying geographical incidence of particular genetic elements common to the whole species. It is also acknowledged that visible characteristics, popularly regarded as major racial pointers, are not inherited in any simple package and that they reflect only a small proportion of an individual's genetical make-up. In short, the term race is applied more often to phenotype - the physical characteristics, such as colour, more frequently found in particular ethnic groups - than genotype - those distinctions which have some biological underpinning. Socially race has a significance dependent not upon science but upon belief. As a consequence it might be argued that as soon as there was a popular recognition that the term 'race' had no scientific base its utility would diminish. Nonetheless, it is a social fact that people see themselves in terms of group. However frail their objective bases, such groups assume social importance and race relations between groups so identified become a social fact irrespective of their lack of any scientific underpinning. Unfortunately the history of racial ideology from Roman slavery through to social Darwinism, imperialism, antisemitism, the British Movement and the National Front represents a continuum of association and practice which remains embedded in much popular thinking about race. As the well documented increase in racial attacks demonstrates, the term 'racialism' is an important and accurate description of social behaviour which has not been undermined by the scientific irrelevance of the description 'race' pertaining to biological, genetic or other objective descriptions relating to the differentiation of human groups.

ETHNICITY AND INTEGRATION

The terms 'ethnic' and 'ethnicity' are frequently used, erroneously it is suggested, as a substitute or euphemism for race and to that extent are subject to the same criticism. More correctly, however, they may be used to incorporate the homogeneity of a group of people in terms not of physical appearance but of religious, social and cultural norms. Consequently reference to ethnicity in relation to Pakistanis, Chinese or Sikhs, for example, may well be relevant in relation to behavioural norms, particularly in determining the nature of local authority provision for the needs of such groups which may in part be determined by such factors. Evidently, however, it is possible, and indeed not uncommon, to make generalised and insensitive value judgments based on a perception of behavioural norms attributable to the ethnicity of a particular group and thereby to restrict or limit access to private or public housing without, of necessity, having any intention to discriminate on racial grounds adversely.

Roy Jenkins, as Home Secretary, introducing the Race Relations Bill 1968, rejected the concept of assimilation in favour of integration - a mutual respect for different cultures and religions. This symbolised Government's recognition that being British and claiming equal opportunity could not and should not be dependent on a conversion to English ethnicity, i.e. the dominant social, cultural and religious norms. But what Governments have meant by integration has been far from static. Some homilies have been offered by Government Ministers on this subject following the demand by some Muslims that Salman Rushdie's <u>Satanic Verses</u> be banned. Speaking to Anglo-Asian Conservatives in Coventry on 14 April 1989, Mr Timothy Renton, the Home Office Minister of State, advised that equal opportunity would remain only an aspiration if ethnic minorities did not choose the route to greater integration (Knewstub, 1989). 'That means making a very real effort to communicate in English, to learn the norms and customs of British life. It means looking outward rather than living introspectively within the confines of a small community.' At the same time that Renton was stressing the desirability of a single school system, the Labour Party published its own policy proposals which accommodated separate Muslim schools. Clearly there is no one view in the host community as to what integration means or how it is best achieved.

Nor is there unanimity amongst the different ethnic minority communities. Modood (1989) has observed that the right-wing 'Become British or go home!' attitude is threatening, coercive and likely to produce the insecure and ill-fitting communities it aims to avoid while in contrast the 'Become black and fight racism' of the left is equally assimilationist to the Asian communities: the choice - become quasi-whites or quasi-blacks - appears stark and unreal. In a 1988 BBC Network East telephone poll on the question 'Should Asians be called Black?' nearly two-thirds of the over 3,000 who rang voted No. Consequently while Afro-Caribbeans and Asians may choose to identify with the Black label as a symbol of mutual oppression against racism and racial discrimination, Asians, in particular, see the danger of cultural submergence and loss of identity in this process. The road to social integration and race equality politics, Modood observed, has to go through ethnicity, not against it: people who feel more secure in their own identities and in some ability to control the pace and nature of change are more likely to adapt with confidence and become genuinely bi-cultural. Western secular individualism, Modood argues, is no less a threat to historical communities than western racism, particularly as in Britain today when such individualism is far more confident and unapologetically interfering.

Ethnicity has an obvious spatial dimension in the provision of housing - such as sheltered housing association provision for the elderly Chinese, Sikh or Afro-Caribbean communities. But choice in housing provision may demonstrate a clash of cultural values which are not readily reconciled. Moreover the law will not be a neutral arbitrator but will reflect inherited values. Conversely if the law is to be certain and predictable it cannot be all things to all people. It may have to seek a median path by eking out a common core but tolerating deviation.

INEQUALITY IN HOUSING

As Smith and Mercer (1987) have observed, the last twenty-five years have witnessed a dramatic expansion of research and writing on the experiences of black people in urban Britain, a large proportion of this work having focused on residential patterns and access to housing. This had often centred on relative disadvantage in terms of a short-lived problem associated with immigration experienced by refugees, migrant labourers and their dependants. Much of this work had depended on the legacy of Park, and the Chicago school of urban geographers, who tended to apply a functionalist and deterministic interpretation to the nature of settlement and social segregation. In such analyses it was often assumed that, in the British context, a process of acculturation and assimilation would dissipate the geographically distinct patterns of settlement over time: thus so-called ghettoes of black concentration would disperse as the need for social support structures waned and opportunities for job and housing mobility increased. By the 1970s and 80s many researchers and commentators began to face the reality that racial inequalities in housing persisted despite the introduction of legislation to combat direct and indirect discrimination and despite the longevity of many so-called immigrant communities. As a consequence theoretical perspectives on race and housing moved from a descriptive concern with settlement patterns and processes of acculturation towards an address of the phenomenon of direct and indirect discrimination based on racial perception. Thus from the 1970s the terms 'structural disadvantage' and 'institutional racism' became part of the common vocabulary and their validity was reinforced by a number of studies, particularly those by PEP and the Runnymede Trust. These showed that, despite the general improvement in housing conditions from the Second World War resulting in few dwellings now lacking the basic amenities and overcrowding being less widespread, black households continue to live in properties that are physically amongst the worst of the housing stock whether in the private or public sector. Brown (1984) demonstrates that although Asian and Afro-Caribbean households have shared in the improved benefits attributable to local authority programmes aimed at slum clearance and rehabilitation of the existing housing stock they remain over-represented in the country's most deprived enumeration districts. Generally they suffer the same extensive relative disadvantage in housing terms as they did ten or twenty years ago.

CAUSES OF INEQUALITY

While the evidence of continuing racial inequality in housing is undeniable its causes are less easy to explain. One basic dichotomy in explanation is between the one extreme of ethnic choice determining patterns of relative disadvantage and racial segregation as opposed to institutional constraints arising from indirect and direct racial discrimination. The former approach attempts to define the issue of the housing experience of Britain's black ethnic minorities as being predominantly a reflection of cultural choice. Such theories, however, have difficulty in maintaining credibility in the face of the evidence that black households have been found paying more rent than whites for lower-quality properties (Doling and Davies, 1982) and that black owner-occupiers have frequently found access to mortgage finance restricted. Access to private housing, as a result, often centred on older properties in the inner city which are not infrequently expensive to maintain and slow to increase in value (Ward, 1982; Karn et al., 1985). Similarly in the public sector studies by Flett, (1979), Phillips (1986), Henderson and Karn (1987) as well as the investigations of the Commission for Racial Equality (Hackney: CRE, 1984b; Liverpool: CRE, 1984a) in tandem with studies of housing association experience (CRE, 1983b; Niner, 1985 and 1987; Dalton and Daghlian, 1989) demonstrate significant disadvantage among black households in this sector.

Cumulatively these studies of housing in both the public and the private sector confirm the conclusions drawn from Brown (1984) in the second chapter but at a more specific level that while it may be difficult to identify the extent to which the genuine exercise of choice by ethnic minority groups in relation to housing tenure and location may limit housing opportunities, there can be no doubt that discrimination both direct and indirect is a more significant factor in explaining relative housing disadvantage. This conclusion is far from new. In 1967 Political and Economic Planning (PEP, 1967) issued a report based on interviews with both white and black people on a series of situation tests carried out in six towns where they focused on opportunities available in the fields of employment, housing and commercial services. In housing it was discovered that out of a total of sixty personal applications to landlords, the West Indian applicant was refused accommodation or asked for a higher rent than the white Englishman on forty-five occasions and was discriminated against in twenty out of thirty inquiries about accommodation through estate agents. In summary the tests revealed that black people's views about the existence of racial discrimination were not exaggerated but were closely related to their own experience or to their knowledge of the experience of others. The findings showed that the groups who were most physically distinct in colour and racial features from the English control group experienced the greatest discrimination and that the group who were culturally most like the English and who sought integration were the most likely to experience rejection. The report established that racial discrimination was a serious and growing problem in Britain and that it related more to colour than to ethnic origin or culture.

GENDER AND RACE

There is strong evidence that the needs and aspirations of black women suffer double jeopardy: they are under-valued, under-researched, under-resourced, but far from understood. Superficially it might be argued that the anti-discrimination legislation in dealing with both race and gender discrimination would secure that black women were no less under-privileged in housing provision than black men and that similarly they were no more discriminated against than white women. The evidence suggests that this is not the case (Institute of Housing, 1987a). Debates on the purview and efficacy of legislation relating to race have been dominated by men, both black and white, while the gender issue has been dominated by white women. Consequently, despite the fact that black women, and young black women in particular, are a growing and significant section of society their voices remain unheard and their experiences largely ignored (Parmar, 1988: 197). Moreover the very segregation of race and gender in legislative provision has, if anything, failed to see the racialised gender roles ascribed to black women as of any significance. As Parmar has argued in the context of young black women, the experiences of living in a racist society are determined by factors of race, gender, age, class and sexuality, and it is a simultaneous operation of these oppressions which shapes their experiences and contributes to their significant lack of power in society. Despite the similarity in the legislative provisions of the Sex Discrimination Act 1975 and the Race Relations Act 1976 and the occasional consultations which have taken place between the Commission for Racial Equality and the Equal Opportunities Commission, the extent to which cooperation has taken place has been limited and piecemeal. The result has been a lack of any concerted strategy to secure that black women's interests in housing, as in other spheres of service provision and employment opportunity, are adequately addressed.

The Institute of Housing (IoH, 1987a) has demonstrated that, despite a decade of equality legislation, women are still disadvantaged in almost all areas of their lives. Furthermore, although the vast majority of local authorities and a number of organisations have now adopted equal opportunity policies, it would appear that in many cases these simply pay lip service to the concept and have had little impact on practice. Part of the reason why the situation of women has been slow to improve can be attributed to the fact that it has often been assumed that women's needs are not distinct from men's and that any policies or initiatives aimed specifically at women are an unnecessary waste of resources. The reality is, however, that the pattern of women's lives is quite different from men's: women carry the main responsibility for bringing up children, for looking after families, for work at home and for looking after people who are ill, disabled or frail. Females are subjected to discriminatory behaviour from an early age: at school they receive different education and at work less training, resulting in fewer skills, low pay and low status jobs. Despite their resultant dependence, in disproportionate terms, upon the public and social services, such services have rarely reflected women's needs in proportion to their number. The 1986 Institute of Housing survey (IoH, 1987a: 6 et seq.) confirmed the generality of the above proposition in respect of housing by demonstrating that of the 7 per cent of housing staff working part-time, 96 per cent were women, that 85 per cent of employees at basic level were women compared to 13 per cent of employees at section head level and, further, that 50 per cent of all male employees as opposed to only 6 per cent of all females were found to be on principal or senior officer grades. As a result, although women do constitute between 43 and 40 per cent of the housing workforce in local authority departments in Britain, their dominance of the part-time, low paid and low status jobs reflects broader social structures in determining that women's ability to shape housing policy is severely inhibited by their abysmally low representation in hierarchies of power.

In respect of women as consumers of public housing the survey concluded that women formed the main client group. One in eight of all families with dependent children are single parents and 90 per cent of these are female-headed. Forty-nine per cent of female-headed one-parent families, compared with only 5 per cent of two-parent families, are reliant upon supplementary benefit as their main source of income. In 1982 some 42 per cent of male-headed one-parent families as opposed to only 23 per cent of those headed by women owned their own homes (IoH, 1987a: 20). Consequently single-parent families, particularly those that are headed by females, are likely to rely on local authorities to provide their housing and council policies in this area will have a profound effect upon the quality of their lives.

In 1985, of those households accepted as homeless by local authorities, 62 per cent contained dependent children and 12 per cent had a member who was pregnant (DoE, 1986a). The 1986 GLC report recorded that female-headed households accounted for two-thirds of those accepted as homeless in London and the majority of these were single-headed families. The increasing use of temporary accommodation for the homeless, particularly sub-standard bed-and-breakfast hotels and short-life property, affects women particularly badly since they are likely to take primary responsibility for child care and domestic responsibilities and are often confined to the home for long periods. In respect of the single homeless the IoH report (1987a: 22) concluded that there are as many, if not more, single homeless women as there are single homeless men although there were far fewer hostels for women than men:

statutory hostel provision for men exceeds that for women by a ratio of 9:1. A survey by Watson and Austerberry (1983), of homeless women living in hostels and calling at women's aid centres, found that the most important reason for homelessness was marital dispute and domestic violence: 15 per cent of those accepted as homeless by local authorities were rendered homeless because of marital breakdown. There is a severe shortage of refuges for women suffering domestic violence and, because of the policies of some councils requiring victims to obtain an injunction, ousting or exclusion order prior to rehousing, there is a risk of further attacks as the whereabouts of women so affected may be known to the past perpetrator of violence.

Ethnic minority women are even more likely than women generally to be poor and to encounter difficulties in obtaining suitable housing and employment. The IoH report (1987a: 23) concluded that whilst the position of women in all sectors of the housing market was disadvantaged compared with that of men, the housing problems experienced by black women were particularly serious. The formal investigation of the CRE into the London borough of Hackney (CRE, 1984b) showed that black households were disproportionately represented amongst the homeless and were more likely than whites to present themselves as literally homeless. Amongst this group it was also found that single parents were over-represented, one-parent families accounting for 27 per cent of white homeless households as opposed to 39 per cent of black homeless households. There is also evidence to suggest that black women in bed-and-breakfast accommodation are particularly vulnerable to racial discrimination by hoteliers (IoH, 1987a: 23).

Cultural factors as well as racial discrimination may also affect opportunities in the allocation process in respect of council housing: Asian women may be precluded from involvement in discussions relating to property and area preferences at the early stages of the housing application process - especially when male housing officers are present.

Most studies on racial harassment (CRE, 1987a, 1987b, 1987c; AMA, 1985, 1987; Home Affairs Committee, 1986; Home Office, 1981, 1989; Gordon, 1986; MacEwen, 1986; Walsh, 1986) do not attempt to isolate gender but the evidence available would suggest that women are particularly vulnerable to racial harassment because they are likely to spend more time in the immediate vicinity of the home and they are also perceived to be physically weaker than

21

men (IoH, 1987a). It also seems likely that because of their dependence on the home the effects of physical attacks upon women are particularly devastating, leading frequently to a sense of imprisonment.

In the private sector there is further evidence of a double disadvantage that black women suffer as a consequence of discrimination by reason of gender and race. The GLC survey of private tenants in London (GLC, 1986b) showed that black women under pension age living alone were twice as likely to be unprotected tenants in comparison with the average for all households in that sector while black women are likely to experience difficulties in securing mortgages with the resultant dependence on high-interest loans from finance houses or deferred purchase schemes.

White male dominance of institutions leads to the projection of negative stereotypical images of Afro-Caribbean and Asian families generally, being associated with the influence of black women in particular. 'Problems' associated with Asian and Afro-Caribbean women and men are traced back to their pathological dependence on black women in their early years: the Afro-Caribbean family is seen as being too fragmented and weak and the Asian family as too strong, cohesive and controlling of its members. Black women are responsible for the numerous 'problems' with which the different welfare agencies are confronted: Afro-Caribbean women are stereotyped as matriarchs or seen as single mothers who expose their children to a stream of different men while Asian women are construed as faithful and passive victims centring their lives upon their religious rituals, family and home (Parmar, 1988: 199). Asian women are seen as failures because of their lack of English, because of their refusal to adopt English eating, dressing and speaking habits and because of their cultural trans- ference of abnormal and idiosyncratic habits to their children. A deterministic view that all social problems relate to the family and its relative strength or weakness may lead to a distinctly ambivalent admiration for the stability and close knit nature of the Asian family in contrast with white families. It is apparent, however, that aspects of cultural identity such as religion, language and dress frequently attract ill-informed and denigratory criticism. Moreover particular facets of culture have not infrequently been depicted as a crude form of female oppression, the media being prone to cite sensational individual cases of Asian girls running away from home. The image of 'innocent' Asian girls being victims of 'backward' and 'tradition-bound' parents forcing them against their will into unwanted marriages has facilitated Central Government's development of immigration laws and policy which seem divisive, sexist and racist. As Parmar (1988) has observed the danger of 'commonsense' images of arranged marriage continuing to dominate official perceptions of the Asian family is that they will systematically distort the delivery of public services and resources to a particular group, namely young Asian women.

Bains, in addressing the professionalisation of ethnicity (Cohen and Bains, 1988: 240), has argued that the growth of Central and Local Government funding for community groups has seen, in tandem, the emergence of professional 'ethnics' as a new intermediary force between black people and the State. With reference to Southall he argues that two communities have emerged: the first, more visible, is created and funded by various State agencies and run by 'career militants' and the second, more hidden one, is populated by the ordinary people who live and work in the area. The first community imagines it stands in a symbiotic relationship to the second - it is serving 'the people' while, in fact, the relationship is largely parasitic. The upper echelon consists of ethnic arts officers, community liaison officers and the like, employed by the local council, the Commission for Racial Equality (CRE) and by implication the local Community Relations Council (CRC) while below them are the black people who set up local organisations, often with the specific intention of obtaining funding. This latter group represents 'community one' which is involved in processes of negotiation for power and resources, with the women in the street, the 'grass roots', being largely bypassed. To the extent that the experience in Southall is likely to be reflected elsewhere this structure ensures not only a filtering of grassroots opinion but also that black women's needs are effectively diluted before the filtering begins to take place. Consequently even where Central and Local Government have established a meaningful dialogue with the professional 'ethnics' it is more rather than less likely that the interests of black women will be largely marginalised or ignored. The fact that this state of affairs is reflected in the inadequate treatment of this issue in this book is neither condoned nor justified. Being dependent on secondary sources, the text reflects the generally pervasive failure to give this issue the prominence it deserves.

However, given the similarity between the anti-discrimination provisions in the Race Relations Act 1976 and the Sex Discrimination Act 1975, there is evidence that the two enforcement agencies, first, recognise areas of common concern and, second, acknowledge that closer co-operation in both promotional and enforcement work has the potential for improving the impact of work in both race and gender issues. To the extent that such cooperation evolves, the issue of black women in housing will be advanced. But such optimism is merely relative. For those who wish to believe that the anti-discrimination legislation is effective, black women constitute an advantaged group. However, so far as this text provides evidence of significant disadvantage in housing on the grounds of race it will in like measure but in silent testimony bear witness to inadequacies in the legislation concerning gender.

IDEOLOGY AND LAW

Some authors such as Richard Quinney (1974) have argued that law itself is a peculiarly western phenomenon - developed in the Judaeo-Christian religion. The concepts of God's law, the law of nature and scientific laws are enmeshed in our intellectual history, and today's concern with the laws of economics and the need for law and order is in keeping with such a conceptual framework. Quinney argues that this western image or ideology of law has been remarkably consistent whether in the popular mind or in academic disciplines. Its leading exponent, Rosco Pound, saw law as a specialised form of social control - to pressure man into upholding civilised society and to deter him from anti-social conduct, i.e. conduct at variance with the postulates of social order.

This consensus model depicts law as reflecting the consciousness of the total society. It is a social institution to satisfy social wants - only the right law can emerge in a civilised society. Pound (1943) states:

> Looked at functionally, the law is an attempt to satisfy, to reconcile, to harmonise, to adjust ... overlapping and often conflicting claims and demands ... so as to give effect to the greatest total of interests and to the interests that weigh

> most in our civilisation, with the least sacrifice of
> the scheme of interests as a whole.

Society is, therefore, depicted as relatively homogeneous and static rather than being characterised by diversity, coercion and change. Moreover, law is not seen as the result of private interests, but as operating outside particular interests for the good of society as a whole. However, if law reflects the particular interests of those who have the power to translate such interests into public policy, the pluralist, consensual model evaporates and law may then be seen as the tool of a dominant group to perpetuate their dominance.

While the creation of new law may be seen as a process aimed at the resolution of conflicts and dilemmas which are inherent in the structure of a particular historical period it does not address the underlying contradictions which may be basic to the interests of different groups (Chambliss, 1977). Consequently a resolution of conflict may not only fail to resolve a contradiction which may emerge in a different form but may spotlight contradictions which were previously less salient or merely latent. Mathieson (1980) has contended that the legal process acts as an ideological filter to emasculate radical proposals for change so that the final legislation does not after all break significantly with dominant interests. The practice of political trimming, stripping down of legislation and the creation of pseudo-alternatives and what he calls 'co-optive co-operation' may all play a part in such a process. Mathieson sees those groups associated with the introduction of radical change as being either absorbed by the State or stigmatised and thereby excluded from effective power and influence for change. In sociological terms the impact of this process has been to inflict a sense of helplessness - at worst despair and at best apathy - in those groups and individuals most seriously disadvantaged by the prevailing ideology, a state Mathieson describes as psychological deprivation.

In short the ideologies of the law, of the professions, of the public interest and indeed of property and housing themselves will have a significant influence not only on new legislative provision in relation to racial discrimination and housing but also on its implementation. Consequently a description of law in the abstract will be deficient: an explanation of what the law is, in statutory terms, must be complemented not only by an explanation of how it is applied in practice but by reference to the climate of ideological opinion which will influence the actions of each participant in the legal process.

IDEOLOGY AND LANGUAGE

The language of ideology and race is a powerful generator of misunderstandings, which has produced much heat and little light. Thus Lord Scarman's rejection of an institutionally racist Britain as a component in the explanation of the Brixton disturbances in April 1981 (Scarman, 1981: 11) and Sharon Atkin's assertion in the lead-up to the 1987 General Election that the Labour Party is a racist institution raised vehement opposition. Self-evidently in forming a view as to the accuracy of such descriptions it is of fundamental importance to ascertain whether the terms employed have a generally accepted meaning. The purpose of examining the meaning of such terms is not purely one of semantics but extends to the functional purpose of labelling. What is meant by this is best explained by illustration. If what Scarman meant by rejecting the concept of institutional racism was that the evident disadvantage suffered by blacks in Brixton and elsewhere was the result of cumulative individual acts of discrimination as opposed to structures, policies and processes which, whether or not with that explicit intention, are nonetheless a product of implicit assumptions concerning race, this conclusion will have implications (the functional purpose) for how racial disadvantage in housing, as elsewhere is tackled. Thus, in the absence of institutional discrimination, where rules - such as those relating to waiting time regarding council house applications - do have an adverse impact on racial minorities, the solution is straightforward - change them. Conversely if such rules are a product of racist ideology which permeates the institutions which are responsible for their formation, implementation and modification, then changing the specific rules without tackling the ideological climate in which they were formed and applied may have only marginal beneficial effects because the underlying racist ideology will find expression in other forms, including the way in which the changed rules are implemented.

INSTITUTIONAL RACISM

As Phillips has observed (in Smith and Mercer, 1987: 128) this concept was first formulated in the United States in stressing the discriminatory effects of institutional rules and procedures which 'reflect and produce racial inequalities in American society' irrespective of individual intentions (Jones, 1972): the emphasis was on the racist consequences of the normal processes of institutional operation as exemplified by the American ghettoes of the 60s. Then the concept was linked, by black political activists, to 'internal colonialism' and the existence of a black underclass. Despite the historical specificity of the concept in the United States, the subordination of black minorities in Britain in terms of wealth, power and status (Sivanandan, 1982; Castles, 1984; Miles and Phizacklea, 1984) is sufficiently clear to permit certain analogies (Kushnick, 1981), including the marginal political and economic status of blacks affecting opportunities in the housing market (Phillips; Smith and Mercer, 1987: 128). Thus after three decades of New Commonwealth (NCWP) settlement in Britain striking inequalities persist across all types of tenure and the declining inner-city reception areas still provide the focus of minority clustering: in some cities segregation is increasing. The option of moving is largely theoretical for many black families suffering the cumulative disadvantage of inadequate education, limited employment opportunities and a growing threat of racial harassment outside established areas of settlement.

However, it is necessary to go beyond a statement of parallel disadvantage of blacks in the States and Britain (albeit in distinct forms of expression) in a search for explanations. Not infrequently the concept of institutional racism has lacked conviction not because of any dubiety about relative disadvantage but because there had been demonstrable weaknesses in linking cause and effect.

In the public sector such linkage depends on two factors. First, the normal processes associated with the selection, rationing and matching of housing need with housing availability must be shown to be discriminatory in effect and, second, it must be shown that such processes are not merely inadvertent but are the concrete expression of racial ideology. The former proposition is now well accounted for. Thus it has been noted that the formal CRE investigation into public housing allocation in Hackney

(CRE, 1984b) and the CRE research report on Liverpool (CRE, 1984a) demonstrate, unequivocally, that it is the policies and procedures adopted (rather than the individual acts of discrimination) which have resulted in blacks being systematically allocated worse housing than whites in like circumstances. Moreover, Henderson and Karn (1987) have shown, in their analysis of public housing allocation in Birmingham, that even where the formal rules should not have resulted in racial disadvantage and there was no explicit reference to race in the application of such rules, blacks fared less well than whites in housing outcome. In that instance class, by reference to socio-economic group, may also have influenced the exercise of discretion by housing officials but this facet did not, in isolation, provide a full explanation of disadvantage.

It is the second factor, however - that the evidence of black disadvantage was the concrete expression of racial ideology - which is most difficult to prove. In Britain there is no legacy of overtly and transparently racist provision in the public sector as there has been in the United States (Greenberg, 1959). As a consequence both in historical and in contemporary provision, taking into account the requirements of the Race Relations Acts, the racial intentions of a particular practice, policy or process will not be apparent, but may only be implied from the circumstances and the effects. The effects, however, are a premise and not a conclusion: there is irrefutable lack of logic in imputing motive or design from effect in respect of a particular practice, policy or process. Nonetheless it may be argued that where, cumulatively, a pattern of racial disadvantage emerges, racial ideological underpinning may provide a 'real' explanation when no other single factor may be imputed as a constant in the various equations analysed. In short the argument may be that where racial disadvantage in the allocation of public housing is demonstrated to be present in a variety of different situations and no satisfactory explanation is evinced in respect of any specific practice, policy or process, then, by a process of elimination, racial discrimination may remain as the sole constant and the only rational explanation.

However, it is seldom possible to eliminate other explanations completely from a particular situation leading to racial disadvantage. Not infrequently, therefore, an alternative partial explanation may be argued. Thus the allocation of blacks to poor housing may be explained, in part, by their proportional over-representation on the homeless persons access route in a particular district which, in turn, is provided with poorer quality council housing. In another situation blacks, because of the immediacy of housing need, may take up a greater proportion of first offers of allocation leaving them at a comparative disadvantage with those who could wait for better offers (by the accumulation of points which would result in improved offers). Individually such explanations of disadvantage in outcome appear to refute any overt intention of discrimination and suggest that the outcome is merely an unfortunate but unintended consequence of a non-discriminatory system. Where, however, these various systems and outcomes are collated and racial disadvantage is seen as a constant, then such partial explanations take on a different perspective - are they not the superficial and immediate explanation in non-racial form of underlying institutional practices designed to perpetuate black disadvantage? Whether such an explanation is tenable does not depend on a master plan of intent: it merely contends that an ideology perpetuates the classification of blacks as a relatively undeserving underclass. Such categorisation may not be expressed in intentional or overt racism but may find itself in subliminal or subterranean attitudes which, often unconsciously, inform the value systems which determine the formation, implementation and modification of the practices, policies and processes of any given institution.

What has been argued, then, in respect of institutional racism, is that while, individually, racial disadvantage in housing as elsewhere, may not be directly attributable to processes which appear to have an ideological rationale, cumulatively such processes are not readily explained otherwise, while acknowledging that overt intention to discriminate may be absent. Such explanation does not imply that individual discriminatory acts do not exist nor that they are of necessity tolerated or condoned. Certainly individual acts are crucially bound up in the reproduction of structures (Giddens, 1979) and may reflect or, in concert, modify the ideology which the structures symbolise but the concept of institutional racism does not imply that all participants within the system need subscribe to, or indeed appreciate, the consequences of specific decisions which it generates for its viability.

The relevance of this debate is twofold. First, an unqualified anti-racist commitment by an individual employed within a system that generates or perpetuates racial disadvantage is likely to have marginal impact on the structures themselves: moreover such an individual, when presented with a non-racial explanation for the adverse impact of policies and procedures on racial minorities, may accept such explanation as both coherent and logical. Indeed such a conclusion may be justifiable not only for the reasons already discussed but also because the cause of such disadvantage may lie outwith the control of his or her employer. Low income, poor communication skills and a failure to optimise choice within housing allocations may all be critical aspects of an applicant's profile on which the allocations officer can have little impact as they reflect a history and process of limited opportunity outside the sphere of influence of the housing department. Any legal provision designed to combat direct and indirect discrimination in housing, therefore, will, to be effective, have to burrow beneath individual motive and superficial explanation to challenge preconceptions and assumptions which, although not expressed in racial terms, may reinforce racial stereotyping and racial discrimination.

Second, the relevance of explicit motive and intention must be open to challenge within the legal process. Evidently if it is argued that institutional racism is to be located in practices, processes and procedures then those responsible for their implementation will frequently be innocent in both senses of the term. Unlawful racial discrimination is a civil but not a criminal offence. Consequently proving innocence or guilt is not an issue before the court. But the relative innocence of the defendant may well be an uninvited intruder into the psychology and psychopathology of the process of adjudication.

It is suggested, therefore, that a sense of guilt for past discrimination or the issue of innocence in respect of institutional racism is a wasteful diversion from the general goal of establishing equality of opportunity and the specific task of administering anti-discrimination legislation.

The CRE investigation into the immigration control procedures demonstrated that racial stereotyping occurred sufficiently frequently to raise serious questions about the ethos in which the work was conducted. The rules and their application created a climate conducive to racial stereotyping, confirmed by an examination of the files where Mirpuris were said to be liars, Moroccans, like Mirpuris, both simple and cunning, Ghanaians like lost and confused children and Nigerians, like Ghanaians, having ambitions out of all proportion to their capabilities and circumstances (CRE, 1983a). If nothing else the CRE report shows that in one Government department the framework of law sustains an ethos of racial stereotyping allowing official discretion to reinforce racial discrimination. Such discrimination on the grounds of nationality is overt but not unlawful in the context of immigration controls. In contrast, to discriminate overtly on such grounds in housing is unlawful and generally recognised as such by public housing authorities.

A STRUCTURAL APPROACH

Nevertheless, if the existence of such stereotyping is demonstrable inferences of discrimination may be more readily accepted as a logical explanation for racial disadvantage. In Britain there was a legacy of a colonial society in which the colonised people were considered inferior (Hill, 1965: 49; Biddiss (ed.) 1979: 189) and an industrial and social structure in which socially mobile whites had abandoned certain jobs and other social positions and black workers had been brought in to fill them. The problems were not those predominantly of racism on the psychological or theoretical level, but questions of structure and inequality (Rex, 1986: 105). As Marx (1962) observed, 'It is not the consciousness of men that determines their existence, but the social existence that determines their consciousness.' Pareto (1963) assumes that much of our social action is of a non-logical kind. We do not first set our goals and then choose the scientifically appropriate means for attaining them. Instead we begin with certain sentiments requiring expression and at the same time as we act we offer a verbal explanation of that act. This contemporaneous explanation (which Pareto terms the 'residue') is subsequently rationalised to give the appearance of logic ('derivations'). Human action may therefore be observed at two levels. One deals with actual behaviour and the other with the verbal justification of that behaviour - a post hoc, ad hoc rationalisation which is likely to be very distant from the sentiment motivating the behaviour itself.

An illustration of this approach may be given by the example of crime and its apparent increase under the Conservative Government from 1979 to 1989. First, this

Government, in the manifestoes relating to the General Elections immediately prior to and during its term of office, stated a commitment to the rule of law and a crack-down on crime. The various police forces have witnessed an increase in resourcing at various levels, including the number in the workforce to meet, or nearly meet, target establishment levels. Generally the police forces are better manned, better paid, better equipped and better trained. So far as law enforcement, therefore, is a factor inhibiting crime one would expect a decline in crime statistics, and an explanation for crime increase will have to be sought elsewhere.

At a psychological and theoretical level explanation may be sought in a waning respect for property and the person, a sense of personal alienation from dominant social values, a breakdown in family and community life. Such explanations, although not always severable from an explanation rooted in structural causes, often take the form of Pareto's derivations. Conversely the clearly structural explanations - poverty, bad housing, poor educational provision and unemployment - have been rejected by Government. It may be argued that increased policing, neighbourhood watch schemes and increased security measures promoted by this Government have focused on crime containment and have not sought to tackle the causes of crime - whatever explanation is preferred - but in any event Government has strenuously denied any links between increased unemployment and crime.

France, which has suffered from rising unemployment, a legacy of inadequate housing and increasing numbers below the poverty level during this period, has assumed that these issues are important structural components affecting the incidence of crime. Critical measures have been taken to intercede at a structural level and crime figures have been substantially and consistently reduced in tandem with the measures taken. Since 1981 France has had no inner-city disturbances or violent confrontations with the police. In 1982 the Bonnemaison Commission's report - sparked by the 1981 disturbances and a survey showing that 88 per cent thought they lived in a 'violent age' - initiated the establishment of a national crime prevention council and 480 local councils. These councils have been involved in renovating dilapidated public housing blocks (HLM) after consultation with local youths who are frequently provided with jobs and training. Councils have funded workshops for the repair of motorbikes, bicycles, scooters and cars on housing estates and even provided some holidays for the young to go sailing, canoeing, swimming and cross-country cycling. In 1985 the Toulouse Council supported 300 to 400 youngsters each day for the whole of the summer period at a cost of £50,000. Michael King has emphasised that crime cannot be isolated from other behaviour, and other changes in French society must play their part. Although the cost of such an approach is substantial the drop in crime rates is impressive. For example the crime prevention council of Chanteloupe les Vignes, one of Paris's northern suburbs, reports that robbery with violence has dropped by 75 per cent, car thefts by half and burglaries by 20 per cent since 1985-86. But the benefits are longer-term. King observed:

> If the French are right about the underlying causes of youth crime, the real benefits of the present policies will be felt five years or more from now (Boseley, 1987).

Since 1981, and Lord Scarman's report on the Brixton disturbances of that year, the incidence of racial violence and attacks has increased - not infrequently in those boroughs which have introduced Equal Opportunity policies and addressed multi-cultural and anti-racist educational provision. Although the impact of such approaches should not be exaggerated and the benefits must be considered longer-term, it is evident that their immediate affect on racial harassment is marginal. Obviously local authorities are not the only or dominant influence in sustaining and counteracting racist ideology. Central Government, the work-place and the media may inform and influence the climate of opinion. However, by analogy with the experience of crime, the hypothesis which emerges is that, to tackle the expression of racism effectively, there must be an attack on the underlying economic and social structures which facilitate such expression. Thus, while the existence of a 'racial' underclass in education, housing and employ-ment may both be a product of and reinforce racial beliefs, more general lack of opportunity in those areas sustains the conditions for scapegoating and the translation of ideology into practice. Evidently these 'structural vehicles' may not be the root cause of criminal or racial mentalities but they are an immediate and tangible explanation for their expression in practice.

CHAPTER TWO

FACTS AND FIGURES

THE PATTERN OF SETTLEMENT

The growth of Britain's black population is essentially a post-war phenomenon: in 1951 0.2 million of people living in Britain were born in the New Commonwealth; this figure had increased to 1.3 million in 1981. The 1981 census, like those before it, included no question on ethnic origin, as opposed to place of birth: the 1981 estimate for the total British population of New Commonwealth ethnic origin (including Pakistan: hereinafter NCWP) whether born here or not was 2.2 million. A 1988 estimate is 2.4 million, or 4.5 per cent of the general population (CRE, 1988e). The largest ethnic minority groups within this number are those of Indian origin (750,000), of Caribbean origin (550,000) and of Pakistani and Bangladeshi origin (500,000) (CRE, 1988e: 13).

The word 'immigrant' is often wrongly used to refer only to black people when in fact white people form the majority of immigrants. The 1981 census shows that nearly 3.4 million people in Britain were born overseas. Well over half of these (1.8 million) are white; over 607,000 were born in Ireland, 153,000 in the Old Commonwealth (Australia, New Zealand and Canada) and about 1.13 million in other countries, including those in Western Europe. The remainder (1.41 million) are brown or black, having been born in the NCWP. It is estimated that 100,000 white people were born in the Indian sub-continent and East Africa while their parents were on overseas work (CRE, 1985a).

Between 1971 and 1983 more people had left Britain than had come in. The net loss of population during this period was 465,000, mainly as a result of emigration to Australia, Canada and New Zealand; in addition a number of emigrants left for the United States of America, South Africa and the EEC countries. In Britain it is estimated that 40 per cent, in 1981, of the NCWP ethnic minority population were born in this country.

One principal reason for the downturn in immigration has been the nature of successive laws passed by Government from 1962 to 1988 to control immigration, the most stringent of these being the 1971 Immigration Act. This Act imposed strict controls on the entry of men seeking work and cumulatively the Act had three general effects:

1. Primary immigration, i.e. men accepted here for settlement on arrival and on removal of time limit, peaked in 1972 at nearly 18,000 and fell to 6,400 in 1983, now being confined to people with those job skills which are still in short supply in this country.
2. Secondary immigration, that is the immigration of dependent women and children in order to effect family reunion, has been more than halved from about 50,000 in 1972 to about 21,000 in 1983.
3. Total immigration from NCWP countries has declined from a peak of 68,000 in 1972 to 27,000 in 1983: it declined to 23,000 in 1988 (HMSO, 1989).

REASONS FOR IMMIGRATION

The reasons for immigration have been crudely categorised as push/pull factors. Post-war reconstruction and the subsequent expansion of the Western European economies, coupled with a fall in the economically active population because of death or injury in the war and an increase in the number of the old and retired created a shortage of labour in Western Europe and migrant labour was needed and recruited in order to meet demand. This labour shortage in Britain in the immediate post-war period was met initially by volunteer workers from Eastern European countries such as Poland but also from Italy and in the mid-1950s to late 1960s from the former colonies of the New Commonwealth countries, mostly the West Indies and Guyana, India, Pakistan and Bangladesh. Concurrently a large number of migrant workers arrived from Ireland in search of work. This pull factor of employment opportunities and potential improvement of living standards in Britain has been contrasted with the push factors, essentially the lack of such opportunities in the immigrants' countries of origin.

AREAS OF RELATIVE CONCENTRATION

Self-evidently areas of relative concentration largely reflect settlement patterns of New Commonwealth immigrants in the 1950s and 1960s before the Immigration Act 1971 severely restricted primary immigration and reflected, generally, labour demand in London and the South East, the employment opportunities in metal manufactures in the West Midlands and the textile industries in the North West.

Approximately 56 per cent of people in Britain of New Commonwealth origin settled in the South East, London having 43 per cent, 15 per cent settled in the West Midlands, 7 per cent in Yorkshire and Humberside, 7.6 per cent in the North West and 6.5 per cent in the East Midlands (CRE, 1985a). The vast majority of NCWP population is therefore located in England (96.8 per cent, with 2.1 per cent in Scotland and 1.1 per cent in Wales. It should be noted, however, that these figures represent persons resident in private households with head of household born in the NCWP. To that extent they will be less reliable figures in relation to those communities of NCWP origin with a history of settlement in this country such as those in Cardiff and Liverpool.

This relatively uneven distribution at regional level is also reflected in the population at district level. Thus in the inner London area ethnic minorities form only 4.5 per cent of the population of the City of London but 30 per cent of Haringey's, with a significant population in Newham and Tower Hamlets. Similarly, with regard to the outer London boroughs relative concentration varies from 3.6 per cent New Commonwealth population in Bromley to 33.5 per cent in Brent, with Ealing, Hounslow and Waltham Forest having ethnic minority populations of 15 per cent or more.

Outside of London, Birmingham, Leicester, Slough and Wolverhampton, for example, have very substantial ethnic minority populations. Major centres such as Edinburgh, Glasgow, Leeds and Sheffield have relatively small ethnic minority populations.

This uneven distribution means that black people, although a relatively small proportion of the national population, are relatively numerous in some localities, and this has often contributed to a very inaccurate perception of the total size of the black community in a town or city and to some distorted perception of the national figures. It also means that in most areas of the country there are very few black people indeed. A high proportion of the white population in Britain live in towns and rural areas that have less than 0.5 per cent of their local residents coming from ethnic minorities. In comparison three-quarters of the black population live in a set of enumeration districts in which only one-tenth of the white population is found (Brown, 1984: 20). If we look more closely at London, Birmingham and Manchester, for example, only 6 per cent of the white population is found in the inner areas of these three cities in comparison with 43 per cent of the West Indian and 23 per cent of the Asian populations. The West Indian population is more concentrated in the inner areas than the Asian with the exception of the Bangladeshis, two-thirds of whom are found in these areas. Just over one-quarter of Asians and West Indians live in the remaining areas of the three conurbations, compared with 13 per cent of whites.

DEMOGRAPHIC CHARACTERISTICS

Age structure

The striking feature of the black population in Britain is that, in comparison with the total population, it is very young. This is a consequence of the youthfulness of past immigrants, who even now are not generally of retirement age (RSRG, 1980: 9). In 1971 all sections of the black population were in comparative terms under-represented in the age group 65 and over. The black population was over-represented in the age group from birth to 4 (11 per cent in comparison with 6.5 per cent) in the age group 5 to 15 (23 per cent, compared with 17.3 per cent). There was more marginal over-representation in the two age groups 16 to 19 and 20 to 26, while there was marginal under representation in the two age groups 30 to 44 and 45 to 64 (CRE, 1985a).

Sex ratios

In the black population there are more males than females and this is a result of past immigration patterns. The shortage of women is particularly noticeable among Asians, especially those from Pakistan and Bangladesh, for whom there were more than two males for every female in 1971. However, there is broad comparability between the sexes for Afro-Caribbeans in all age groups. The arrival of the comparatively few dependent female relatives (fiancees, wives and children) still overseas should help to resolve this sex imbalance in respect of the Pakistan and Bangladeshi populations. It is expected that the strict control of primary immigration of males, coupled with the continued fall in the fertility rates of ethnic minority women towards the general rate, in time will result in the ethnic minority age structure reflecting that of the general population and the sex imbalance will even out as it has appeared to do amongst the West Indian and Indian populations.

Household structure

Amongst the issues studied by the third PSI survey (Brown, 1984) in relation to ethnic minorities was household structure. This survey found that 90 per cent of West Indian and white households contain a single family unit where the figure for Asian households is lower, at about 80 per cent; 18 per cent of Asian households contain two family units and over 3 per cent contain more than two (op. cit., Table 14). With regard to household size, counting both adults and children, the average household sizes amongst the ethnic minorities are larger than amongst whites, although there is a complex pattern of ethnic differences. The average number of people in white households is 2.6, for Asians it is 4.6 and for West Indians 3.4. These averages are the products of very different distributions: while 20 per cent of white households comprise a single person only 13 per cent of West Indian households and 5 per cent of Asian households contain one person. At the other end of the scale there are only 3 per cent of white households with more than five people while the figures for West Indian and Asian households are 11 per cent and 29 per cent respectively. Within the Asian group, the Pakistanis and Bangladeshis have very much larger households than the others. Only 31 per cent of white households contain children while the figure for West Indians is 57 per cent and the figure is much higher for Asians at 73 per cent: nearly one-third of the Asian households have more than two children resident, compared with 12 per cent of West Indian households and 5 per cent of white households. The PSI survey also found that in the households that contain children the average number of children is almost the same for whites and West Indians (1.6 and 1.7 respectively) but much higher for Asians, 2.6.

If the household structure is analysed in a vertical fashion, that is, by the number of different generations resident, then the PSI survey demonstrates that the Asian and West Indian households will more frequently contain a combination of adults and children than the white households, and black households comprising only people over 54 are relatively rare. Despite the small proportion of black people aged over 54, households containing all three age groups are more common amongst blacks than whites (Asians 9 per cent, West Indians 4 per cent, whites 2 per cent).

The PSI survey also examined the horizontal extension of household structure, that is, for example, two brothers living with their wives and children without a resident parent. This structure was found to be more common amongst West Indians (8 per cent) and Asians (16 per cent) than whites (4 per cent). Amongst Asians it was more common amongst Pakistanis, Bangladeshis and Sikhs than others.

With regard to lone-parent households, compared with white and Asian, a high percentage of West Indian households consist of a lone parent with children under 16. Of all the West Indian households, 18 per cent are of this type in comparison with 4 per cent for Asians and 3 per cent for whites. Almost one-third (31 per cent) of all West Indian households with any children are headed by a lone parent, compared with 5 per cent and 10 per cent of Asian and white households with children (lone-parent families within larger households are excluded from this calculation).

MOVEMENT, TENURE AND QUALITY

House movement and changes in geographical distribution

The PSI survey in 1982 found that the age of the head of household is very strongly related to the likelihood of moving house and that within age groups the rate of movement of black households has equated with that of the white population (Brown, 1984: 56). The survey found that within a five-year period from 1977, 44 per cent of West Indian households, 47 per cent of Asian households and 32 per cent of white households had moved at least once. The Bangladeshi population had the highest rate of movement. The PSI survey found that despite a limited movement between 1971 and 1980 towards dispersal, more black people

- and a higher percentage of black people - are now living in the areas of high ethnic minority concentration. A preliminary comparison of the distribution changes between the 1974 PEP survey (Smith and Whalley, 1975) and the 1982 survey, in the restricted geographical area of the former survey, suggests that the overall changes in distribution have affected Asians and West Indians to a similar extent. This finding is different from those of a study of movements between the 1961 and 1971 censuses, which showed the indices of residential segregation from white people were, in that period, increasing for Asians but decreasing for West Indians (Peach et. al., 1975). However, the 1982 survey indicated that West Indian households who had moved out of areas of relatively high concentration were more likely to move to areas that were previously exclusively white; 12 per cent of West Indian movers compared to 8 per cent of Asian movers are now in wards that in 1971 had concentrations of immigrants of 0.5 per cent or lower.

Housing tenure

Of white households, 59 per cent are owner-occupiers, 30 per cent rent from the local authority, 7 per cent rent unfurnished dwellings and 2 per cent rent furnished dwellings in the private sector while 2 per cent occupy housing associations' property either rented or on a co-ownership basis. Of West Indian households 46 per cent are council tenants, 41 per cent are owners, 8 per cent live in housing association dwellings, while 6 per cent rent privately either furnished (3 per cent) or unfurnished (3 per cent). Amongst Asian households 72 per cent owned their own house, 19 per cent rented from the council, 2 per cent from housing associations and 6 per cent rented in the private sector either unfurnished (4 per cent) or furnished (2 per cent).

Amongst Asian households the level of owner-occupation is low for Bangladeshis (30 per cent) and very high for Sikhs (91 per cent): 53 per cent of Bangladeshis are in council tenancies while only 6 per cent of Sikhs are in this form of tenure (Table 2.1).

Housing tenure and household type

The 1982 PSI survey (Brown, 1984: Table 34) demonstrates that while some of the ethnic variation in tenure patterns is

Table 2.1: Tenure patterns in 1982 (%)

Type of tenure	White	West Indian	Asian	Indian	Pakistani	Bangladeshi	African Asian	Muslim	Hindu	Sikh
Owner-occupied	59	41	72	77	80	30	73	67	73	91
Rented from council	30	46	19	16	13	53	19	24	16	6
Privately rented	9	6	6	5	5	11	5	6	8	3
(Furnished)	2	3	4	4	4	8	4	5	6	1
(Unfurnished)	7	3	1	1	1	2	1	1	1	1
Housing association	2	8	2	2	1	4	2	2	3	*
Other	*	-	*	*	-	*	1	*	1	*
Base:Households (weighted)	2694	1834	2851	1150	751	277	604	1339	748	520
(unweighted)	2305	1189	1893	726	518	197	411	937	481	349

* <0.5%

Source: Brown, 1984: Table 29.

related to the differences of households types in the different ethnic groups, the essential tenure characteristics of the Asian, West Indian and white samples are reproduced even after separating the different household types. Thus, for almost every comparison, owner-occupation is most common amongst Asians and least common amongst West Indians. Council tenancy is most common amongst West Indians and least common amongst Asians. In all household types West Indians are more commonly found in housing associations than Asians or whites. White households nearly always have a larger proportion in privately rented dwellings than Asian or West Indian households.

New household tenure patterns

The 1982 survey included an analysis of tenure patterns amongst new households which concluded that they were becoming even more ethnically polarised (see Brown, 1984: 69, Table 30). Half the new households of West Indian origin were council tenants, 13 per cent were in housing association accommodation while 28 per cent were owner-occupiers. For Asians the figures were 14 per cent, 1 per cent and 78 per cent respectively. In the private rented sector the West Indians represented 8 per cent and the Asians 5 per cent of their respective new households, which contrasts sharply with the 17 per cent figure for white new households. In comparison with existing households within each ethnic group, new households show first for whites a diminution in council house renting (21 per cent from 31 per cent), and an increase in private furnished renting (11 per cent from 2 per cent), second for West Indians a diminution in owner-occupation (28 per cent from 43 per cent), an increase in housing association renting (13 per cent from 5 per cent), council renting (49 per cent from 45 per cent) and furnished private renting (6 per cent from 2 per cent) and third for Asians a diminution in council renting (14 per cent from 20 per cent) and housing association renting (1 per cent from 2 per cent) and an increase in owner-occupation (78 per cent from 71 per cent). While the increase in private renting by whites and West Indians, to a lesser extent, may reflect a pattern of the younger households deferring owner-occupation until later, when incomes and savings make this transition easier, the lack of earlier analysis of new household tenure patterns in the 1960s and 1970s makes analysis speculative. Moreover while the right to buy in respect of council housing and housing association tenancies may, in theory, facilitate a transition to owner-occupation by the West Indian community there is no evidence that this option has been taken up more voraciously than by the white tenants from 1980, the year of its inception. Accordingly the polarisation by ethnic origin of new household tenures is likely to persist over time despite some movement expected to owner-occupation, particularly in respect of white households.

Although there are geographical variations in the pattern of housing tenure and these have some effect on the differences between the housing tenure of whites, Asians and West Indians, the differences in tenure pattern between ethnic groups persist when regions, cities and smaller areas are considered separately. However, there are indications that in some types of area the ethnic differences are not of the same character as those evident in the national figures. In the conurbations the characteristic tenure pattern of West Indian households is very marked; the proportions of council tenants and households in housing association dwellings are much higher than those amongst white households but in the rest of the country the levels of council tenancy amongst whites and West Indians are identical, at 30 per cent, although there is still a large proportion of West Indians in housing associations. The proportion of Asians in council tenancies is very low nationally, but it is about the same as for whites in the inner-city areas. There is, however, a high level of owner occupation amongst Asians in all areas (Brown, 1984: 70).

The PSI survey noted that one of the most startling differences evident in the inner-city areas is between the proportion of whites and blacks in privately rented dwellings. Nearly a quarter of white households in the inner areas are private tenants, compared with 7 per cent of Asian households and 3 per cent of West Indian households. This is as much a demonstration of how the white population differs between the inner cities and elsewhere as it is of the difference between whites and blacks in the inner cities.

Housing tenure and job levels

Amongst the white population there is a straightforward correlation between job level and likely tenure. The higher the job level the higher the proportion of owner-occupation. Although it is also true that a larger proportion of Asian council tenants is found amongst the lower socio-economic groups and a higher proportion of owner-occupier West Indians is found amongst the higher socio-economic groups there is no visible straightforward relationship between job level and tenure of black households. The Asian communities generally have a remarkably high level of owner-occupation, irrespective of job level; this factor not only indicates that income has little direct correlation with tenure but also that tenure is not, in the Asian community, a useful indicator of the quality of housing occupied. While this is especially marked in relation to the Hindu and Sikh communities it also applies to the Moslem groups with the exception of the Bangladeshi community, the most recently settled in the UK, over 50 per cent of whom are in the public rented sector.

Housing quality and income

Irrespective of race, income will evidently be an important determinant of housing quality, although not, as noted above, of housing tenure. However, studies of housing in both the private sector (PEP, 1977; PSI, 1984) and the public sector (Henderson and Karn, 1987) demonstrate that ethnic minorities get poorer quality housing (by reference to type, size, space and amenities) than whites for the same expenditure. Accordingly to gain equivalent housing, in the private sector, ethnic minorities must either spend a greater proportion of their income on housing or must have greater incomes. Clearly the negotiation of equal access to quality housing in the public sector by paying more is not an available option. While it is possible for individuals to buy their way out of racial disadvantage in housing, the nature of cyclical disadvantage and evidence of continuing racial discrimination and disadvantage in employment make such a possibility in private housing of marginal interest. The 1986 Labour Force Survey figures, which were based on an average of survey results for 1984, 1985 and 1986 (Department of Employment, 1988) demonstrate that the overall unemployment rate among black people was 20 per cent compared with 10 per cent for whites. For those in the 16 to 24 age group, the group predominantly forming new households and entering the housing market for the first time, the unemployment rates were 32 per cent for blacks and 17 per cent for whites: in the case of Pakistanis and Bangladeshis the rate was 43 per cent for this age group.

These differentials were unaffected when young black and white job-seekers were matched for age, sex, qualifications and experience. The CRE, by reference to a study undertaken by the Midlands Careers Service, quotes figures in its 1987 Annual Report (1988a: 39) which, in alluding to the fact that missing out on YTS employer-run schemes may result in missing out on permanent job prospects, indicates that a substantially higher proportion of whites (45 per cent) enter YTS schemes on reaching 16 than Asians (16 per cent) and a marginally higher proportion of whites do so than West Indians (41 per cent). In respect of the last figure, however, there is evidence that West Indians receive a proportionately higher allocation of Premium Placements than employer-led provision. Whether or not this reflects a scarcity of the latter in some inner-city areas, the consequences are disadvantageous in respect of employment opportunities. For those in employment Brown (1984: 213) shows that the median weekly earnings of Asian and West Indian men is 85 per cent of whites for England and Wales and while regional variations occur the median earnings of the former are consistently lower than those of the latter (variation: 70 per cent (North West) to 91 per cent (London)).

CONCLUSIONS

With reference to the longer-term implications of this profile for housing, certain tentative conclusions may be drawn. First, while the original patterns of settlement may clearly relate to job opportunities and 'replacement labour', and have effected structural unemployment because of the decline in particular sections of the economy, particularly manufacturing and metal industries, housing tenure, quality and cost, as well as aspects of dispersal and concentration, have been influenced by other factors. Moreover the characteristics of ethnic minority groups, so far as they are distinguishable by age, sex ratio and household size, have shown a distinct movement towards the 'domestic' norm over a period of time. Because, during this period of change, there has been little if any comparative improvement in terms of housing quality, cost or indeed location, such distinctions fail to explain, to a substantial extent, disadvantage in housing by ethnic origin. While socio-economic group is one determinant of housing choice and housing quality, it likewise fails to explain such disadvantage and the additional cost of private housing begs any alternative explanation to that of racial discrimination made unlawful by the race relations legislation, even where differences attributable to locational factors are taken into account. For this reason 'ethnic choice' in the sense of choosing to pay more for less desirable housing can be discounted as a principal factor in determining ethnic disadvantage in housing quality. In addition to racial discrimination in the private sector there is clear evidence of racial disadvantage and racial discrimination in the public sector.

Moreover not only are ethnic minorities disadvantaged in the job market by suffering higher rates of unemployment, for those employed income levels are appreciably less than for the rest of the community: geographical location and educational attainment do not explain such discrepancies and lack of equal opportunity in job training schemes illustrates continuing racial discrimination in employment.

On the evidence of this profile alone, racial discrimination appears to be a significant direct factor in determining housing opportunities in both the public and private sectors and, because of the lack of bargaining power of ethnic minorities through discrimination in employment and income, ethnic minorities will suffer further, indirect, disadvantage in the housing process. The constancy of such disadvantage between surveys demonstrates, in the absence of any other plausible explanation, that the race relations legislation - principally the Acts of 1968 and 1976 - has had little if any effect on the housing circumstances of Britain's black communities.

CHAPTER THREE

DISCRIMINATION AND THE LAW

ISSUES AND APPROACHES

It was not until the 1960s that race became a major public issue in British politics. It was, however, immigration which became the focus of attention rather than the question of racial equality. In the eighteen months before the passage of the Commonwealth Immigrants Act 1962 over 200,000 black immigrants had arrived in Britain - almost as many as in the five years 1955 to 1960 and marginally fewer than black immigrants entering the country between July 1962 and the end of 1967. This period was marked by a sharp shift in public opinion towards immigration: in 1962, a few months before the passage of the Act, 62 per cent of the public favoured controls and 23 per cent favoured free entry but by April 1968 the corresponding figures were 95 per cent and 1 per cent respectively. While the diversity of reasons for this shift in public opinion has been well documented (Rose, 1969; Lester and Bindman, 1972; Foot, 1965), one constant theme has been, in the politics of race in Britain, the search by political leaders from the major parties to depoliticise this issue by papering over party differences (McKay, 1977).

By the 1960s both major parties subscribed to the view that immigration should be controlled because immigrants placed great strain on employment and housing. It was only the Labour Party, albeit with substantial ambivalence, which supported the idea of anti-discrimination legislation. By the time that Britain's first civil rights law, the Race Relations Act 1965, had been passed, for many, blacks had become synonymous with immigrants. Moreover prior to the Local Government Act of 1966 (Section 11) the general laissez-faire attitude of Central and Local Government meant that virtually nothing had been done to cater for the problems that many immigrants experienced in housing, employment and education. As Burney (1967) observed, in the absence of Central Government direction and incentives, many local authorities, frequently in areas of high immigration and Labour controlled, behaved as though blacks did not exist: 'most Labour controlled councils made a habit of resolutely ignoring immigration, to the extent of, wherever possible, ignoring immigrants'.

So far as access to housing was recognised as an issue of concern in relation to ethnic minorities, a resolutely colour-blind approach was advocated whereby the real need was perceived to be to alleviate the housing shortage and to provide for those in greatest need. There should be no attempt to discriminate positively in favour of such minorities to remove the racial disadvantage which they had experienced (Home Office, 1965).

At the time of the 1965 Race Relations Act the majority of blacks resident in the United Kingdom were immigrants and were likely to experience or to have experienced difficulties in the following areas:

1. Problems associated with settlement and establishment.
2. Differences in the dominant form of household structure and size being met by shortage in the private and public housing sector.
3. Cultural, social, linguistic and religious differences -the ethnic character of the group concerned - forming distinct needs and aspirations from those of the host community.
4. The experience of ethnocentricism by the largely monocultural white majority resulting in racial disadvantage through expectations of social, linguistic and religious 'acculturation' to the dominant norms of the host community.
5. Racism - the experience of direct and indirect discrimination through racially motivated behaviour by the white community.

It may be artificial to attempt a rigid isolation of these factors, given that they will interrelate more or less significantly in particular situations. Nonetheless, to the extent that they inform both opinion and practice, reference to the law alone may result in an inadequate or misleading explanation of social behaviour and consequently in a distortion of the relevant merits or demerits of legislative provision in achieving the goal of racial equality in access to housing.

Britain legislated against racial discrimination at almost the same time as the United States. Although the British experienced nothing equivalent to the American racial troubles of the 1960s, there has been considerable disillusionment with the 1965, 1968 and 1976 Race Relations Acts; the disturbances in Brixton, St Pauls, Handsworth and Toxteth in the early 1980s would appear to question either whether the legislation in its present form is capable of effecting significant change towards equality of opportunity or alternatively whether the law, in whatever form, is capable of securing the goals which it sought to achieve.

McKay (1977: 13) in his comparative critique of British and American housing legislation in the 1960s and mid 1970s suggested that in general British and American civil rights commentators have been divided. On the one hand there are those who dismiss the laws as at best irrelevant and at worst a deliberate attempt by white elites to manipulate actual or potentially rebellious blacks into political quiescence. On the other hand there are those who believe the legislation has failed not because it is doomed to failure, but because it has been inadequately formulated and/or implemented. Crudely the radicals, eager to construct theories explaining the discrepancy between expressed goals and performance, have focused on linking the alleged existence of an elite consensus, or of institutionalised racism, to the subordinate position of non-whites without a thorough examination of the nature of the enforcement of law itself. Conversely, the liberal approach has been to focus on the details of the administrative process or on the nature of discrimination in isolation from structural, political and economic forces. The one, then, studies theory divorced from practice while the other studies practice but fails to locate it in any theoretical structure. Moreover, the difficulty that social scientists have experienced in accommodating racial inequality and conflict within broadly based theories of class stratification and conflict has resulted in such theories lacking an empirical base compatible with the evidence available. It has been suggested, even at a purely theoretical level, that there is often failure to produce a theory worth testing in that the principal assumptions were refuted, in part or in whole, by a cursory examination of empirical evidence. Thus Katznelson (1973) claimed that in both the United States and Britain there had been a conscious attempt of white elites to depoliticise the race issue through the creation of 'buffer institutions' (mainly community relations agencies) whose function had been not to improve race relations and reduce discrimination but to isolate black aspirations and reinforce institutional racism. This theory appears to be at one and the same time a confirmation and a refutation of conflict management and suffers from such inconsistency: it is difficult to reconcile the purpose of political elites in removing race from the political arena to defuse conflict situations with the practice of setting up buffer institutions with the function of frustrating demands to the extent that they are then expressed in the formulation of political organisations outside the accepted political arena.

However, as Philip Mason, a former Director of the Institute of Race Relations, has observed, the Government of the day, the Civil Service and Parliament while frequently insensitive, often complacent and pompous, often reluctant to make any change and sometimes legally pedantic have not been evidently cynical on this kind of policy. In contrast, in his view, it has always been taken for granted that it was against the interests of this country that there should be a section of society which is easily identified, which is discriminated against and which is alienated from Government and the majority (Brier and Axford, 1975, referred to in McKay, 1977: 15). Moreover, the development of the legislation on race relations (virtually ignored by Katznelson) and the ability of some buffer institutions, such as the Home Affairs Sub-committee on Race Relations and Immigration, operating from 1979 to 1988, to expose Government inaction on a number of important issues and to influence change in departmental policy in respect of urban aid, health and social service issues, Civil Service monitoring, training and Section 11 funding (Nixon, 1986) would suggest, if nothing else, that Katznelson's analysis is far too simplistic to sustain any continuing credibility.

It is clear that the nature of social structures in western democracies has undergone significant change. A sense of local community identity has largely evaporated as a reflection of increased mobility. The separation of home and work and of home and leisure activities and the fact that the average household changes its living accommodation once every five years have, particularly in the large urban conurbations where the majority of black people live, lessened the relevance of integrated local networks of interaction between the individual and his domestic location to the extent that most people would now have difficulty in defining the boundaries of their local community. The symbols of community structure, the church, the school, the football team, the pub and the cinema have, for a variety of reasons, lost much of their local relevance and significance if not their very existence. Predominantly our remaining sense of community is framed by a complex interaction with different groups for different purposes which lack common spatial identity. Our spacial reference points have become blurred at the level where personal identity and a true sense of community are meaningful. Our sense of spatial identity, conversely, has probably increased at the level where political decisions are made - the district, the county or region and the national, whether Northern Ireland, Wales, Scotland and England or the United Kingdom. Such changes may well have repercussions on the receptivity of local communities to cultural diversity. While a strong community structure may reject newcomers as well as accept and thereafter protect their interests, the very act of rejection would have exposed conflict and defined issues requiring to be addressed. Weak local structures, in their inability to absorb or reject cultural diversity, will both camouflage patent and latent conflict and result in newcomers building alternative and potentially insular structures of community support frequently provoking antagonism, perhaps partially emanating from envy, in the host community. Self-evidently the term 'newcomer' so far as it equates with first-generation immigrants is, by now, an inappropriate description for black British citizens but these processes of change have been evident for at least thirty years, the period which has witnessed by far the greatest immigration of black residents in the United Kingdom.

This period has also witnessed the growth in the power of the mass media and its centralised and restrictive base. Social Trends (HMSO, 1987b) confirms that television watching absorbs a phenomenal amount of our time. Its impact is strengthened by the absence of a local community reference point which might refute or qualify its powerful and selective messages. The lack of personal interaction between most whites and blacks within a strong local community is likely to have two consequences. First, a particular message - such as the continuing fear of being swamped by black immigration - is not challenged by a knowledge of black experience within a given locality - effectively denying the relevance of such knowledge.

Second, issues are depersonalised and the not infrequently inhumane responses by Central Government (such as the provision of offshore accommodation for immigrants, including dependent relatives, in atrocious living conditions while their status is being verified) and by Local Government (such as the provision of deplorable temporary accommodation for homeless Bangladeshis by Tower Hamlets) frequently fail to provoke indignation and effective opposition in the communities affected.

Given that such influences are more or less extraneous to the legal process in respect of anti-discrimination provision it becomes apparent that a description and critique of that process provides a very partial view of the nature of racial discrimination and the prospects for legislation affecting substantial social change.

Despite this we must judge legislation and its enforcement on its own terms. The extent to which extraneous influences provide a context for legislation is a truism which Government accepts and the objectives of legislation assume that there are countervailing forces with which it has to contend. While we may not expect a Government to be omniscient we do expect it to be honest, well informed and intelligent. But when Governments have consistently juxtaposed race relations with immigration and, in respect of the latter, have demonstrated, at least to the satisfaction of many critics, that they have shown dishonesty, ignorance and wilful disregard for rational analysis it does raise serious questions regarding their bona fides in respect of the other side of the equation, race relations: it may be naive, therefore, to assume that all is what it is said to be.

RACE DISCRIMINATION AND THE COURTS

General

As Bindman and Lester (1972) have observed, cases of racial injustice have come before the courts in three different situations, each involving conflicts because of the movement of members of an ethnic or national group from one country to another. The traffic in African slaves was perhaps the earliest and most dramatic legal issue which came before the courts either because merchants sought to enforce agreements for the sale of slaves or because masters sought forcibly to remove unwilling slaves from England to the colonies. Second, a diversity of courts, including the Privy Council, were involved in determining legal issues involving discrimination against Chinese and Japanese migrants to Canada and, third, problems of discrimination have arisen in relation to the immigration of Jewish refugees and black Commonwealth citizens.

As an aftermath of the Spanish wars black slaves first appeared in Britain at the end of the sixteenth century and by the end of the eighteenth century about one-third of the British merchant fleet was engaged in transporting 50,000 negroes a year to the New World. By this time the slave trader was not only socially respectable but his business was a recognised route to gentility and was officially approved by the Board of Trade, the navy, and the nobility (Davis, 1966: 154). It would appear that both French and English colonial law assumed that the slave had essentially the attributes of personal property, and like a horse or cow could be moved, sold or rented at the will of his owner (op. cit.: 248). In Chambers v Walkhouse (1693) 3 Salk. 140 negroes were described as merchandise and compared with muskcats and monkeys. By this time the courts had already decided that slavery was legal in England because negroes were infidels who therefore had no entitlement to enjoy the rights enjoyed by Christians. In 1749 Lord Hardwicke opined from the Bench in Pearne v Lisle (1749) Amb. 75 that a slave did not become free on coming to England nor did he become free by being baptised and further that any master might lawfully force his slave to return with him from England to the plantations. This view conflicted with that of the Lord Chancellor in Chanley v Harvey (1762) 2 Eden. 126 where he stated that a man became free as soon as he set foot in England and that a negro could maintain an action against his master for ill usage as well as having a writ of habeas corpus if restrained of his liberty.

Eventually, and most reluctantly, in the case of James Somersett (1772) 20 St. Tr. 1 slavery was pronounced to be unlawful in England - 'Whatever inconveniences, therefore, may follow from the decision I cannot say this case is allowed or approved by the law of England; and therefore the black must be discharged.' While Sommersett's case was decided thirty-six years before Parliament prohibited the slave trade and sixty-two years before it abolished slavery in the colonies (in 1834) its importance, as Bindman and

Lester have observed, has been vastly exaggerated. In essence slavery had not been proscribed but its continued existence had been allowed in attenuated form and only the worst abuse - the forcible removal of negroes back to the colonies - rendered unequivocally unlawful despite the fact that it undoubtedly continued.

It was left to Lord Stowell, who described himself as an abolitionist, in the Slave Grace case (1827) 2 St. Tr (NS) 273 to point out the evident hypocrisy of Sommersett's case, stating that it went no further than the extinction of slavery in England as it was:

> unsuitable to the genius of the country ... the air of our island is too pure for slavery to breath in it. How far this air was useful for the purpose of respiration during the many centuries in which two systems of villeinage maintained their sway in this country, history has not recorded. The arguments ... do not go further than to establish that the methods of force and violence which are necessary to maintain slavery are not practicable upon this spot.

He rejected the argument that colonial slavery was against public policy, recalling first that it had been 'favoured and supported by our own courts which have liberally imported to it their protection and encouragement' and second that it had been continually sanctioned by Acts of Parliament.

In holding that temporary residence in England did not destroy the status of slavery, which revived on a slave's return to the colonies, he sought to place the onus firmly on Parliament to take legislative action against slavery in the colonies rather than through public conscience to outlaw slavery through the back door by means of judicial decision. Although Parliament eventually passed an Act in 1833 abolishing slavery in the colonies, as late as 1860 the courts would still refuse to invalidate a contract made by a British subject for the sale of slaves in Brazil because the possession of slaves was lawful in that country.

Turning to the second issue, that of discrimination against Chinese and Japanese Canadians, on four occasions the Privy Council had to decide whether it was lawful for the Provincial Government of British Columbia to practise racial discrimination against Chinese and Japanese Canadians. In each instance the area of concern addressed by the Privy Council was whether or not racially discriminatory legislation passed by the British Columbian legislature encroached unlawfully upon the exclusive power of the Dominion legislature, under the British North America Act 1867, over aliens and naturalised persons. In Union Colliery Co. of British Columbia v Bryden (1899) AC 580, a shareholder of the colliery company sought to rely upon a provincial Act which barred 'Chinamen' from employment in underground coal workings because he was of the view that the company's employment of Chinese in responsible positions would be a source of danger to others in the mine. The Privy Council, in deciding that the provincial statute was unlawful because it applied to aliens or naturalised citizens who were under the exclusive authority of the Dominion Parliament, took care in refusing to comment on the merits of the discriminatory statutes since the courts had 'no right whatever to enquire whether their jurisdiction [i.e. that of the legislature] has been exercised wisely or not'. Thus they chose to remain silent on the issue of principle as to whether or not the Canadian Constitution should be interpreted, in the absence of Dominion legislation to the contrary, so as to apply equally to all citizens living in Canada, without unfair discrimination.

In Cunningham v Tommey Homma (1903) AC 151, Tommey Homma, a naturalised British subject of Japanese origin, was excluded from the electoral voting register in Vancouver under another British Columbia statute which provided that 'any native of the Japanese Empire ... not born of British parents' and 'any person of Japanese race, naturalised or not' was to be denied the right to vote. The Privy Council reversed the decision of the Supreme Court of British Columbia and held that Tommey Homma had properly been denied the franchise despite the fact that the provision referred to 'naturalisation' which was within the exclusive competence of the Dominion Parliament. In this instance the Privy Council had managed to interpret the expression 'political rights' in the Canadian Naturalisation Act as not including the right to vote in provincial elections. This latter decision was followed in Brooks-Bidlake and Whittal v Attorney General for British Columbia (1923) AC 450, which concerned licences issued by the Minister of Lands for British Columbia to cut timber on Crown property on condition that 'no Chinese or Japanese shall be employed in connection therewith'. This restriction had been held invalid by the provincial Court of Appeal in 1920 but in 1921 the provincial legislature passed a statute providing that this restriction had the force of law. As Brooks-Bidlake, the licensee, employed both Chinese and Japanese workers, it sought a ruling from the courts that it was entitled to employ such workers despite the terms of the licence. Amongst other things, the company relied upon the Japanese Treaty Act of 1913 which had been passed by the Dominion Parliament to give effect to a treaty under which Japanese subjects were guaranteed the right to equal opportunities in employment in Canada. The Privy Council decided that the provincial legislature had acted within its constitutional powers because the pith and substance of the statute concerned the management and licensing of public property in British Columbia rather than the employment of aliens and naturalised persons. However, a year later, in Attorney General for British Columbia v Attorney General of Canada (1924) AC 203, the Privy Council changed course by determining that the discriminatory statute was invalid because it violated the principle laid down in the Japanese Treaty Act. Apparently what was raised in the latter case was 'a wholly different question' because the employment of Chinese labour was not in issue.

As Lester and Bindman (1972) observe, taken together the four cases cannot rationally be reconciled. The qualities, prized in the common law, of certainty, predictability and consistency are demonstrably absent, as are considerations of equality and justice. In 1947, two years before the Canadian Parliament abolished the right of appeal to the Privy Council, that body was asked to determine an appeal, Co-operative Committee of Japanese Canadians v Attorney General of Canada (1947) AC 87, in which three Orders in Council made in December 1944 (after the war had ended) by the Governor General under the War Measures Act 1914 were under challenge. These Orders authorised the deportation to Japan first of Japanese nationals resident in Canada and second of naturalised and natural-born British subjects 'of Japanese race' together with their wives and children under 16 years of age. The Privy Council rejected the argument that the expression 'of Japanese race' was too vague to be capable of being applied to ascertained persons as well as the argument that no emergency existed which would justify such action being taken under the War Measures Act in confirming the validity of the three Orders.

These cases illustrate the difficulty of determining to

what extent the Privy Council and indeed the House of Lords feel it appropriate to take on board the issue of public policy in developing the common law. To the extent that public policy may be relevant to the judiciary in coming to a determination the courts would be bound to acknowledge that the development of common law is a dynamic process as public policy itself is seldom static. This view was reflected by the judgment of Lord Pearson in Addie v Dumbreck (1929) AC 358 where he observed that the formulation of the duty of occupier to trespass is plainly inadequate for modern conditions, and its rigid and restrictive characteristic had impeded the proper development of the common law in this field. 'It has become an anomaly and should be discarded.' Stevens (1979: 622) has argued that the last twenty years have witnessed a greater willingness to see the courts as an integral part of the Government process, but that change should not be over-estimated. Politicians and political scientists writing about the English scene still normally ignore the courts as an integral part of Government, and the wide acceptance of the distinction between law and policy (closely associated with the idea of the rule of law) still allows the judiciary and even the legal profession a measure of immunity from criticism, which judges and lawyers in other nations might well envy. Self-evidently an emphasis on logic, certainty and predictability downplays discretion and creativity but is a price which many feel the courts are bound to pay. A former High Court judge, attacking the 1966 Practice Statement, which enabled the House of Lords to overrule its previous decisions, observed that judges

> should refrain from broad statements of principle and from obiter dicta. They should also be scrupulous to apply the law as it exists even if they think it be wrong or unfair or unjust and should resist the temptation to twist the law to conform with their sympathies or theories, as a proper instrument for the reform of law is Parliament, aided where necessary by the Law Commission, a Law Reform Committee or Royal or Departmental Commissions' (Sir Henry Fisher, quoted in Stevens, 1979: 622).

Racial discrimination in the UK

Given this ambivalence towards the legitimacy of applying public policy in the development of common law generally, it is not surprising that the history of judgments relating to the application of common law, prior to the first race relations legislation in 1965, lacks any clear-cut espousal of principle. With few exceptions when they have addressed issues relating to private persons discriminating against other private individuals, the courts have not been prepared to invalidate such discrimination, indirectly giving it the force of law. In <u>Weinberger</u> v <u>Ingles</u> (1919) AC 606 the court took the view that Hugo Weinberger, a naturalised British subject for thirty years but who had been born in Germany, had not been unlawfully discriminated against when the Committee of the Stock Exchange, of which he had been an exemplary member for twenty-one years, had refused to re-elect him. The decision followed pressure from the Stock Exchange Anti-German Union who had objected to his re-election on the sole ground that he had been born in Bavaria. Even where the courts have found against the discriminator the public policy element has often been ambiguous. Thus in <u>Scala Ballroom (Woverhampton)</u> v <u>Ratcliffe</u> (1958) 3 All ER 220 the Musicians Union, which had many black members, was held to be entitled to advise its members not to play in the Scala Ballroom, which, from its opening, had denied entry to black people. The court in indicating, although <u>obiter</u> to the judgment, that the ballroom proprietors were entitled to maintain the colour bar 'in their own business interests' did not outlaw racial discrimination. The court merely allowed the Musicians Union to act on its own initiative to protect its members' interests.

This non-interventionist character of the common law, as portrayed by the courts, also made it incapable of combating racial discrimination in the field of housing. Thus prior to the legislative provision in the 1968 Race Relations Act, because there was no common-law duty to do business with anyone, the courts would not intervene to help someone who was refused housing accommodation or the services of an estate agent solely on the grounds of race. However, there were, potentially, at least two major exceptions to this proposition. First, in England and Wales, conspiracy might be involved leading to an action for damages where two or more people agreed to prevent someone of a particular race or colour from obtaining a house or flat where the purpose was not primarily to protect their own interests but to inflict harm on the victim. Second, where property transactions are bound by restrictive covenants, when there is an expressed reference to the consent of the owner or landlord which should not be arbitrarily withheld, then, at common law, the courts would regard it as arbitrary to withhold consent on grounds of race or colour (Mills v Cannon Brewery Co. (1920) 2 Ch 38, and see generally Lester and Bindman, 1972: 58 et seq.). The law in this regard, however, is far from clear-cut: thus in Schlegel v Corcoran (1942) IR 19 the sale of a dental practice to a Mr Gross had been refused on the grounds that he was Jewish and that the practice might therefore develop a Jewish complexion. Mr Justice Duffy asserted

> that caprice is not the right word for antisemitism, which far from being a peculiar crotchet, is notoriously shared by a number of other citizens, and, if prejudice be the right word, the antagonism between Christian and Jew has its roots in nearly 2,000 years of history and is too prevalent as a habit of mind to be dismissed off-hand.

Although this was an Irish case and was held under the equivalent of the Landlord and Tenant Act where withholding of consent to the transfer of rooms should 'not be unreasonably withheld' the court was clearly of a mind that racial discrimination was not unreasonable in this context. This interpretation may be contrasted with the views expressed by Sir Hartley Shawcross in 1949:

> A clause in a lease or other agreement discriminating between different classes of His Majesty's subjects on the grounds of colour may well be void under the existing law as being contrary to the rules of public policy upheld by the English court.

Earlier reference to public policy has already suggested that such optimism may well be misplaced but, although not strictly in the housing sphere, the concept of 'public trust' flowing from the edict Nautae Caupones appears to have developed in parallel strands in England and Scotland to ensure that innkeepers should not arbitrarily discriminate against bona fide travellers. The Scottish case Rothfield v North British Railway Co. (1920) SC 805 confirmed that a person who had been refused lodgings by the innkeeper on racial grounds could obtain a remedy.

The innkeeper's duty is a rare example of a common-law obligation to give equal treatment and consequently its practical value is limited. It is also important to note that the definition of an inn excluded lodging houses, boarding houses, private residential hotels, ale houses, houses of entertainment and restaurants.

Conclusions

In summary, then, although this examination of common-law provision prior to the enactment of race relations legislation in 1965 has been cursory, it is suggested that the illustrations are sufficiently typical of the approach of the courts to indicate the inchoate nature of the development of public interest and public policy in the common law on matters of racial discrimination. Thus in the sphere of housing a person would be unlikely to obtain any assistance from the courts if he were refused a house or flat in the private sector which was available and which he could afford. Similarly if he were denied the services of an estate agent or an estate agent imposed the discriminatory practices of his client no remedies would be available. Conversely, however, as a matter of public policy he might have the opportunity of seeking redress from a local authority housing department which had barred access to council housing solely on the grounds of his race or who had discriminated against him in respect of the nature or quality of housing on this ground. In addition the common law might read in a requirement that a decision should not be made arbitrarily or capriciously in connection with restrictive covenants affecting the transfer of property and like requirements might be placed on innkeepers, who have a public trust in relation to providing accommodation and facilities for all bona fide travellers.

The common law in this area also serves to illustrate a more general dilemma faced by the courts in relation to their creative law-making role. As a generality judges prefer to see their role as that of the logical and rational interpreter of legislation and common law. In the name of public policy, this might permit an element of creative not to say judicious development. However, this should not be seen as an instrument of radical reform.

The relevance of this view is not confined to the pages

of history, for it will be evident with the passage of new legislation which the courts are required to interpret that the judges will bring with them this apprehension of innovation and this scepticism concerning their role as interpreters of public policy. Moreover it is also clear that the courts reflect the attitudes and prejudices of the general public, and in the absence of any training which might qualify such prejudices, it cannot be assumed that new legislation will work in tandem with a new and more sympathetic approach from the courts.

WHY LEGISLATION?

As Lester and Bindman (1972) have observed the practical effect of racial discrimination and its disastrous cost to the economic, social and moral well-being of society are an essential element in the arguments of those who have advocated the use of legislation. Lord Boyle expressed the purpose of legislation as follows:

> It is not to try to enforce moral attitudes by law. I think that I am more opposed than most people to using the criminal law in order to enforce morality as such. But race relations is not just a matter of private morality - it is a major issue affecting what John Stewart Mill would have called the 'public domain'. Racial discrimination as a practice is both wasteful and divisive, in a manner that quite transcends taste and prejudices. It is also extremely insidious (Lester and Bindman, 1972: 85).

Despite Lord Boyle's rejection of legislation intruding into private morality, his reference to its place in the 'public domain' appears to be an unconvincing attempt at a rigid severance of the two. Clearly law has a public education facet: it may attempt to change social norms from a practice of discriminating and tolerating discrimination to that of isolating the discriminator because he or she has become a non-conforming law breaker. The importance of this facet was illustrated by evidence given by the president of Levitt & Sons Inc., a major American property developer, to a Congressional committee concerned with Title IV of the Civil Rights Bill 1966 which outlawed racial discrimination in housing (Street et al.,

1967). In his evidence the president referred to the fact that the company was building at more than a dozen locations in four eastern states, in Puerto Rico and in France. In all these locations, with the exception of Maryland, which then had no anti-discrimination housing law, the company operated a non-discrimination policy. In the non-discriminatory or 'open occupancy' communities which were established he reported that he knew of no single racial disturbance, no outbreak of violence, no picketing and an absence of social tension.

The opponents of anti-race discrimination legislation seldom go so far as to justify discrimination in itself although a lecture by Lord Radcliffe in 1969 which referred to Britain's black communities as guests and 'a large alien wedge' carrying with their colour 'a flag of strangeness and all that strangeness implies' appeared not too distant from that view. He went on to observe that the two key words, 'prejudice' and 'discrimination', did not carry any association of moral ill-doing and suggested that the law should only be used in situations in which the moral issue was generally regarded as being beyond doubt. He argued that race relations could not be founded on such a vague concept.

> The conduct of human life consists of choices, and it is a very large undertaking indeed to outlaw some particular grounds of choice, unless you can confine yourself to such blatant combinations of circumstances as are unlikely to have any typical embodiment in this country. I try to distinguish in my mind between an act of discrimination and an act of preference, each time my attempt breaks down.

He did, however, make a distinction between the Race Relations Act 1965, which outlawed discrimination in a place of public resort, and the Act of 1968, which made discrimination by employers and landlords, for example, unlawful, in the following terms:

> [There is] an immense difference in principle between imposing conditions on persons whose calling excludes freedom of choice, or preferences between members of the public, such as inn keepers, restaurant keepers and common carriers and persons such as employers and landlords who

are bound by the nature of their position to exercise preference in selection (Radcliffe, 1969).

This reference to freedom of choice was reflected in the speech of Quintin Hogg MP (as he then was) during the third reading debate on the 1968 Bill when he said:

> I recognise that individual liberty involves the liberty to do what other people regard as wrong. I would fight for that liberty even where it was unpopular to do so ... I am, however, bound to add this proviso to that doctrine. Where individual acts add up cumulatively to a large scale social injustice, it is at least for consideration whether Parliament ought to deal with them. Where one is dealing with services offered to the public ... and more particularly where those services are offered to a section of the public by one of the great sources of supply - the great landlords, the great trade unions, the great businesses or the greater providers of transport - I begin to think that one is dealing with something where, if the individual right to the service is infringed, a serious source of injury might take place. In at least two of the areas ... those of housing and employment, there is positive evidence that injustice on a serious scale may be taking place. I cannot therefore say that it is inherently and necessarily an infringement of the liberty which I respect to attempt to legislate about it (9 July 1968, Hansard (HC), Vol. 768, Col. 473-4).

Lord Radcliffe's opinion raises the spectre of the 'thought police' and an apprehension about legislating in the realm of morals, particularly where there is no clarity of public opinion or where public opinion is out of sympathy with what is being proposed. With regard to the former it is important to note that legislation on discrimination in this country is designed to affect racist behaviour and not racist opinion unless the result of the latter is to the tangible detriment of the sufferer in a specified area covered by the legislation. Similarly it is difficult to attribute much credence to concerns regarding legislation having some moral imputation. As exemplified by the Abortion Act 1967 and the Housing (Homeless Person) Act 1977 the law will inevitably reflect social value judgments: what is of concern therefore is not the fact that the law will reflect moral judgments but rather the extent to which such judgments are felt to be of sufficient weight to justify legal controls. It is Quintin Hogg's speech, therefore, in alluding to competing claims between the freedom of the individual and the protection of a section of the public from arbitrary discrimination which is more central to the debate on the efficacy of race relations legislation.

However, it is important to appreciate that such a debate does not take place in a historical, social, economic or legal vacuum. Together these factors create an ideology of contemporary law which affects not only the formulation of legislation and its revision but also its interpretation and implementation.

DISADVANTAGE AND DISCRIMINATION IN HOUSING

Before examining UK anti-discrimination legislation it is necessary to outline some of the complexities which are involved in analysing disadvantage and discrimination: in practice, what is the nature of the problem which the legislation is to tackle?

Ward (1984) has argued that ethnic preferences in housing have to find expression within the constrained opportunities in a particular locality. The highly divergent patterns of housing achievement among minorities, particularly those from south Asia, in urban areas in Britain suggests that discrimination, while it originates in much broader forces in society, is shaped by local circumstances, which fashion the incentive to impose differential terms, the opportunity to do so, the form that discrimination takes and its objective consequences. Such a view has important consequences with regard to the analysis of patterns of ethnic minority housing on a large scale. Because ethnic preferences, economic circumstances, housing opportunities and settlement and mobility patterns have been shown to be factors which affect housing tenure, location and quality, any attempt to isolate and quantify discrimination can be successful, if at all, only when the analysis is undertaken in a particular locality: and by definition the experience of that locality will not be, of necessity, reflected in the experience elsewhere. For this reason it is particularly problematic in coming to conclusions about the extent of racial discrimination exercised by any of the participants in the housing process. However, because public housing is subject to more uniform processes of generation and control, this form of tenure is more accessible in any attempt to identify discrimination as a causal factor in the experience of relative disadvantage by ethnic minorities. But, as Henderson and Karn (1987) have observed, it is not possible to understand the racial and social patterns which are produced by council housing allocation without consideration of the racial and social patterns produced by the operation of the numerically dominant private market. With this qualification, certain observations can be made about the public sector in relation to access, quality of housing and segregation.

With regard to access, in 1967 a study of six local authorities (PEP, 1967) found that racial minorities were effectively excluded from access to council housing via the waiting list because they failed to meet certain residential qualifications. This conclusion concurred with that of Rex and Moore (1967) in their work on Birmingham. Some blacks were coming into council housing via slum clearance programmes, but on the whole they were not concentrated in the areas scheduled for redevelopment but in adjacent multi-occupied property. PEP issued a further study of racial minorities in council housing in 1975 (Smith and Whalley, 1975) which, in examining the internal administrative data from ten local authorities, concluded that racial minorities in each were at a disadvantage in gaining access to council housing. The explanation was predominantly ascribed to residential qualifications. In addition it was found that a number of West Indians and Asians had moved into the owner-occupation of cheap old property; the exclusion of owner-occupiers from waiting lists thus prevented them from subsequently entering council housing other than through slum clearance. In one of the local authorities studied it was found that racial bias in access was operating specifically against Asians although no satisfactory explanation for this outcome could be found.

With reference to the quality of the property allocated Smith and Whalley concluded that the allocation of council housing tends to work to the disadvantage of minority groups in a variety of ways for a variety of reasons, including the priority system, the nature and location of the stock available, the acceptance of offers through desperation and a differential choice of areas by whites and blacks. Poor communication and explicit racial bias were also identified as factors. In the same year the Runnymede Trust (Lomas with Monck, 1975) published an analysis of the 1971 Census Small Areas Statistics for London which highlighted the segregation of black households on the oldest inner-city flatted estates, identifying specifically estates owned by Lambeth, Newham, Wandsworth and the GLC. A study of GLC housing allocation (Parker and Dugmore, 1976) the following year suggested that the following factors were critical:

1. The disproportionate number of black homeless families.
2. The poor knowledge that blacks had over the council housing system.
3. Acceptance of a first offer more readily than whites.
4. Preference for inner-city locations where the property available was worse.

However, the report went on to acknowledge that these factors did not properly account for the differential between black and white tenants and reluctantly concluded that racial bias was occurring. Further studies on Lewisham, Islington, the GLC, Nottingham, Bedford and Liverpool all confirmed this pattern of relative disadvantage in relation to quality of property allocated.

With regard to racial concentrations, studies of the private sector had already shown up the growth of areas of segregation (Rex and Moore, 1967) but an explanation had been sought in the combination of recent arrival in Britain, the weak purchasing power of black households in the private market and racial exclusion both through the market mechanism and through the activities of vendors, landlords and their agents. Given the absence of these factors in council housing allocation, it was assumed that when substantial numbers of Asians and West Indians began to be rehoused in council properties these concentrations would tend to be evened out (Smith and Whalley, 1975: 113). However, reports already referred to on the GLC, Lewisham, Islington, Bedford, Nottingham and Liverpool, as well as the CRE investigation and research regarding Hackney and Liverpool respectively and a more recent study of Edinburgh allocations (Thwaites, 1986; Taylor, 1987) have all confirmed that this assumption is false. The fact that such concentrations are happening in the worst quality estates and that allocations have not reflected black preference in the same way that they have reflected white preference challenges a view that cultural choice is a significant factor in determining the housing situation of ethnic minorities in the public sector. Nonetheless these studies also demonstrate the need to have some conception as to what is meant by racial disadvantage and racial discrimination.

As Henderson and Karn (1987) have observed, one way of conceptualising racial discrimination is to see it as a product of racially prejudiced individuals and, therefore, the behavioural consequence of rigid dogmatic and irrational thinking about race. The studies referred to, however, have demonstrated that racial disadvantage has not predominantly been the product of such explicit prejudice but has rather reflected culturally sanctioned rational responses to struggles over scarce resources (Wellman, 1977: 4). Evidently racial discrimination is not the only culturally sanctioned response in that judgments on class and gender also have ideological structural underpinning. Consequently an explanation for racial disadvantage will not be found in a search for the overt discriminator who wears his or her prejudice on a garment displayed for public view.

It has been noted that while the manifestations of racial disadvantage are well documented the causal factors are much less readily located. It is suggested that this is one factor, and no doubt a significant factor, compounding the difficulties of implementing and policing legislation which renders both direct and indirect discrimination unlawful.

HOUSING POLICY AND THE LAW

Although Government by 1975 (Racial Discrimination, Home Office, 1975: 5) had decided that legislation 'is the essential precondition for an effective policy to combat the problems experienced by the coloured minority groups and to promote equality of opportunity and treatment', it was also recognised that legislation was not and never could be a sufficient condition for effective progress towards equality of opportunity. A wide range of administrative and voluntary measures are needed to give practical effect to the objectives of the law. While it is not the purpose of this book to focus on general housing policy and its effect on ethnic minorities resident in Britain in relation to their housing experience, the law cannot be viewed in isolation and will reflect Central Government policy in respect of its design and in respect of its implementation. Moreover in an area such as public authority housing, Central Government policy while not immediately responsible for council housing has, through fiscal policy, design standards, loan criteria and the promotion of clearance and rehabilitation both at the individual and area level, established the ground rules and determined the pace in which housing change may take place at the local level. In the 1975 White Paper the Government stated that it fully recognised 'that the policies and attitudes of central and local government are of critical importance in themselves and in their potential influence on the country as a whole' (Home Office, 1975).

As noted earlier, the first major policy issue on race which attracted Government attention related to immigration and, given the continuing relevance of immigration, most recently in connection with the requirement of Commonwealth citizens from black countries to have visas, Government's conflict with the courts over the refugee status of Tamils from Sri Lanka and the controversy over the 1988 Immigration Act, the issue of immigration controls and Government policy which dictates them may be seen as an important barometer of Central Government opinion. Further the ideologies which have underpinned immigration control have had a direct impact on ethnic minority communities settled in this country, particularly those from the Asian sub-continent and East Africa whose opportunities for marriage and for family reunion have been severely restricted by such policy: Governments of both Labour and Tory persuasion have consistently argued that the issue of immigration control and race relations must be taken in tandem. At first sight the impact of immigration controls on housing and discrimination legislation may appear marginal but this view merely demonstrates the gap between the white public perception of relevance and black experience. The importance of this divergence is its impact on Government policy, which, it is contended, has reflected white prejudice in an ideological commitment which denies the reality of black experience.

In so far as that ideological commitment is demonstrated to have contemporary relevance two factors emerge. First, studies of the Immigration Acts of 1962, 1968 and 1971 have established that the racial bias in controls from the 1960s up to the 1988 Act may be seen as a continuum. Second, in describing the discriminatory impact of controls on a broad section of the Asian community, in particular, it will be clear that the credibility of Central Government policy in anti-discrimination measures in all areas, including housing, is severely threatened. It is difficult to underestimate the importance of this facet in shaping the black communities' response to Central Government initiatives. In relation to housing, therefore, the black communities' response to anti-discrimination legislation in relation to allocation systems, for example, will be shaped not only by their experience in the housing market but also by their experience of other Government policies where stated goals and intentions have been denied by the reality of personal experience. It is also worthy of note that in describing the generality of impact of Government policy whether in the area of immigration controls or in that of housing, there is a danger of obscuring the personal devastation of particular case histories which, when reflected on a broad level, will cumulatively create a scepticism if not antagonism towards new initiatives and the explicit objectives behind them.

But a mere acknowledgment by Government that the law in itself will be inadequate to effect change but must be supported by complementary policy initiatives cannot be judged on its face value. Clearly it is necessary, particularly in the context of housing, to evaluate whether policy statements are more than rhetoric: to the extent that policies, apparently designed to support the legislative provisions, are inconsistent, ill resourced, lack evaluation and fail to set specific or attainable objectives, they will rebound not only on the Government responsible for their conception but also on the local authorities and other agencies through which they are expected to be implemented. In this respect the issue of motive, goodwill or intention, while no doubt informing policy adoption and review, is secondary to performance. Conversely, however, given the nature of institutional discrimination in respect of its diversity of expression generally and, more particularly in the context of housing, its different forms in different localities, together with the lack of direct Central Government control over local discretion, it may also be argued that however well structured and resourced policies may be, their effective implementation will depend on a number of significant externalities: these include structural economic change and employment opportunities and public attitudes towards ethnic diversity, class and community identity. This proposition does not suggest that such externalities are not influenced by and on occasion constitute a reflection of Government policy but equally Government cannot exercise absolute control over social and economic change.

It must also be recognised that the various studies, to which reference has been made in this chapter, have focused on racial disadvantage and discrimination in the distribution process, in other words, who gets what and why from the existing supply of resources, including public and private-sector housing. However, if institutional bias is identified as the key determinant of outcomes in distribution then it is to be expected that it will be equally significant in the processes of production. In housing do residential planning, housebuilding and rehabilitation reflect the needs of ethnic minority groups to the same extent as those of the rest of the community and, if not, is the legislative provision capable of tackling this issue?

Clearly the processes of housing production, so far as they disadvantage ethnic minorities, have an indirect impact and are likely to stem from practices reflecting institutional or indirect discrimination. Assessment of such processes depends to a greater extent on concepts of normative justice - or equality measured by results - rather than direct discrimination which may be redressed by formal justice - equality measured by the absence of explicit discrimination on racial grounds. However, before attempting to analyse the effect of the Race Relations Act 1976 on housing - which, for the first time encompassed both concepts - it is necessary to map the progression of the anti-discrimination legislation and to state what it now is, or appears to be, as a necessary context for the discussion of the more specific areas in housing to which it has been applied.

CHAPTER FOUR

ACTS I AND II:
THE RACE RELATIONS ACTS 1965 AND 1968

THE RACE RELATIONS ACT 1965: AN OUTLINE

The Labour Party manifesto for the 1964 General Election contained a commitment to legislate against racial discrimination and incitement in public places and, at the beginning of 1964, the Shadow Cabinet and the Society of Labour Lawyers were both asked to draft proposals which would give effect to this commitment. The Shadow Cabinet's working committee, under Sir Frank Soskice, confined the scope of legislation against racial discrimination to public places, thereby ignoring some of the provisions contained in Lord Brockway's Bills which would have banned discriminatory leases or provided for the withdrawal of licences from discriminators or the bringing of civil proceedings against them. It was the view of the working committee that racial discrimination should be a criminal offence and prosecutions would be brought only with the consent of the Attorney General or, in Scotland, the Lord Advocate.

The Committee on Racial Discrimination established by the Society of Labour Lawyers under the chairmanship of Andrew Martin QC, while recognising the importance of racial discrimination in employment and housing similarly confined the focus to discrimination in public places.

Meanwhile a group of Labour Party supporters, who had been impressed by North American evidence that anti-discrimination laws were more likely to be effectively enforced by administrative machinery than by proceedings in the criminal courts, presented proposals to the Prime Minister in late November 1964. They suggested a citizens' council should be created with powers to investigate discrimination on the grounds of race, colour, religion, sex and national origin in the fields of education, employment, housing, insurance, credit facilities and the administration of justice. It would publish findings and make recommendations and attempt to end discrimination by persuasion and conciliation. About this period a multi-racial Campaign against Racial Discrimination (CARD) was established after meetings between representatives of the main immigrant groups. Its legal committee developed these proposals and presented them to a meeting of the organisation in February 1965, where they were approved. The resolution approving the recommendations agreed that the scope of the legislation should include employment, housing and the provision of commercial services and that a statutory body should be established with power to investigate, conciliate, hold hearings and make legally enforceable orders against those found to have breached the law: the object of the legislation, it was agreed, should be to alter conduct and provide individual remedies rather than to punish.

The first reading of the Race Relations Bill was on 9 April 1965: under the Bill racial discrimination was to be made a criminal offence, punishable by a maximum fine of £100, if practised in hotels, public houses, restaurants, theatres, cinemas, public transport or any place of public resort maintained by a public authority. Discrimination in employment was ignored and the only reference to housing was a clause dealing with discriminatory restrictions on the disposal of tenancies. There was no reference to a conciliation agency and no civil remedy was available for those harmed by racial discrimination. The publication of the Bill met critical acclaim by a number of newspapers, by CARD and by the Society of Labour Lawyers, who issued a further report recommending the extension of the Bill to discrimination in housing, employment and certain credit and other service industries and the establishment of a conciliation commission. The Shadow Home Secretary, Peter Thornycroft, who had had meetings with the representatives of CARD at which the principle of enforcement by conciliation had been urged, moved the Opposition amendment to the second reading. The Opposition opposed the Bill because it would 'introduce criminal sanctions into a field more appropriate to conciliation and the encouragement of fair employment practices while also importing a new principle into the law affecting freedom of speech' (Lester and Bindman, 1972).

During the three weeks before the second reading and the committee stage, the Home Secretary drafted amendments to his own Bill: criminal sanctions were to be dropped in cases of discrimination and retained only for racial incitement. A Race Relations Board with local conciliation committees would investigate complaints of discrimination, settle disputes and generally secure compliance with the law. Only where this conciliation process had failed and the Board considered it likely that discrimination would continue would there be reference to the Attorney General or Lord Advocate, who might then seek an injunction or interdict to prevent further acts of discrimination being committed. This change of heart largely appeased the Opposition, who had seen the conciliation process, in tactical terms, as a softening of the Bill rather than as a strengthening of the process whereby discrimination would be diminished in public places. Furthermore Soskice refused to widen the Bill's coverage in any way and was, therefore, assured of carrying his Bill with Conservative support before the Standing Committee. Indeed the damage that was done was to Labour Party unity: seven of the thirteen Labour MPs on the Standing Committee openly expressed their view that the Bill was inadequate.

The Race Relations Act 1965 was brief: it contained eight sections and one schedule. In addition to providing for racial discrimination in a place of public resort being unlawful it established the Race Relations Board and Conciliation Committees (Section 2), set out proceedings in England and Wales (Section 3) and in Scotland (Section 4), made provisions relating to incitement and public order (Sections 6 and 7) and contained provisions dealing with racial restrictions on the transfer of tenancies (Section 5). Section 5 comprised the following:

> 5(1) In any case where the licence or consent of the landlord or of any other person is required for the disposal to any person of premises comprised in their tenancy, that licence or consent shall be treated as unreasonably withheld if and so far as it is withheld on the ground of colour, race or ethnic or national origins: provided that this subsection does not apply to a tenancy of premises forming part of a dwelling house of which the remainder or part of the remainder is occupied by the person whose licence or consent is required as his own residence if the landlord is entitled in common with that person to the use of any accommodation other than accommodation required for the

purposes of access to the premises.

5(2) Any covenant, agreement or stipulation which purports to prohibit the disposal of premises comprised in a tenancy to any persons by reference to colour, race or ethnic or national origins shall be construed as prohibiting such disposal except with the consent of the landlord, such consent not to be unreasonably withheld.

5(3) In this section 'tenancy' means a tenancy created by a lease or sublease, by an agreement for a lease or sublease or by a tenancy agreement or in pursuance of any enactment; and 'disposal', in relation to premises comprising a tenancy, includes assignment or assignation of the tenancy and subletting or parting with possession of the premises or any part of the premises.

5(4) This section applies to tenancies created before as well as after the passing of this Act.

It could be argued that Section 5 marked a significant improvement on the common-law provision in relation to restrictive covenants and conditions. In reality it ignored all the major problems faced by Britain's black population in relation to discrimination in both public and private housing.

THE RACE RELATIONS ACT 1965:
PROPOSALS FOR CHANGE

The Labour Party's opposition to the 1962 Commonwealth Immigrants Act related the appeasement of racial prejudice through racially discriminatory controls to an adverse impact on race relations rather than the reverse and to the stimulation of demands for increased control. By August 1965, however, the Labour Party, now in government, had largely come to accept the Opposition view that the larger the number of black people who were admitted to Britain the more racial prejudice was likely to increase: this was evident in the White Paper on Immigration from the Commonwealth (Home Office, 1965) which dealt first with immigration policy and second with departmental proposals to tackle the problems of integration. The latter section appeared to relate difficulties experienced in housing, education and health to the presence of black immigrants resulting in strong opposition not only in respect of the immigration provisions but also in respect of the assumptions underlying the apparent need for control to protect service provision. Obviously the small size of the Government's parliamentary majority, the prospect of another General Election in the near future and the electoral unpopularity feared to have resulted from the passage of the Race Relations Act were factors in Government's shift of view but that these factors were able to prevail demonstrated, first, a lack of ideological commitment by Government and, second, its ability to marginalise the influence of those who saw this shift in terms of blatant political opportunism.

Despite this lack of Central Government conviction, the appointment of Roy Jenkins in December 1965 to replace Sir Frank Soskice as Home Secretary injected new hope into the prospects of extending the 1965 Act to both housing and employment. Mark Bonham Carter was appointed by Jenkins to be the first chairman of the Race Relations Board, still vacant at that time, and he made it a condition of appointment that he should be able to put the case, after a year's operation, for the extension of the scope of the Act and a revision of the Board's powers (Rose et al., 1969: 520).

In July 1966, at Roy Jenkins's personal instigation, two studies were commissioned jointly by the Race Relations Board and the National Committee for Commonwealth Immigrants (NCCI), the voluntary forebear of the Community Relations Commission. The first study was undertaken by Political and Economic Planning (PEP, later to become PSI), to evaluate the nature and extent of racial discrimination in Britain, and the second was an inquiry under Professor Harry Street to examine overseas experience regarding anti-discrimination legislation and to recommend amendments to the 1965 Act in the light of such analysis.

At a conference in February 1967, convened by the NCCI, on racial equality in employment, Roy Jenkins indicated that if the conference itself, the PEP study and the Race Relations Board's report 'show a clear case that legislation is needed and would be helpful - we shall not shirk the issue'. Both reports were issued two months later. The former revealed that the extent of racial discrimination in Britain ranged from substantial to massive and that if left to itself the problem was likely to grow worse, while the latter concluded that:

> the law has an essential part to play in dealing with this difficult and explosive area of human relations ... So long as the law is unequivocal and unambiguous it need not be in any way oppressive as we believe our experience shows. There is no reason to suppose that employers of labour or that those who sell houses are in any respect less law abiding than publicans; nor why they should be any less susceptible to the processes of conciliation backed by an ultimate legal sanction (Lester and Bindman, 1972: 128).

It was not until October 1967 that Professor Harry Street's committee published its report. This report emphasised that the scope of the law had to be wide enough to cover the most damaging forms of racial discrimination: in employment - recruitment, training, promotion, redundancies, conditions of work and membership of trade unions; in housing - the sale or letting of both private and local authority property; and in a great variety of commercial and social facilities and services. The report endorsed the philosophy underlying the anti-discrimination provisions in the United States and Canada, where the primary purpose of the law was to create the climate of opinion which would obviate discrimination rather than seek out and punish discrimination which had occurred. As a consequence they concluded against the use of criminal enforcement (Street, et al., 1967).

In July 1967 Roy Jenkins announced in the House of Commons that the Government had decided that the 1965 Act should be extended to deal with racial discrimination in employment, housing, insurance and credit facilities and that public places would be given a wider definition than under the then present legislation (Hansard (HC) Vol. 751, Col. 744). This statement coincided with the publication of a report by a Labour Party working party on race relations which recommended that where the Race Relations Board had failed to conciliate, cases should be referred to an independent race relations tribunal whose members would have specialist knowledge of industrial relations, housing, Local Government and race relations. This report was unanimously adopted at the annual conference of the Labour Party in October 1967.

Cumulatively these four reports had, by October 1967, cleared the path for the introduction of a comprehensive Bill.

For housing the most significant development was Government's acceptance that the Bill should cover discrimination in the controversial area of the sale and letting of private housing together with the allocation of local authority housing stock. However, the Government refused to give the Board the wide range of enforcement powers recommended by the Street report, e.g. the power to subpoena; the power to obtain a court order to suspend the disposal of property pending the investigation of a complaint; the power to obtain a court order requiring the discriminator to rectify the act of unlawful discrimination by, for example, offering a house or job which had been refused on the grounds of race or colour. Furthermore the alleged victim was not given rights to proceed on his own behalf, the Board having the sole right to bring legal proceedings.

Despite a somewhat turbulent passage on 25 October 1968 the Bill received the royal assent and a month later its provisions entered into force. As Bindman and Lester have noted (1972: 148) the Bill's enactment reflected the success of a campaign which had had to overcome significant disadvantages. Britain's racial minorities did not have significant voting power (at that time there was not one black MP in the House of Commons); the black community consisted mainly of immigrants whose claim to equal treatment was inevitably made with less insistence and understood with less sympathy than had they been indigenous inhabitants: there was little public awareness of the nature or extent of racial discrimination; the ideal of equality whether embodied in a written constitution or elsewhere was not a foundation stone in the fabric of British justice; the black lobby had suffered from internal fragmentation; the Government did not have an unequivocal ideological commitment towards equal opportunity (as demonstrated by the passage of the Commonwealth Immigrants Act 1968); powerful lobbying groups such as the trade unions and CBI were either defensive or resistant to the proposals in the Bill;; and it could be anticipated that the political losses far outweighed the gains in its passage. What had been critical to the Bill's success, however, was not only the skilful lobbying by interested pressure groups but also Roy Jenkins's personal commitment to the achievement of racial equality and his willingness to lead public opinion in that direction.

THE RACE RELATIONS ACT 1968: AN OUTLINE

The Race Relations Act 1968 largely repealed the Act of 1965 with the exception of the provisions relating to restrictions on the disposal of tenancies (previously referred to) and public order (Section 6).

Definition

Unlawful discrimination was defined in terms of the attitude and intention of the discriminator and not in terms of the group identity of his victim. The Act applied to discrimination on the grounds of colour, race or ethnic or national origin. None of these terms was defined. This is less critical than may appear, as what matters is whether people have been treated less favourably on such grounds, rather than whether they really belong to groups of such a description (but see Lester and Bindman, 1972: Ch. 4). The term 'colour' is relatively straightforward in that it applies to discrimination against someone solely because of the colour of his skin, whether black, brown, yellow or white: although the term 'race' may have no objective scientific meaning it would encompass a situation where someone is discriminated against because he is a black African. The term 'ethnic group' has been defined as one of a number of breeding populations, which populations together comprise this species of homo sapiens, and which individually maintain their differences, physical or genetic and cultural, by means of isolating mechanisms such as geographical and social barriers (Mongagu, 1964: 25). However, it is social organisation rather than biological description which is to be emphasised in relation to the use of the term 'ethnic'. With regard to 'national origins' it was envisaged that it would include discrimination not only against someone because he or his forebears were born in a particular country but also because he was of a particular nationality or citizenship. However, in Ealing London Borough Council v Race Relations Board ex p. Zesko (1971) 1 QB 309, where a local housing authority adopted a rule that a condition of

78

acceptance on its waiting list for council house accommodation was that 'the applicant must be a British subject within the meaning of the British Nationality Act 1948' and refused to place a Polish national on the list, the High Court determined that it was not unlawful to discriminate on the grounds of present nationality rather than national origin.

In the decision of the United States Supreme Court in Plessy v Ferguson 136 US 537 (1896) it was held that 'separate but equal' facilities satisfied the constitutional requirement of 'equal protection of the laws' and this remained intact until the decision in Brown v Board of Education 348 US 483 (1954), where the Supreme Court decided that separate educational facilities were inherently unequal even if the facilities were equal in all material respects. Section 1(2) provided that 'segregating a person from others' on racial grounds is 'treating him less favourably than they are treated'.

Scope

Following the general definition of discrimination in Section 1 the Act was followed by specific definitions in four of the situations to which it applied (Sections 2 to 5). Thus under Section 1 a person discriminates against another if, 'on the ground of colour, race or ethnic or national origins, he treats that other in any situation to which Section 2, 3, 4 or 5 ... applies, less favourably than he treats or would treat other persons'. These sections dealt respectively with unlawful discrimination in (1) the provision of goods, facilities and services; (2) employment; (3) trade unions and employers' and trade organisations; (4) housing accommodation and business and other premises.

Housing: general

The last section, dealing with housing (Section 5), was as follows:

(5) It shall be unlawful for any person having power to dispose, or being otherwise concerned with the disposal, of housing accommodation, business premises or other land to discriminate -

(a) against any person seeking to acquire any such accommodation, premises or other land by refusing or deliberately omitting to dispose of it to him, or to dispose of it to him on the like terms and in the like circumstances as in the case of other persons;

(b) against any person occupying any such accommodation, premises or other land, by deliberately treating him differently from other such occupiers in the like circumstances; or

(c) against any person in need of any such accommodation, premises or other land by deliberately treating that person differently in respect of any list of persons in need of it.

At the time of the passage of the Act it was clear that the housing market then, as now, was characterised by serious and persistent shortages of residential accommodation. Such shortages may create powerful financial incentives for the property owner to discriminate on racial grounds. First, estate agents or large property owners may seek to exclude black people from obtaining accommodation in a particular area from the fear that their presence would have a detrimental effect on house values and potentially cause white families to panic-sell, leading to a fall in property values. In such an eventuality a developer or property company may purchase houses cheaply with the expectation of reaping significant profits on later sales to black families. In the United States this practice was known as 'blockbusting' and the Street report (Street et al., 1967: para. 134.9) recommended that it be penalised in the Race Relations Act. At the time of the Street report, however, there was very little evidence that this specific exploitation of 'blockbusting' was taking place in the same fashion or to the same extent as in the United States. Thus the discrimination which was taking place in the housing market was characterised by social rather than economic behaviour on a large scale. Although it did not impair the general utility of the Act, certain provisions, in reflecting the experience elsewhere and assuming or anticipating problems in housing on a like scale in this country, failed to reflect the reality of the experience of discrimination by blacks in the United Kingdom.

As both Cullingworth (1969) and Burney (1967) noted, the opportunities for exploiting housing shortage in the private sector were exacerbated by difficulties experienced by black applicants to local authority housing from the application of residential rules.

Rex and Moore (1967): 274) identified six types of housing situation:

1. Outright ownership.
2. Ownership with mortgage.
3. Council tenant.
4. Tenant of a private house.
5. Outright ownership but where the owner has purchased with a short-term loan and is compelled to let rooms in order to meet obligations regarding repayment.
6. Tenancy of rooms in a lodging house.

Rex and Moore in their study of Birmingham concluded that black immigrants were most likely to find themselves in the latter two categories in the neglected 'twilight areas' or 'zones of transition', where the amount of available private rented accommodation was shrinking but which were not at that time being considered for redevelopment by the local authority. Although it was clear that both the cycle of transmitted deprivation and the influences of discrimination in related areas such as employment and education would influence housing opportunities it was considered that a housing class or sub-class of black immigrants had been created and that the model constructed, although no doubt a simplistic one, as Rex (1973) conceded, was a useful analytical tool in determining the social opportunities for black families in improving their housing conditions. For the 1968 Act to be effective with regard to housing, it was therefore necessary not only to reduce discrimination in the situations in which racial minorities then were typically housed but also to promote equal opportunities for them to move into better residential accommodation and for these to be accompanied by complementary programmes and measures designed to diminish urban poverty (see Rose et al., 1969). As a starting point, however, it was important that anti-discrimination measures were sufficiently broad to ensure that no important sector of the housing market was left untouched.

The 1968 Act, in its application to 'any person having power to dispose, or being otherwise concerned with the disposal, of housing accommodation ...', covered not only a person with a freehold or leasehold interest in property but also a tenant who sub-lets, a housing manager controlling allocation or an estate agent involved in the disposal of premises. Moreover the Act was not confined to the principal participants but by the inclusion of 'those

81

otherwise concerned' would cover agents both formal and informal who had an influence on the disposal of premises. An employer would be liable for the acts of his employees as a principal would be liable for the acts of his agent. Both local authorities and estate agents would be liable as persons concerned with the disposal of housing accommodation and also, by reason of Section 2 dealing with the provision of facilities or services to the public or a section of the public, as proprietors of boarding houses or hotels. Aiding, inducing or inciting another to discriminate was made unlawful by Section 12 and thus any attempt to coerce a property owner or estate agent to discriminate would itself be unlawful. The display or publication of advertisements indicating an intention to discriminate (whether or not such discrimination was unlawful) was also made unlawful by the Act (Section 6(1)). While there had to be a deliberate intention to discriminate on any of the unlawful grounds the intention would still be unlawful if directed not against the would-be purchaser or tenant but his or her association with another on the basis of that other person's colour or race, etc.

The first case heard under the 1968 Act, Race Relations Board v Geo. H. Haigh & Co. (1969) 119 New L.J. 858 illustrates the fact that a person would not be liable for refusing to dispose of accommodation unless it were available, but immediate availability is not a requirement. The defendants in developing a housing estate in Huddersfield entered into agreements with prospective customers to build a complete house and lease it for a 999 year period. They had refused to enter into such an agreement with the complainant on the ground of his colour but had contended that as no house had been completed which was then available for sale they were not persons 'having power to dispose of housing accommodation'. The court held that there was nothing to prevent an owner from letting or selling a house which would not be available for some time and therefore rejected this argument along with the contention that an incomplete house was not housing accommodation.

In addition to being unlawful to refuse outright to dispose of premises, whether by sale, lease or licence, it was also unlawful to refuse to allow someone to acquire it on equal terms or in like circumstances. The acceptance of discriminatory instructions from a property owner would render an estate agent or accommodation bureau liable both as agent for the owner and also as principal in discriminating in the running of a business.

Although the Act was directed at any person having power to dispose, or otherwise concerned with the disposal, of housing accommodation, etc., in the reference in Section 5(b) to discrimination by such a person against anyone occupying such accommodation by deliberately treating him differently from other occupiers in like circumstances discrimination in management was clearly covered. Thus although a landlord may agree to let property to someone on a non-discriminatory basis the landlord would still be liable if thereafter he were to treat that person differently from other tenants on racial grounds.

Similarly, by Section 5(c), affording different treatment to a person in respect of any list of persons in need of accommodation was designed to prevent local authorities from discriminating in their treatment of people on a waiting list for council housing, although it is clear that such discrimination would also infringe Section 2, which applies to the services of local authorities. The provisions of Section 5 were not, however, confined to local authorities, so, for example, an estate agent or university lodgings bureau would be obliged to maintain lists of those seeking accommodation, or in the case of a building developer those seeking to purchase accommodation, on a non-discriminatory basis. Thus in Race Relations Board v London Accommodation Bureau (1971) the bureau admitted that it had acted unlawfully in refusing to supply a black Ugandan student with a list of accommodation which it had made available generally to white applicants.

The provisions relating to restrictive covenants found in the 1965 Act were not repealed by the 1968 Act. It should be noted, however, that such restrictive covenants in so far as they discriminated on racial grounds were only rendered unlawful in relation to leases and tenancy agreements and did not apply when contained in agreements for the sale of freehold properties. The validity of such restrictions at common law, as noted above, was far from clear and the omission of such a provision is difficult to understand or justify. It should also be noted that where racial restrictions on leases and tenancy agreements were rendered unlawful by the 1965 Act the law was confined to rendering their implementation unenforceable rather than declaring them to be void as proposed by the Street Committee (Street et al., 1967: 89). There was and is no sanction therefore against the

83

landlord who continues to impose racial restrictions in his leases in the hope that his tenant will believe them to be legally binding. Nonetheless it could well be argued that a landlord, in this situation, would be seeking to induce his tenant to discriminate unlawfully and would therefore incur liability under Section 12 of the 1968 Act (now Section 31 of the 1976 Act). The Race Relations Board was given power under Section 23 of the 1968 Act to apply to the courts to have the discriminatory clause expunged by appropriate revision of the contract but only after an investigation into an alleged act of discrimination. Confusion about the validity of such restrictions was increased by the different tests used in the two Acts to determine whether a landlord was exempted from their scope. The 1965 Act exempted a landlord living on the premises in which there was not normally accommodation for more than two households other than his own, and sharing accommodation other than means of access or storage accommodation with his tenants.

Housing: local authority

Local authority housing departments throughout England and Wales and Scotland have an extraordinarily broad impact on the lives of a significant proportion of the population. As Cullingworth has observed (1969: para. 54) about a quarter of the 1.5 million households who move each year are housed by local authorities. Their powers, however, are not restricted to the public sector in the sense that their broad statutory responsibilities in relation to demolition, acquisition, rehabilitation, urban renewal and planning development extend beyond the individual property to a broad area basis. As the Street report (1967: 131-2) observed the result of this complex set of provisions is that, in practice, over large areas of their housing activities, local authorities have an uncontrolled discretion regarding their obligation to rehouse and their powers of providing housing. The relevant minister does not record information about how these powers are exercised by various local authorities and, if an impact on this discretion was to be made so that the elimination of unlawful discrimination in housing became a practical prospect, it would be necessary for Central Government to develop a more comprehensive supervisory role to ensure that such discretion was not abused. The Street report proposed that the Minister should be empowered to review and amend the allocation rules of local authorities so that the needs of applicants could be fairly weighed. Where the Board was of the view that a particular code operated unfairly it could make representations to the minister to seek amendment. A similar recommendation was made with regard to access channels for both housing and rehousing which would prevent, for example, the disproportionate allocation of poor property to those displaced by development or demolition schemes. Moreover, in recognising the past imbalances in local authority housing allocation due, in part, to the imposition of residential qualifications, the report observed that the authority concerned ought not to be prevented from issuing a fair policy for the purpose of reducing or eliminating imbalances between different areas by positive discrimination in favour of members of a particular minority group. It was supposed that a scheme, for such a purpose, would be submitted to the Race Relations Board and, if approved, the execution of the plan would not then constitute unlawful discrimination. The report also recognised the difficulty which stemmed from certain local authorities omitting from clearance schemes those areas with a concentrated black population in order to avoid rehousing them but it made no specific recommendations to prevent the practice; the remedy would have to lie in political accountability. The general approach of the Street Committee's report was to make local authority housing more uniform and open to public scrutiny while at the same time giving administrative powers to Central Government through the appropriate minister, and the Race Relations Board to monitor practice and direct remedies where unfairness resulted. Although the authors of the report recognised that their recommendations would entail radical innovations in Local Government practice they doubted whether effective control of unlawful discrimination in the critical area of council housing would be achieved merely by the prospect of individual legal remedy through the Race Relations Board. Such a view was echoed in the Cullingworth report, issued a year after the passage of the 1968 Act, which commented on the discriminatory effect of some local authority policies and practices in housing. Black people, being largely newcomers, are affected by residential qualification. Again because newcomers tended to take furnished accommodation they were particularly affected by an interpretation of rehousing obligations which excluded those in furnished tenancies. These, and similar generally applicable rules 'have the affect (though not the purpose) of discriminating against coloured immigrants'. Black immigrants, like any other group from cultural backgrounds which are 'strange' to housing visitors, may also tend to be unfavourably treated. The report observed that black people often found the council house system incomprehensible and shrouded in mystery, underlining the need for better publicity, better public relations and for a housing advice service. In order to improve the service the report recommended that local authorities should maintain records of black residents. In addition to enabling the housing authority to monitor the extent to which black people had been allocated disproportionate quantities of poorer quality housing such records would have been instrumental in enabling the Race Relations Board in terms of the 1968 Act to secure local authority compliance with the requirement not to discriminate in the housing service. Both the Street and Cullingworth reports recognised that record keeping might have the effect of discouraging not only racial discrimination but other kinds of unfair treatment in relation to those in need of housing. As noted above, however, the 1968 Act, while rendering direct discrimination in housing allocations unlawful, did nothing to address practices, policies or procedures which had an adverse but indirect impact on black housing provision in both the public and private sector.

Mortgages

The PEP report (1967: 78-81) demonstrated that discrimination had been widely practised in relation to the provision of mortgage finance for house purchase. Clearly local authorities, building societies, banks and insurance companies provide the vast bulk of mortgage finance for house purchase and racial discrimination by such bodies on a large scale would have a dramatic affect on the opportunities of minorities to become owner-occupiers and to purchase housing of their choice in their area of choice. Direct discrimination by such bodies would be covered by Section 2 of the Act, which was concerned with the provision to the public or a section of the public of facilities or services. However, there was no uniform administrative structure with regard to local authorities in their provision of mortgage finance (or indeed to building societies and insurance companies and banks). Consequently the policing task of the Race Relations Board was made particularly difficult in this area. Moreover, the practice of redlining and applying particular conditions and requirements which, although not intentionally discriminatory, would have an adverse impact on racial minorities was not rendered unlawful in terms of the 1968 Act.

Exceptions

In the same way that the provisions of the 1968 Act applied equally to local authority and private housing so the exceptions avoided such distinction. Section 7 of the Act exempted discrimination against any person with respect to the provision or disposal of any residential accommodation in any premises if at the time of the disposal:

1. The premises are 'small premises'.
2. The person having power to provide or dispose of the accommodation (referred to as the 'landlord') resides or intends to reside on the premises.
3. There is on the premises, in addition to the accommodation occupied by the landlord, 'the relevant accommodation' shared by him with other people residing on the premises who are not members of his household (Section 7(1)).

The exception, then, applied to residential accommodation, whether a private house, hotel or boarding house, but would exclude commercial tenancies. These provisions are, generally, reflected in Section 22 of the 1976 Act.

A further exemption from the Act applied to a person who owned and wholly occupied property, whether the interest was one of freehold or leasehold: he could discriminate in its disposal unless the services of an estate agent were used or there was a discriminatory display or publication of an advertisement. 'Estate agent', by reason of Section 7(8) was given a broad definition to include a person who, 'in connection with the disposal' in the course of his trade, business or profession, brings together or takes steps to bring together the owner and the prospective purchaser, or who acts as an auctioneer. A similarly broad definition is now found in Section 78(1) of the 1976 Act. Consequently if the owner-occupier instructs his solicitor to act on his behalf in finding him a purchaser, a common practice in Scotland, where solicitors act as estate agents, he or she may not discriminate in effecting the sale: the nature of the exemption is of very limited application since the vast majority of sales by owner-occupiers are inevitably through professional third parties or result from advertising or by means of a 'For Sale' notice. This exemption provision, known as the 'Colorado compromise' as it had been modelled on the Colorado Fair Housing Statute, is recognition of the practical impossibility of enforcing the law in respect of private transactions conducted on a personal basis to the exclusion of third parties and any form of advertising.

A further provision, also borrowed from experience in the United States, relates to exemptions with regard to charitable housing trusts and associations. First, nothing in the 1968 Act was to be construed as affecting a provision in any charitable instrument which came into existence after 25 November 1968 (the date upon which the 1968 Act came into force) and which conferred benefits on persons of a particular race, particular descent or particular ethnic or national origin (Section 9 (1)(a)). In respect of any provision in a charitable instrument which came into effect before that date the Act preserved the legality of any action, whether otherwise discriminatory, taken in accordance with such provisions. As a consequence any new charitable trust might be created to confer a benefit on a particular racial or ethnic group while a provision, existing before the Act came into effect, might discriminate against any group on racial grounds. These provisions have been recast in Section 34 of the 1976 Act. Now any provision which confers benefits by reference to colour alone is void.

Investigations

The Street Committee report (Street et al., 1967: para. 144.4) attached importance to the Board's positive or affirmative role in administering the Act. Thus as well as a reactive responsibility in relation to the investigation of complaints referred to it the Board should be empowered to conduct investigations for the purpose of uncovering unlawful discriminatory practices and taking action to eliminate them. By Section 17 of the Act the Board was given power to conduct investigations even though no formal complaint had been received in a situation either where someone had alleged discrimination or 'for any other cause'. However, before conducting an investigation it was necessary for the Board to have 'reason to suspect that an unlawful act has been done'. Therefore, although the Board had been given a wide discretion to investigate suspected unlawful conduct, it could only exercise this power, in most cases, if the act of discrimination involved an identifiable victim. This power to conduct investigations itself was circumscribed not only by the requirement to have reason to believe that a specific act of discrimination had occurred but also by the Board's lack of power to compel the disclosure of relevant evidence from those suspected of unlawful conduct. In housing matters the Board could investigate a complaint or delegate such an investigation to the relevant regional conciliation committee.

The Race Relations Board and conciliation committees

The Race Relations Board and local conciliation committees were established by the Race Relations Act 1965. The 1968 Act enlarged the functions of the conciliation process from being concerned only with the removing of public wrong to the further task of securing redress for the individual victim. The 1968 Act (Section 14) provided for a Board comprising a chairman and not more than eleven other members appointed by the Secretary of State. The Board had the power to appoint such conciliation committees as it considered necessary for assisting it in the discharge of its functions. The Board, under Section 15, had the duty, along with a conciliation committee, to receive any complaint made to it within two months of the act complained of. This period could be extended at the Board's discretion where 'special circumstances' warranted. The Board itself had power to conduct an investigation into any complaint received by it or by a conciliation committee and the Board could direct a conciliation committee to refer a complaint to itself for investigation. In conducting an investigation the Board or a conciliation committee had to make such inquiries as they thought necessary with regard to the facts alleged in the complaint and thereafter form an opinion as to whether or not an act of discrimination had taken place. On forming such an opinion it was necessary for the Board or conciliation committee to 'use their best endeavours' first to secure a settlement of any difference between the complainant and the respondent, and second to obtain a satisfactory written assurance against any repetition of the act considered to be unlawful or the doing of further acts of a similar kind by the party against whom the complaint was made.

In the event of either the Board or a conciliation committee failing to achieve a satisfactory settlement and undertaking it was then in the Board's discretion alone as to whether or not to pursue legal proceedings in the County Court or in the Sheriff Court in Scotland (Section 19).

The Board's powers and functions had, therefore, expanded considerably from the provisions of the 1965 Act: it had originally consisted of a chairman and two members and its functions were limited to appointing local conciliation committees and considering complaints of discrimination which the local committees had been unable to resolve. Under the 1968 Act it retained the duty to appoint conciliation committees to help to discharge its functions locally but it no longer delegated the investigation and conciliation of all complaints to its local committees. Should either a local conciliation committee or the Board fail to resolve a complaint following an investigation it would then be for the Board rather than the Attorney General to decide whether to bring proceedings in the courts and for it alone to conduct such proceedings. The role of the Attorney General and, in Scotland, the Lord Advocate was confined to the prosecution of cases of incitement to racial hatred, which remained under Section 6 of the 1965 Act.

By 1971 there were sixty-eight members of the Board's staff operating either from the Board's headquarters in London or from the location of one of the nine conciliation committees. The nine conciliation committees covered the Metropolitan area (three), Scotland, Wales, the North, the West Midlands, Yorkshire, and the East Midlands. Each part of Great Britain came within the jurisdiction of one of these committees whose local office was manned by at least one conciliation officer employed by the Board with secretarial support staff. Under the 1965 Act the conciliation committees were involved in the process of investigation but under the 1968 Act, with the increase of professional staff, the committees were generally concerned with considering written statements and documentary evidence rather than with interviewing witnesses. The work of the conciliation committees, therefore, focused on hearing all representations concerning a complaint and in forming an opinion as to whether or not unlawful discrimination had taken place.

The Community Relations Commission

The 1968 Act established the Community Relations Commission for the purpose of (a) encouraging 'harmonious community relations' and (b) coordinating on a national basis the measures adopted for that purpose by others. It was also charged with the responsibility of advising the Home Secretary on any matter referred to the Commission by him and to make recommendations to him on any matter which the Commission considered should be brought to his attention (Section 25). 'Community relations' was to mean 'relations within the community between people of different colour, race, ethnic or national origins'. As with the Race Relations Board, the Commission had a network of local committees - Community Relations Councils - whose main function, although not a statutory one, was to work in co-operation with the local authorities on various aspects of community relations. The Community Relations Commission and its local councils were given no statutory power to investigate complaints in terms of the 1968 Act. Their duties were confined to promotional activities. Consequently there was a conscious division between the responsibilities of the enforcement agency, the Race Relations Board, and the conciliation committees, and the promotional activities assigned to the Community Relations Commission and its local councils.

THE RACE RELATIONS ACT 1968: AN ASSESSMENT

Introduction

The obvious starting point for assessing the 1968 Act is to ask whether it was effective within its own limitations. In relation to housing, had unlawful racial discrimination diminished between 1968 and 1976, when a new Act was introduced? This question is not easy to answer. A significant diminution in the number of complaints over this period would suggest that the Act had stemmed unlawful discriminatory practice. But such a conclusion, in itself, is difficult to sustain. Complaints may increase, not because discrimination is increasing, but because of increased public awareness of a potentially effective remedy: conversely they may diminish, not because discrimination has diminished, but because the investigatory process and potential remedies are seen, generally, to be ineffective and not worth the effort. Even if we can provide a satisfactory answer to the first question - has unlawful racial discrimination diminished? - it may be difficult, if not impossible, to demonstrate a causal link between such a conclusion and the presence of legislation. Social, cultural, economic, political, administrative and other 'non-legal' factors may have played a more significant role in changing the incidence of racial discrimination and/or social behaviour than the threat of legal sanctions.

Beliefs and experiences of discrimination

McKay (1977: 86) referred to the PEP reports (PEP, 1967; Smith, 1976) on racial discrimination in concluding that, during the 1960s, the most common cases of discrimination involved landlords discriminating against blacks.

Table 4.1 shows a significant decline in both the belief in discrimination and the experience of discrimination by West Indians and Asians between 1967, over a year before the 1968 Act's inception, and 1974. The decline in the former for the West Indian sample was 14 per cent and the latter 18 per cent while the figures were 8 per cent and 10 per cent respectively for the Asian sample. If beliefs and experience of discrimination are directly linked to legislation then, with the passage of the 1976 Race Relations Act providing more comprehensive controls on discrimination, one would expect to see, at least, no worsening of beliefs and experience by 1982. Ironically, the belief that there are landlords who discriminate had decreased between 1974 and 1982 by 2 per cent in respect of both West Indians and Asians but the personal experience of discrimination increased very dramatically for both groups - 28 per cent and 27 per cent respectively. These figures suggest that to the extent that there is a direct correlation between the formal legislation and the incidence of discrimination there are more significant factors which determine personal experience. It is difficult to identify factors which determine personal experience. Further it is difficult to reconcile a decrease in the belief that private landlords discriminate with a substantial increase in the personal experience of discrimination in this sector. One explanation might be found in changes of landlord type or geographical location effected for substantial numbers of each group over the period resulting in a substantial proportion occupying accommodation where discrimination was not experienced and which was sufficiently isolated in social terms from other occupiers, whose tenants had experienced an increase in discrimination, to ensure that the beliefs of the former were not affected by the experience of the latter. It is also possible to conjecture that the diminution of overt discrimination effected by the 1968 Act and reinforced by the 1976 Act has had an impact on beliefs in the sense that discrimination, in being less obvious, is more directly correlated to personal experience than previously but neither hypothesis is readily tested.

If the public's perception of discrimination has been attuned to the legal definition by which is meant that 'discrimination' in 1974 was understood as meaning direct discrimination while in 1982 it was understood as encompassing, in addition, indirect discrimination, then the sharp decline in personal experience of discrimination between 1967 and 1974 and the sharp rise from 1974 to 1982 may suggest that the 1968 Act did, after all, have a direct impact on personal experience and the apparent failure of the 1976 Act reflected the fact that it promised much more than it was able to deliver. Whatever the interpretation put on these figures, in themselves they are highly ambiguous.

Table 4.1: Discrimination by private landlords: belief and experience (%)

	West Indian			Asian[1]		
	1967	1974	1982	1967	1974	1982
1. There are landlords who discriminate	88	74	72	49	41	39
2. Personal experience of discrimination by private landlords	39	21	49	17	7	34

[1] The Asian figures for 1967 and 1974 are 'averaged' for Indian and Pakistani responses.

Source: McKay (1977), Brown (1984)

Complaints to the Race Relations Board

Similarly, an examination of complaints received and investigated by the Race Relations Board between 1968 and 1978 is bound to beg more questions than it answers. Table 4.2 shows a remarkably consistent pattern of complaints (700 to 1,000 per annum) throughout this period, with little appreciable change. The only significant feature is the increased percentage of findings of discrimination, 12.6 per cent in 1969-70 and 23.7 per cent in 1976 - the first and last complete years of investigation when the Act was fully operable. The figures are dominated by employment cases (Section 3) and the number of housing cases, falling either within Section 2 or Section 5, are not separated from other complaints falling within these sections but they ran at between 120 and 140 per annum, 30 per cent of which resulted in opinions of unlawful discrimination (Gribbin, 1978).

The Board received 'several dozen' (McKay, 1977) complaints concerning local authority housing between 1970 and 1976, a period coinciding with increased West Indian and Asian access to council housing (Smith, D., 1976) and a significant decrease in the private rented sector, generally. The extent of contact between minorities and private landlords had decreased sharply since 1966, partly because of the spread of owner-occupation, partly because of the increases in the numbers of blacks renting from black landlords and partly because of the eligibility requirements of local authorities being met by blacks either through slum clearance or through the general waiting list. As McKay observed, it is obvious from the pattern of council housing allocation documented between 1969 and 1976 that very little had changed from the previous period. An increasing number of minority families were housed but black families continued to be housed predominantly in the inner city and frequently in the inferior estates. While the 1968 Act made no distinction between private and public-sector housing clearly a policy of dispersal of black applicants, effectively restricting black choice, was unlawful. But both Birmingham and Lewisham applied quotas to allocations, leading the Board, in its 1974 report, to issue a special memorandum on local authority housing stating that compulsory dispersal was illegal (RRB, 1975: 42-3). The fact that no subsequent Race Relations Board investigations (under Section 17) or CRE investigations (under Section 49) have focused on formal dispersal policies suggests, to put it no more firmly, that the Act had some affect in stemming such discriminatory practices. But the success of individual complaints was very limited: only a handful of authorities were found to be acting illegally, in two instances because they treated immigrants unfairly: in Ealing London Borough Council v RRB ex p. Zesko (1971) 1 QB 309, the refusal of the London borough of Ealing to register foreign nationals was ultimately upheld by the House of Lords ('national origins' not encompassing current nationality) and in the other case concerning Wolverhampton, the council dropped a rule requiring non-UK-born applicants to wait longer by substituting one requiring applicants to have lived in the UK for ten years. Although this replacement rule was dropped after further investigation (RRB, 1972) and before the introduction of the concept of indirect discrimination by the 1976 Act, both cases illustrate the facility with which the intentions of the legislation could be circumvented and the significant lack of 'willing' complainants from the large number of those who had suffered from unlawful discriminatory practices.

In June 1971 the Board gave oral evidence to the Select Committee on Race Relations and Immigration in connection with its inquiry into housing. The Board drew attention to the fact that no one had complained against Wolverhampton's waiting list rule although the investigation brought to light more than 200 victims. Moreover only one person complained about the London Accommodation Bureau's practice of withholding accommodation lists from black applicants when several hundred had been adversely affected.

In the Board's first report on the 1968 Act (RRB, 1969), it was noted that some 'immigrants' and 'immigrant organisations' lacked confidence in the Act, creating a vicious circle: lack of confidence causing a reluctance to complain, resulting in fewer complaints being sustained, leading in turn to a further decrease in confidence. The Board bemoaned the widespread ignorance of the Act's provisions and the less than enthusiastic support provided by Community Relations Councils (CRCs) in referring complaints to the Board. The fact that complaints, which were seen as the tip of the iceberg, and the CRC referrals had not increased substantially during the period of the Act suggests that, in so far as the 'vicious circle' explanation holds good, despite attempts at increased and improved

Table 4.2: Race Relations Board, completed cases 1968-78

(a) Year	(b) Discrimination	(c) No discrimination	(d) Other	(e) Total	(f) Percentage of discrimination viz. (b) as % of (b) + (c)
1. Employment (Section 3) cases, other than those disposed of by industry machinery, disposed of by year and outcome					
1968-69	2	35	42	79	5.4
1969-70	34	466	125	625	6.8
1970-71	38	414	128	580	8.4
1971	35	301	97	435	10.4
1972	62	277	128	467	18.3
1973	43	221	102	366	16.3
1974	56	302	116	474	15.6
1975	84	318	116	518	20.9
1976	59	356	134	549	14.2
1977	28	244	88	360	10.3
1978	5	38	7	50	11.6
2. Sections 2 and 5 cases disposed of by year and outcome					
1968-69	2	35	58	95	5.4
1969-70	69	246	135	450	21.9
1970-71	92	275	113	480	25.1
1971	49	161	71	281	23.3

Table 4.2 Race Relations Board, completed cases 1968–78 (continued)

1972	89	190	127	406	31.9
1973	64	162	119	345	28.3
1974	138	221	161	520	38.4
1975	79	240	152	471	24.8
1976	128	247	152	527	34.1
1977	71	124	90	285	36.4
1978	6	7	1	14	46.2

3. Sections 2, 3 and 5 cases disposed of by year and outcome

1968–69	4	70	100	174	5.4
1969–70	103	712	260	1,075	12.6
1970–71	130	689	141	1,060	15.9
1971	84	462	168	716	15.4
1972	151	467	255	873	24.4
1973	107	383	221	711	21.8
1974	194	523	277	994	27.1
1975	163	558	268	989	22.6
1976	187	603	286	1,076	23.7
1977	99	368	178	645	21.2
1978	11	45	8	64	19.6

NB: The figures for 1969–70 and 1970–71 cover the twelve-month periods 1 April to 31 March. Those for 1971 to 1978 are for calendar years. There is an overlap for the period 1 January 1971 to 31 March 1971.

Source: CRE Annual Report 1978, HMSO, London, 1979

public relations by the Board, the issue of credibility continued to impede the utility of the legislation.

Advertisements

One aspect of the Act which the public did quickly become aware of, and respond to, was Section 6, which dealt with discriminatory advertisements. The report of the Committee on Housing in Greater London (Milner-Holland, 1965) referred to the committee's own survey of private accommodation conducted in August 1964 which showed that black people were specifically barred from over a quarter of all the lettings (1,258): notices accounted for 78 per cent, newspapers for 13 per cent and agents and bureaux 9 per cent of such exclusion. Another survey was conducted for the committee between December 1963 and June 1964 which showed that the proportion of lettings where the vacancy would 'be both advertised and a coloured tenant considered was only 4 per cent'. Clearly discriminatory advertisements were widespread before the 1968 Act.

In the Race Relations Board's report for 1969-70 (RRB, 1970: 53) a comparison was made between advertisements for accommodation to let in the Hackney Gazette and North London Advertiser in July 1968 and July 1969. In the former eighteen of the adverts demonstrated an intention to discriminate on racial grounds, in the latter none. On 2 February 1970 a survey of notice boards was undertaken, also in Hackney: on twelve boards advertising accommodation there were thirty-six advertisements, only one of which was discriminatory, expressing a preference for Europeans.

Conclusion

Gribbin (1978: 100), a previous employee of the Race Relations Board and now of the Commission for Racial Equality, observed in 1978 that 'it is now common knowledge how inadequate the old legislation [the 1968 Act] was in practice', the PEP reports of 1975 and 1976 confirming the suspicion that housing complaints received by the Board did not reflect the true extent of racial discrimination. But give the fact that the intervention of the Race Relations Board was inevitably spasmodic and that other agencies such as housing associations, local authorities, accommodation agencies, building societies and estate agents were in a position to be far more effective in ensuring equal opportunity than anything the Board itself could do, the Act should not be judged solely by reference to complaints and investigations. The most important consequence of the passage of the 1968 Act, referred to in the Board's report for 1971-72 (RRB, 1972: 23) was to stimulate those and other bodies to fulfil their social responsibilities. The success of the Act in this area is, perhaps, more a matter of opinion than of fact. Certainly it provided the framework within which the Board and its conciliation committees and the Community Relations Commission and the local Community Relations Councils could negotiate with national and local bodies in both the private and the public sectors to reflect such social responsibilities in promoting equality of opportunity. But its weaknesses in scope, procedure and policing meant that the obdurate and recalcitrant were faced with insufficient sanctions to secure widespread change. The good may have become better, the bad may, on occasion, have been chastened, if not harnessed, but the indifferent largely retained their indifference.

Of course what the Race Relations Board did and how it did it are as much a reflection of personality and attitude as of law and procedure. Katznelson (1973) has argued that the Board and its sister agency, the Commission, were colonial buffer institutions reflecting a paternalistic view of race relations. Black British were faced with the stark choice of cooperation with essentially unrepresentative and undemocratic white-dominated institutions established to depoliticise race and manage social unrest or to opt out. Because the former, whether by complaining to the Board or by participation in a local Community Relations Council, emasculated direct participation in political change, it was always going to be viewed with scepticism if not cynicism by black activists. While the Board may not have shared Katznelson's view of its role as a buffer institution, it shared the view that complainants were reluctant to pursue complaints, creating the 'vicious circle' referred to previously. Whether the 1968 Act and the institutional arrangements designed to give it effect were seen by black victims of racial discrimination in housing as a reflection of white power manipulation and control or as merely ineffectual is, perhaps, less important than the composite effect - lack of credibility. It may have been that black political leaders, at that time, advocated not dissimilar arrangements but the fact that they played little part in the framing of the legislation was bound to be reflected in its reception and their perception of its credibility. Consequently, just as the success of that Act cannot be gauged purely from an analysis of complaints but must be judged in the context of its social impact, stretching beyond the formal controls and sanctions, so its failings cannot be attributed, solely, to its formal weaknesses in scope and procedure, but must also be located in the perception of the black community of its credibility and relevance.

THE RACE RELATIONS ACT 1968: PROPOSALS FOR CHANGE

The Government in its White Paper Racial Discrimination (Home Office, 1975) set out its proposals for change which were to become the 1976 Race Relations Act. In conducting this review it had been assisted by information provided by the Race Relations Board, the Community Relations Commission, the Runnymede Trust, Political and Economic Planning, organisations within the community, voluntary bodies dealing with different aspects of race and community relations, as well as individual experts. In particular, it acknowledged the help of the Select Committee of the House of Commons on Race Relations and Immigration which, since 1968, had conducted a number of inquiries and produced a series of detailed reports and recommendations on some of the major aspects of race relations and immigration (education, employment, housing, police/ immigrant relations, the problems of coloured school leavers and the control of Commonwealth immigration). In addition its report on the organisation of race relations administration had been published on 21 July 1975.

Inevitably, Central Government linked its review of the existing race relations provisions with immigration. The White Paper intimated Government's intention to undertake a comprehensive review of the citizenship laws in such a way as to enable the control of future immigration to be seen as effective and flexible but free from any racial discrimination. Government expressed the hope that it would then be possible to bring to an end the acrimony, controversy and uncertainty which 'have hampered our capacity as a society to deal with the problems of race relations' (para. 3). In making reference to patterns of immigration and settlement, the White Paper referred to an excessively high proportion of black people living in relatively more deprived inner-city areas: these areas of housing stress, in other words were disproportionately black (para. 6). It went on to observe that the latest figures suggested that the housing conditions of the black population had hardly improved in the last ten to fifteen years and the proportion of those who lived in overcrowded conditions or who were forced to share the basic amenities was higher than for the population at large. 'Coloured people are grossly over-represented in the private furnished rented sector, where conditions are worst and insecurity greatest, and significantly under-represented in the council housing sector.' In referring to racial disadvantage present in housing, employment, educational facilities and environmental conditions, the White Paper acknowledged that it was a Government's duty to prevent these morally unacceptable and socially divisive inequalities from hardening into entrenched patterns.

The White Paper then advised that Government had decided that the first priority in fashioning a coherent and long-term strategy to deal with the interlocking problems of immigration, cultural differences, racial disadvantage and discrimination was to give more substantial effect to what it had already undertaken to do: to strengthen the law already on the statute book. Legislation, it was said, is the essential precondition for an effective policy to combat the problems experienced by the black minority groups and to promote equality of opportunity and treatment. 'To abandon a whole group of people in society without legal redress against unfair discrimination is to leave them with no option but to find their own redress.' In recognising the nature of structural disadvantage the White Paper went on to refer to the fact that merely to increase the scale of resources had, by itself, no effect on the unequal and frequently inequitable allocation of the increased resources. Indeed, it observed, such a policy may, by increasing the differentials, exacerbate the very problems it was intended to solve. An effective strategy, then, to deal with the problems of deprivation and disadvantage must of necessity attend both to the scale of resources required and to the equitable allocation of the increased resources. Racial disadvantage most often occurred in the context of generalised disadvantage and could not be realistically dealt with unless there were mechanisms for correcting the maldistribution of resources. The White Paper went on to acknowledge that the

Government, in providing new legislative provision as a first priority, had an equally onerous responsibility to ensure that the policies and attitudes of both Central and Local Government were addressed. New strategies had major implications in relation to resource allocation which would be considered in the context of the then current major public expenditure review.

In respect of the 1968 Act, the White Paper referred to one important weakness as being the narrowness of the definition of unlawful discrimination. It intimated the intention to introduce the concept of indirect discrimination which would prohibit practices which might be fair in a formal sense but discriminatory in their operational effect. In summary it also intimated the following proposals for change:

1. The removal of the obligation of the then Race Relations Board to investigate each individual complaint, which was seen as hampering the crucial strategic role of identifying and dealing with discriminatory practices and encouraging positive action to secure equal opportunity.
2. The strengthening of the power of the enforcement agency to initiate formal investigations, to compel the attendance of witnesses and the production of documents and other information for the purposes of an investigation.
3. The power to require unlawful discrimination to be brought to an end.
4. The opportunity for individual complainants to pursue complaints in the courts, whether or not assisted by the enforcement agency.
5. Provision for compensation to reflect injury to feelings as a consequence of an act of direct discrimination.
6. An extension of the enforcement agency's power to seek injunctions and interdicts.
7. An amalgamation of the Race Relations Board and the Community Relations Commission; the latter had been seen to be handicapped by an uncertainty of aim since its inception and by the overlap of its responsibilities with those of the Board.

In referring to the White Paper Equality for Women (HMSO, 1975b) Government reiterated its intention therein stated to 'harmonise the powers and procedures for dealing with sex and race discrimination so as to secure genuine equality of opportunity in both fields'. In respect of housing, apart from the structural changes intimated above, and the extension of the concept of discrimination to include indirect discrimination, the proposals for change affecting housing were not radical. The only substantive alteration intimated was the specific provision making it unlawful for a landlord to discriminate against a prospective assignee or sub-lessee of a tenancy by withholding consent to the assignment or subletting.

Undoubtedly the most innovatory aspect of the White Paper was its recognition of the concept of indirect discrimination whereby the collective interests of a group might be disadvantaged by the imposition of apparently non-discriminatory provisions, which in race relations law represented what Kamenka and Tay describe as the shift from gemeinschaft to gesellschaft law (see Lustgarten, 1980: 11). Although this movement from individual to collective rights was not unique and is evidenced in pre-capitalist times, the individualism of contemporary law, particularly within the UK, has run counter to such recognition: thus the provision of housing by local authorities for the 'working classes' was abolished after the Second World War due to the questioning of the contemporary relevance of such description and implicitly because the ascription of class was no longer an immutable determinant of opportunities nor a fixed reference point for privileges and advantages. Conversely, however, colour was an involuntary and unalterable characteristic and the studies by PEP and the Runnymede Trust, for example, had demon-strated, beyond peradventure, that the disabilities imposed by the white majority on those who were black became a real determinant of an individual's life chances. However, given that the proposals contained in the White Paper did not include an attack on discrimination based on class, which, although not immutable, contained propensities for inherited encyclical deprivation and disadvantage, it was apparent that the concept of indirect discrimination, so far as it could be related to class in the sense of economic and social status, would throw up contradictions which would not readily be resolved. Thus a building society's refusal to lend to those resident in the inner cities by reason of a demon-strable correlation between residence and income might, apparently, fall within the definition of indirect racial discrimination where black people formed a disproportionate element of the population so resident, but in so far as such discrimination was based on the income of such residents, or indeed the potential capital value of the real assets being acquired, such discrimination might well be justifiable in terms of the Government's proposals and therefore lawful. As a result unless an individual complainant could seek a radial path to the centre of where unjustifiable discrimination was occurring on the basis of race the cyclical wheels of deprivation would continue to turn to his disadvantage.

As Lustgarten (1980: 9) has observed racial inequality is not therefore equated with other forms of stratification even though in the experience of most blacks they may be inseparable; in James v Valtierra 402 US 137 (1971) the Supreme Court of the United States rejected the argument that differentials on the basis of income or wealth were discriminatory against racial minorities. Other criteria may, however, be called in to question, for example, seniority, test results or length of residence: where such criteria are considered justifiable the fact that they heavily disadvantage black groups will be held to be irrelevant in law. Where a particular criterion is found unjustifiable then the beneficiaries of its abandonment, although disproportionately black, may well be a numerically greater number of whites. Thus in the United States the experience of the 'open admissions' programme of the City University of New York introduced in 1970 for the purpose of broadening the criteria for acceptance to facilitate the entrance of black and Hispanic students had the effect of benefiting a large proportion of white students who had also hitherto been denied admission on the same 'not otherwise justifiable' grounds (New York Times, 24 December 1978). In summary then the Government's purpose in intimating the intention to introduce legislation to outlaw indirect discrimination was not to secure that blacks were not disadvantaged by the application of unjustifiable criteria, whether in employment, housing or the provision of services, but merely that where such criteria had the demonstrable effect of a disproportionate impact a remedy would be available to the individual black complainant.

The White Paper referred to the Government's duty to prevent morally unacceptable and socially divisive inequalities from hardening into entrenched patterns but clearly there were limits on how Government saw intervention taking place; there was no provision for positive action to remedy the accumulated disadvantage through patterns of previous discrimination; where the law was going to prove capable of intervention such intervention would be piecemeal; the penalties imposed by way of compensation were restricted to individual complaints rather than class actions: there were no financial penalties, generally, in respect of indirect discrimination. Cumulatively, the economic impact of these proposals was unlikely to demand from major employers or housing authorities, for that matter, an immediate and comprehensive appraisal of existing practices to ensure that indirect discrimination was not occurring.

CHAPTER FIVE

ACT III: THE RACE RELATIONS ACT 1976

INTRODUCTION

In replacing the Acts of 1965 and 1968, the Race Relations Act 1976 effected four main areas of change. First, it extended the meaning and scope of discrimination. Second, it changed the whole institutional structure, including the replacement of the Race Relations Board and Community Relations Commission by one body, the Commission for Racial Equality. Third, it altered the procedural arrangements for pursuing a complaint by allowing individual access to courts and tribunals and, fourth, the Act altered the criminal offence of incitement to racial hatred.

With regard to the meaning and scope of discrimination, the definition now includes indirect discrimination (Section 1(1)(b)) as well as discrimination by victimisation (Section 2). The grounds on which discrimination is unlawful are extended to cover discrimination on the grounds of nationality (Section 3), thus negating the effect of the Ealing decision. Contract workers (Section 7), partnerships (Section 10) and various professional and training bodies (Sections 12 to 15) as well as clubs with a membership of twenty-five or more (Section 25) are all brought within the scope of the Act. Although it can be said for most purposes that the entire law on race relations is contained within a single Act, and the cases interpreting it, a number of exceptions remain.

First, it is evident that the principles of nationality and/or citizenship and immigration law run counter to those of the Race Relations Act. Obviously both nationality and citizenship law are designed to discriminate in favour of citizens. Immigration law will reflect the discriminatory provisions of the law relating to citizenship or nationality and may also in itself discriminate either directly or indirectly on racial grounds. The boundary between these provisions and those of the Race Relations Act is drawn by simply exempting the former from the latter in the Act itself.

Second, in private international law cases, particularly those relating to matrimonial issues, questions relating to nationality, citizenship, etc., may arise but will not, as a generality, come within the ambit of the Race Relations Act.

Third, in so far as the Race Relations Act involves the use of tribunals and courts the law relating to race relations will include the procedural rules and requirements which regulate their conduct, jurisdiction and practice.

Fourth, European Community law, which does not explicitly forbid racial discrimination as it does sexual discrimination, is directly applicable and enforceable as part of UK law. As part of the principle of free movement, the EC provisions forbid, with certain exceptions, discrimination in employment against any Community national but the relationship between the law of the EC and the Race Relations Act has not been judicially considered. It remains a possibility, however, that the principles of equal opportunity relating to housing embodied for example in the European Convention on Human Rights and Fundamental Freedoms may become part of municipal British law in so far as such principles may be adopted by the European Court as reflecting obligations imposed on the member States.

As important as these qualifications to the general impact of the Race Relations Act 1976 is the existence of the Sex Discrimination Act 1975 (SDA), which is drafted largely in identical fashion to that of the race relations provisions. The SDA was seen as a model because it incorporated lessons from the experience of the 1968 Race Relations Act. As a consequence cases arising under the SDA will frequently provide useful precedents for cases under the Race Relations Act (and vice versa). Despite this, judges on occasion have sought to provide distinctly different meanings to identical phrases in the two Acts on the basis that the purposes are distinguishable: it is not always safe to assume that a precedent under one Act will have binding effect on the interpretation of the other.

A more tenuous reference point is the European Convention on Human Rights and Fundamental Freedoms, which provides that the various rights protected under it are secured without discrimination on grounds of race, colour, language, religion, national origin or association with national minorities. It is important to appreciate, however, that these rights and protections are not part of UK law and are held only against States and not against individuals. In a number of instances, however, they have been used by the British courts as interpretive devices and, perhaps more importantly, they can be pursued by individual petition to the European Commission on Human Rights and, in certain instances, on to the European Court itself. Despite the fact that neither decisions of the Court nor opinions of the Commission or the Council of Ministers are legally enforceable, it is not uncommon for international pressure to result in a particular State adopting legislative amendments in order to comply with such opinions and judgements. The first report from the Home Affairs Committee 1979-80, Proposed New Immigration Rules and the European Convention on Human Rights (HC 434), illustrates this potential.

Other international obligations, for example those created by the International Labour Office and the United Nations as well as the Council of Europe, while similarly not being part of UK law and, unlike the European Convention on Human Rights and Fundamental Freedoms, not providing any right of individual petition, may nonetheless influence domestic legislation should the courts choose to consider such conventions and agreements of assistance in their deliberations.

Lastly UK courts may, on occasion, make reference to cases from foreign jurisdictions although they can never be more than persuasive: it was of course Griggs v Duke Power Co. 401 US 424 (1971) which was a formulative influence on the definition of 'indirect discrimination' in the 1976 Act.

THE ACT IN SUMMARY

As with the 1968 Act, the 1976 provisions create a statutory tort or delict of unlawful racial discrimination which may be committed in particular circumstances which are defined and cover most aspects of employment, education, housing, the provision of goods, facilities or services, and advertising. Some discrimination may be contractual, for example, instructing (Section 30), and aiding (Section 33) or the actual terms of a contract itself (Section 72), including one of employment (Section 4(2)). A number of exceptions exist and a very limited amount of 'positive discrimination' is permitted in relation to training.

The individual believing him or herself to be the victim of the delict or tort may, generally, pursue the case either before an industrial tribunal in respect of employment or otherwise through a designated County Court or Sheriff Court. The remedies normally associated with tort or delict are available under the Race Relations Act and extend to solatium, or damages in respect of hurt feelings. While legal aid may be available in respect of actions through the County Court or Sheriff Court, it is not available, as with other tribunal proceedings, in respect of actions relating to employment although initial advice may be available under the respective schemes. Because the Commission for Racial Equality is empowered to assist applicants at all levels from initial advice to representation and instructing counsel it has the effect of operating as an ad hoc legal aid scheme.

In addition to these powers, the CRE has the power to carry out 'formal investigations' into the activities of individuals, companies, partnerships, local authorities, etc., and this would include investigations into areas such as employment and housing. Such investigations may result in 'Non-discrimination Notices' which are ultimately enforceable by court orders and may require those acting unlawfully to alter or cease from particular practices as specified. The CRE, alone, may take enforcement proceedings against racially discriminatory advertisements and against persons instructing, inducing or pressuring others to discriminate unlawfully on racial grounds. The CRE may issue codes of practice in relation to employment (Section 47) which require the approval of the relevant Secretary of State; this code came into effect on 1 April 1986. The Housing Act 1988 (Section 137) amends Section 47 to enable the CRE to issue codes of practice in respect of rented or licensed housing: an extension of the CRE's 'housing code' powers is expected (Clause 135, Local Government and Housing Bill 1989: Inside Housing, 1989).

Unlawful discrimination

Three tests are applied to determine whether a person, corporate or otherwise, has committed an unlawful act of discrimination on racial grounds:

1. There must be an act of discrimination in terms of Section 1 (Racial Discrimination), Section 2 (Victimisation) or Part IV (Other Unlawful Acts).

2.   Such discrimination must be on the grounds of race as defined in Section 3.
3.   Such racial discrimination must, to be unlawful, fall into one of the situations specified in Part II (Employment), or Part III (Other Fields) and not be specifically or generally exempt (Part VI).

The Act recognises three principle types of discrimination: direct discrimination, segregation and indirect discrimination.

Direct discrimination

First, discrimination is defined by Section 1(1)(a) in the following terms 'a person discriminates against another ... if, on racial grounds ... he treats that other less favourably than he treats or would treat other persons'. This is readily illustrated by the case of Owen and Briggs v James (1981) IRLR 133 where a 'coloured' English girl was interviewed for a job and turned down because of her colour. More frequently, however, admissions of discrimination are absent and denials are made. Section 65 of the Act enables a questionnaire to be sent to the respondent (SI 1977, No. 842). Failure to respond or equivocation may result in inferences being drawn by the tribunal or court.

As a generality the burden of proof lies upon the applicant and this in practice may be very difficult to discharge. In Moberley v Commonwealth Hall (1977) IRLR 176 (an SDA case) the Employment Appeal Tribunal (EAT) held that where the applicant and respondent are of different sexes and an act of discrimination has occurred a case exists to answer; this approach was adopted by the EAT in Khanna v MoD (1981) IRLR 331 under the Race Relations Act but was qualified by the EAT, which went on to suggest that where the evidence conflicted, rather than think in terms of the burden of proof shifting, the tribunal should decide on the balance of probability. Nonetheless there is some recognition of the difficulties inherent in proving such a case, as illustrated by James above, where it was determined that the tribunal was entitled to stop proceedings at the end of the applicant's case only in exceptional circumstances.

In these circumstances, where a prima facie case of unfair treatment has been established, in the absence of evidence to the contrary the tribunal may be justified on the balance of probabilities in determining that unfair racial discrimination was present. In many cases, however, a tribunal must rely heavily upon inferences, and these may be drawn on occasion, from statistical data, a practice not disapproved in James (as above) where the Court of Appeal upheld the EAT, which affirmed the need to be bold in drawing inferences. Generally, however, the courts' view of statistical evidence is one of apprehension and they have even discouraged its use, as in the case of Jalota v IMI Kynoch (1979) IRLR 313.

The dominant view in relation to motive or intention is that it is required in respect of direct discrimination while, by reason of the fact that a condition or requirement may have been motivated by other considerations, it is not a necessary component of indirect discrimination. Whereas negligent direct discrimination is conceivable it is more difficult to imagine a direct act of discrimination which was made unconsciously or without intent. It may be, however, that the unconscious prejudice of an individual results in direct discrimination without the individual being immediately aware of what he is doing. It may also be possible for the individual to deliberately discriminate on the grounds of perceived ethnic preference but, unconsciously, apply racial stereotyping in determining how that judgment was to be made; thus a housing officer who sought to influence housing allocations by steering Muslim applicants to areas near a mosque would appear to be discriminating on religious grounds. However, if his judgment was based on the ethnic origin of the applicant, e.g. he was from Pakistan, the result might be unconscious direct discrimination (as well as unlawful indirect discrimination by the imposition of a condition or requirement).

In direct discrimination, race itself must be more than a mere background fact (Seide v Gillette Industries (1980) IRLR 427) but need not be a 'causa sine qua non', as illustrated by the case of Din v Carrington Viyella (1982) IRLR 281. That case, however, is difficult to square with the judgment of the Court of Appeal in James (as above) where it was held that a lawful reason does not efface an unlawful one. It is also necessary to distinguish between purpose and intention. Thus in the Din case (as above) although the discrimination was clearly intended to be on racial grounds it was for an ulterior purpose, in order to avoid industrial action. But this, of itself, was no defence to an act of discrimination.

Segregation

In accordance with the definition in the Act segregation is not strictly speaking a separate type of discrimination but a form of direct discrimination (Section 1(2)). Because it is a form of direct discrimination the question of intent may arise: it may be possible therefore that word-of-mouth hiring may result in segregation but, race not having been a consideration with regard to the practice, no act of unlawful direct discrimination would have occurred. Subject to the practice of word-of-mouth recruitment, or placement on a housing association waiting list, constituting a requirement or condition, this may form indirect discrimination, although the Court of Appeal's decision in Perera v Civil Service (1983) IRLR 166 may make it difficult to prove.

Indirect discrimination

A fundamental aspect of the 1976 Act, as opposed to the provisions in the 1968 Act, was the inclusion of indirect discrimination in certain instances as being unlawful. The legal draughtsman adopted, in statutory form, the essence of the US case Griggs v Duke Power Co. 401 US 424 (1971). Here the court held that the use of employment tests or educational qualifications as prerequisites for hiring and promotion fell within the statutory proscription of 'discrimination' unless those prerequisites had 'a manifest relationship to the employment in question'. Burger CJ. said:

> The Act prescribes not only overt discrimination, but also practices that are fair in form, but discriminatory in operation ... good intent or absence of discriminatory intent does not redeem employment procedures or testing mechanisms that operate as 'built-in headwinds' for minority groups and are unrelated to measuring a job capability. (401 US at 431 (1971))

In summary Section 1(1)(b) provides that a person alleging indirect discrimination must prove four things:

1. That the respondent applied 'a requirement or condition';
2. such that, although applied indiscriminately, the proportion of persons of the victim's 'racial group' who

can 'comply' with it is 'considerably smaller' than the proportion of persons not of that group who can comply with it;

3. that it was to the detriment of the victim because he could not comply with it; and

4. that it cannot be shown by the discriminator to be justifiable irrespective of race, colour, nationality or the ethnic or national origins of the person to whom it is applied.

It will be apparent that in many of the indirect discrimination cases race would not be in the mind of the discriminator at all. Thus in Mandla v Dowell-Lee (1983) IRLR 17 a prerequisite of entry to a school was not wearing headgear. This applied to everyone but adversely affected Sikhs more than others because of their religious requirement to wear a turban. Given that Sikhs were considered to be a racial group for the purposes of the Race Relations Act and that no adequate justification was produced (albeit standards and discipline were offered) it was found that application of this prerequisite was unlawful.

The scope of indirect discrimination is potentially enormous. Prerequisites held to be unlawful have included, in race relations cases, the following:

Not living in central Liverpool (Hussein v Saints Complete Home Furnishings (1979) IRLR 337).

Not being bearded (Singh v Rowntree Mackintosh) (1979) IRLR 199).

A degree from a British or Irish university (Bohon-Mitchell (1978) IRLR 525).

In addition the following SDA cases may constitute possible precedents in respect of the Race Relations Act:

Minimum age (Turton v MacGregor Wallcoverings (1979) IRLR 244 CA).

Maximum age (Price v Civil Service Commission (1978) IRLR 3).

Certain seniority (Steel v Post Office (1977) IRLR 288).

Being childless (<u>Hurley</u> v <u>Mustoe</u> (1981) IRLR 208).

Being part-time (<u>Holmes</u> v <u>Home Office</u> (1984) IRLR 299) but see <u>Kidd</u> v <u>DRG (UK)</u> (1985) IRLR 190.

Minimum strength and height (<u>Thorn</u> v <u>Meggit Engineering</u> (1976) IRLR 241).

Being a certain grade of employee (<u>Francis</u> v <u>British Airways Engineering</u> (1982) IRLR 10).

From the above it will be seen that the area in which indirect discrimination has been explored most fully by the courts has been that of employment and consequently the extent to which certain conditions or requirements constitute unlawful indirect discrimination in relation to housing are less well established although the principles are fairly clear.

1. <u>'Condition or requirement'</u>: although any sort of written or oral aptitude or other test is capable of being a condition or requirement the decision in <u>Perera</u> requires that such a condition or requirement be absolute and not one which can be traded off against others, or to which exceptions can be made. Apparently, then, the statement of a preference, although it may directly affect access to housing, improvement grants or any other service provided by a housing department, now appears to fall outwith the absolute prerequisite test introduced by this decision.

2. <u>'Racial group'</u>: the provisions of indirect discrimination require reference to the complainant's racial group but it is clear that the comparison need not be between everyone falling within his or her racial group but between the 'qualified pool' (see <u>Price</u> (above), an EAT, SDA decision). However, in <u>Orphanos</u> v <u>Queen Mary College</u> (1985) IRLR 349 the House of Lords insisted, albeit <u>obiter</u>, that a particular racial group has to compare itself with all those who are not members of that group: in this case a Cypriot had to be compared with all non-Cypriots 'consisting of all persons ... of every nationality from Chinese to Peruvian inclusive'.

3. <u>'Can comply'</u>: in <u>Price</u>, the EAT decided that the identical wording of the SDA meant 'can in practice comply' and not 'can conceivably comply'. The Court of

Appeal doubted this interpretation in the Mandla case but the House of Lords confirmed the Price approach.

4. 'Considerably smaller': there appears to be a technical problem which may occur when none of the racial group of which the victim is a member can comply with a requirement. In this instance it might be argued that the test of 'considerably smaller' is unhelpful, zero not being a proportion of anything. In fact this arose in Wong v GLC (unreported: EAT, 524/79) which had been brought as a case of indirect discrimination although the requirement or condition claimed was being 'white'. But it is in keeping with the 'open texture' of the language which may be employed in such statutory provision to say that 'considerably smaller' includes 'infinitely smaller'. The case of Percy Ingle Bakeries v CRE followed the interpretation in the Wong case but the argument was rejected in the SDA case before the EAT, Greencroft Social Club and Institute v Mullen (1985) ICR 796. In the latter case a female employee of an exclusively male club (protected by Section 34 of the 1975 Act) was refused a hearing following suspension on the grounds that she was not a member. Although all female employees would be excluded from membership the condition or requirement was found to constitute indirect discrimination: 'It would in our view run counter to the whole spirit and purpose of the ... legislation if a requirement or condition which otherwise fell within the definition ... was held to lie outside the legislation if the proportion was so negligible as to amount to no women at all.' Of greater relevance, however, is how much smaller the proportion has to be to constitute considerable.

5. Justifiable: the Act gives no definition of the term 'justifiable' although there was much debate in standing committee as to whether the alternative 'necessary' would have been preferable. Cases have generally focused on employment matters where the test of economic preference has gained credibility as sufficient to demonstrate justification. There are no reported cases in relation to housing but clearly a less stringent test in relation to justification might suggest that administrative convenience (as opposed to demonstrable inconvenience in relation to any other procedure) might be sufficient to satisfy the courts that a particular allocation procedure, for example, was 'justifiable'

despite its adverse impact on certain ethnic minority groups. The preference given to 'sons and daughters' in Tower Hamlets allocations constitutes indirect discrimination: whether it is 'justifiable' awaits judicial decision (CRE, 1988c; see also Smith, 1989a).

The test of justifiability established in Ojutiku and Oburoni v MSC (1982) IRLR 418 ('Acceptable to right-thinking people as sound and tolerable reasons') has been subject to increasing criticism. In sex discrimination cases some indication of change towards a test of necessity was evidenced in the case of Bilka Kaufhaus Gimlott v Weber von Hartz (1986) IRLR 317, and by Lord Keith's obiter remarks in Rainey v Greater Glasgow Health Board (1987) IRLR 26 HL that the 'justification' defence under the SDA should be interpreted in the same way as the defence under the Equal Pay Act. However, in Hampson v Department of Education and Science (1988) IRLR 87, the EAT refused to accept the argument that, as the SDA and RRA are materially the same in their drafting, the approach in Bilka should apply (that an indirectly discriminatory requirement must be 'necessary' for the employer to escape liability) and, read with Rainey, should overrule the much more lenient test in Ojutiku. The EAT felt bound to apply Ojutiku unless and until it was specifically overruled: justification was not raised in Rainey and Ojutiku was not cited in the House of Lords.

Currently, then, through the influence of the impact of European law on sex discrimination cases but not on race discrimination cases, there is the prospect of a duel test of justification, with the more onerous test of necessity being applied solely to sex cases. Although the House of Lords refrained from offering any guidance on justifiability in Mandla v Dowell-Lee, in Orphanos v Queen Mary College (1985) IRLR 349 it held that it meant 'capable of being justified', which suggested to the editor of the Industrial Relations Law Report that the law prohibiting indirect discrimination would not be worth the paper it was written on (Gregory, 1987: 45). Moreover, the determination that 'justifiable', by the House of Lords in Mandla, was a question of fact, may leave the decision of the original tribunal unchallengeable, where the correct test has been applied.

6. Detriment: the applicant must not only demonstrate unlawful discrimination whether of a direct or indirect

nature but also that it has, in the latter instance, resulted in some tangible detriment. It may be argued that this requirement is to prevent 'public interest actions' but more importantly the term, which again lacks interpretive guidance, may be used by the courts to restrict a remedy where discrimination has taken place. While the case of MOD v Jeremiah (1980) QB 87 illustrated (per Brightman LJ) that the term, 'detriment' might include a simple disadvantage even where subjectively felt, this case focused on the interpretation of Section 4 of the 1976 Act, which also refers to detriment but in the context of employment. It is likely that that interpretation would be applied to Section 1(1)(b)(3)) in respect of indirect discrimination. In the SDA case Holmes v Home Office the EAT rejected the argument that two detriments had to be proved, one under each of the equivalent sections, and attempts to introduce the de minimis principle in direct discrimination under the SDA have been quashed by the Court of Appeal in Gill and Coote v El Vino Co. (1983) 2 WLR 155.

Standard of proof

Given that racial discrimination is a civil rather than criminal offence the standard of proof required is the balance of probability. Nonetheless, it is a widely accepted fact that the most difficult obstacle for a person alleging discrimination to surmount is to prove that he or she is actually the victim of discrimination (SCOLAG, 1988: 93) and in so doing to establish the relevance of demonstrating patterns of discrimination, implicitly, by reference to statistical data. Generally such data has been to little avail since, to obtain a remedy, it would have to be established that the litigant was the victim of a specific discriminatory act. However, in West Midlands Passenger Transport Executive v Singh (1988) IRLR 186 CA the Court of Appeal held that statistical evidence, so far as it established a discernible pattern, might be indicative of discrimination in respect of the specific act alleged. The court indicated that where such evidence suggested a pattern of discrimination against a racial group, then, in the absence of a satisfactory explanation to the contrary, it was reasonable to infer discrimination against a member of that group in respect of a specific act. Indeed statistical evidence of discrimination against a group, for example in promotion, may be more persuasive of discrimination in a particular case than previous treatment of the applicant, which, at a personal level, may not have been motivated by discriminatory intent. To the extent that this approach is adopted by tribunals and the County and Sheriff Courts, it may prove of considerable assistance to those alleging discrimination. It has also qualified the decision in Jalota v IMI (Kynoch) (1979) IRLR 313 so far as it was there argued that statistics had no probative value other than as to the employer's 'credit' and were consequently inadmissible - a clearly perverse contention. This qualification is of critical importance in housing allocation, where patterns of disadvantage, demonstrable by statistical evidence, may be the only evidence of indirect discrimination.

Victimisation

Somewhat unusually the Act also defines as discrimination the victimisation of a person because that person has, for example, asserted his rights under the Act (Section 2(1)). Victimisation arises where, in any of the situations in which the Act applies, a person (the discriminator) treats another person (the person victimised) less favourably than he treats, or would treat, other persons on the ground that the person victimised has done (or intends to do, or is suspected of having done or of intending to do) any of the following:

1. Brought proceedings against the discriminator or anyone else under the Race Relations Act.
2. Given evidence or information in connection with proceedings brought under the Act by another person against the discriminator or anyone else.
3. Otherwise done anything under, or by reference to, the Act in relation to the discriminator or anyone else (e.g. by giving evidence or information to the Commission for Racial Equality in the course of a formal investigation).
4. Alleged that the discriminator or anyone else has committed an act which (whether or not this is expressly stated) would constitute a contravention of the Act.

Nonetheless where the conduct of the person victimised involves the making of an allegation, the less favourable treatment on account of the allegation is not victimisation for the purposes of the Act if the allegation was false and was not made in good faith. It should also be noted that victimisation occurs only where the reason for the victimisation was one of the racial grounds set out above.

In Kirby v MSC (1980) ICR 420 it was held that the disciplining of a Job Centre clerk by his employer because he had passed information to the local Community Relations Council concerning unlawful acts of employers dealing with the Job Centre did not constitute unlawful victimisation on the grounds that the employer normally disciplined employees for passing on confidential information to third parties when this was not in conformity with agreed procedures. The difficulty with the Kirby case is that it would appear to justify a policy of general victimisation by an employer where his employees pass on information for the purposes of preventing the recurrence of unlawful acts or otherwise, irrespective of the objective 'justifiability' of the action concerned. It would seem to have been possible to interpret the phrase 'less favourably than ... other persons' otherwise than contrary to the intention of the Act (as the EAT admitted its interpretation was) since a failure to act in the way that Kirby did had the potentiality of putting a person in the position of unwillingly aiding discrimination and liable therefor under Section 33 of the Act. That view is now vindicated by Aziz v Trinity Street Taxis (1988) IRLR 204 CA where the Court of Appeal clarified the purpose of Section 2(1) as ensuring that an individual should not be deterred from doing any of the 'protected' acts by fear of victimisation. The court found that the making of secret tape-recordings by Mr Aziz for the purpose of establishing whether or not his treatment was racially motivated was a protected act under Section 2(1)(c). However, the court went on to require that the victimisation (in this case dismissal following the protected act) would only be established if the perpetrator had 'a motive consciously connected with the race relations legislation'. If this is to mean that an employee who is dismissed for bringing a race relations complaint is not victimised if the employer shows that the employee would have been dismissed for bringing a complaint under any legislation, then the Kirby grip has not lessened its clasp.

Vicarious liability

By virtue of Section 32 of the Act employers and principals are liable for acts or omissions done on their behalf (in the course of employment, and with or without approval in the first case, and whatever the type of authority in the second). The defence available under Section 31(3) that 'reasonably practicable' steps were taken to avoid the act or type of act found discriminatory is available only in employment cases. As a generality therefore a local authority will be liable for the unlawful act of its employees and, by reason of Sections 20 and 21, the individual employee who has acted unlawfully, being either a person concerned with the provision of goods, facilities and services or a person who has power to dispose of premises will also be directly liable for his act.

Instructing or procuring discrimination

Section 30 of the Act renders liable any person who instructs, procures or attempts to procure another person to do an unlawful act under Parts 2 or 3, that is, all unlawful acts concerning which an individual can complain, provided in any case that the person either 'has authority over' the other or the other 'is accustomed to act ... in accordance with' the wishes of that person. Although this section does not figure largely in the reported cases, it would be possible, for example, for a housing association to instruct a housing authority not to refer black applicants to it under arrangements for referral. In this instance either the housing association would constitute a person 'having authority over' or alternatively the housing authority would be 'accustomed to act in accordance with their wishes'. Proceedings may be brought by the CRE alone (Section 63).

Pressure to discriminate

In addition to instructing or procuring discrimination, it is also unlawful to put pressure on another to induce or attempt to induce an unlawful act under Parts 2 or 3 of the Act. Like Section 30 the provisions of Section 31 can only be used at the suit of the CRE. This section would encompass tenants putting pressure on a housing authority (as in the Tower Hamlets case). The provision also makes it clear that it is no defence that an attempted inducement was not made to the actual person having authority to act provided it was made 'in such a way' that he was likely to hear of it. In the case of CRE v Powell and Birmingham District Council (EAT 33/85) it was held that the innocent passing on of information that a garage discriminated against black MSC trainees in refusing to allow their placement contrary to Section 13 of the Act did not constitute inducement. The EAT referred to the case of CRE v Imperial Society of Teachers of Dancing (1983) ICR 473 where it was held that the statement of a preference by an employer to a careers officer in a local school, to the effect that the school should not send anyone coloured, as that person might feel out of place, would constitute inducement. Here inducement had been interpreted to mean 'to persuade or to prevail upon or to bring about'. The EAT, however, made a distinction in the Powell case in asserting that simply giving information was not an attempt to persuade. In the case of Camden Nominees Ltd v Forcey (1940) 2 All ER I before Simons J it was held that the advice given was of such a nature (was of a character obviously intended to be acted upon) that it was for all practical purposes equivalent to persuasion; but, if the matter be merely advice (in the ordinary sense of that word), this would not constitute inducement. It is suggested that the Powell decision give rise to real difficulties in the sense that the practical consequences of giving information about a discriminatory employer, etc., are likely to perpetuate acts of unlawful discrimination whether or not this was the intention of the advice proffered.

Aiding unlawful acts of discrimination

Knowingly aiding an unlawful act is, in itself, made unlawful by Section 33. Unlike the provisions relating to inducement or procurement Section 33 cases can be brought by an individual. A personnel officer who had discriminated in recruitment would not be liable under Section 4, not being the employer, although the employer would be rendered liable by Section 32 of the Act. The personnel officer, in such a case, would become liable by Section 33 in aiding an unlawful act despite the fact that the personnel officer, himself or herself, would appear to be directly responsible for the act of unlawful discrimination.

A defence is offered by Section 33(3) that the person alleged to be aiding unlawful discrimination was acting in reliance upon a statement made to him by the primary discriminator that '... the act which he aids would not be unlawful and it was reasonable for him to rely on that statement'.

Racial grounds

Section 3(1) of the Act declares 'racial grounds' to mean any of the following, namely colour, race, nationality or ethnic or national origins, and racial groups are defined accordingly.

The term 'colour' has seldom posed particular problems to either tribunals or courts in the interpretation of the Act. Thus potential victims of direct discrimination have been described as white, black or coloured (Hussein) and not infrequently an alternative to racial group such as 'Sikh' in addition to 'coloured' has been employed (see Virdee v EEC Quarries (1978) IRLR 295.

In respect of 'race' we have already noted the inherent lack of objectivity in relation to such a description but it has been employed in a number of cases such as Baywoomi v British Airways (1981) IRLR 431 where the complainant was described as 'by birth an Arab of Aden', and it was added, 'we think that for the present purposes, that it is a sufficient definition of his racial group', and 'the applicant is also dark skinned, but we do not consider that his colour is of any relevance to his racial group for present purposes'.

'Nationality' means 'legal nationality', which equates with 'citizenship' (see Section 78). In terms of the British Nationality Act 1981 the term would clearly include not only a British citizen but also a 'British Overseas Citizen' as well as a national of a British Dependent Territory. Doubt might occur, however, in respect of a 'British Protected Person' or someone who was a UK national for the purposes of the EEC or even an EEC national. In practice, however, the term, 'national' has not given rise to difficulties of interpretation.

'Ethnic origins' fell to be interpreted in Mandla v Dowell-Lee (1983) IRLR 17. In this case Sikhs were held to be not merely a religious group but also an 'ethnic' group. Two criteria were applied: first, the group must have a long shared history and, second, a separate cultural tradition. Moreover it was held to be relevant to take into account a common geographical origin or descent from a small number of common ancestors; a common language, though not necessarily one peculiar to the group; a common religion different from that of neighbouring groups or of the general community surrounding; and being a minority or being an oppressed or dominant group within a larger community. It would therefore be possible to be a member of an ethnic group by reason of marrying in or by reason of conversion. In applying such a broad indicative test the question of the biological distinctiveness of the group was not of the essence. This approach was confirmed by Dawkins v Property Services Agency (1989), unreported (CRE, 1989f) where the tribunal found that Rastafarians constituted an ethnic group.

Although religion as such is not a racial ground, religious discrimination may be indirect racial discrimination. Thus discrimination against practising Jews, for example, might constitute indirect discrimination against Jews as an ethnic group, as defined in Mandla above. Similarly discrimination against Catholics in parts of the UK might fall into indirect discrimination against those of Irish ethnic origin. The test of ethnic origin might separate those from Northern Ireland and the Republic not only on religious grounds but also on grounds of such origin.

Clearly there may be some overlap between 'ethnic origin' and 'national origin'; following the decision of the majority of the House of Lords in Ealing London Borough Council v Race Relations Board ex p. Zesco (1972) 1 All ER 105, national origins exclude current nationality. Although determined under the 1968 Act (and indeed leading to the inclusion of 'nationality' under the 1976 Act) the test applied to 'national origin' in that case still subsists. Lord Simon, while not providing a definition, suggested that it meant descent from a 'nation' by reference to 'national spirit' and to 'tradition, folk memory, a sentiment of community'. Whether the existence of a nation State at any time in a group's history is a prerequisite for the term 'national origin' to apply is not clear: the alternatives of race or ethnic origin may be sufficiently broad to prevent unreasonable exclusion from 'racial grounds'.

In Section 3(1) there is provision that 'references to a person's racial group refer to any racial group into which he falls' and in Section 3(2) that 'the fact that a racial group comprises two or more distinct racial groups does not prevent it from constituting a particular racial group for the purposes of the Act'.

Provision of goods, facilities and services

As mentioned above, it is unlawful to provide goods, facilities or services to the public or a section of it (whether for payment or otherwise) in a discriminatory fashion, whether by way of refusal or by way of a provision not in the normal manner or on the normal terms (Section 20). Section 20(2) of the Act gives examples of the facilities and services mentioned. These include the following:

1. Accommodation in a hotel or boarding house.
2. Facilities in banking and insurance.
3. Facilities in education.
4. Facilities for entertainment, recreation or refreshment.
5. Facilities for transport or travel.
6. Services of any profession or trade, or any local or other public authority.

It should be noted that these provisions are similar to those in the 1968 Act, previously mentioned, and follow closely the provisions in Section 29 of the Sex Discrimination Act 1975. The list is 'self-evidently not intended to be exhaustive' (Applin v RRB (1975) AC 259, at 291 per Lord Simon (on the 1968 Act). On the question of the application of these provisions the courts have determined the following:

1. The Inland Revenue provide a service - by giving relief from tax, for making repayment or giving tax advice (Sanjani v Inland Revenue Commissioners (1981) QB 459).
2. Credit facilities in a department store are encompassed by this section (under Section 29(1) of the 1975 Act, Quinn v Williams Furniture Ltd (1981) ICR 328).
3. Local authority fostering was a service covered (Applin v RRB) but is now expressly exempted.
4. The Home Secretary (and thus the 'Immigration Service' is not providing a service in granting leave to enter or remain in the UK (R v Immigration Appeal Tribunal ex p. Kassam (1980) 1 WLR 10).
5. Discrimination in the grant of permission to use facilities or services (as opposed to their provision) is not unlawful (Amin v Entry Clearance Officer, Bombay (1983) 2 All ER 864).
6. Provisions relating to development control and the

development plan process will constitute a service, if not by reason of the 1976 Act as originally enacted, then by reason of the amendment included in the Housing and Planning Act 1986 (Section 55 of that Act providing a new Section 19A of the 1976 Act).

The Act expressly includes a reference to local or other public authorities (Section 20(2)(g)) and this was not thought to be problematic (see Lester and Bindman, 1972: 280) as it would include quangos and in any event was exemplary rather than taxative. A major issue arising under the 1968 Act was whether or not the service was being provided 'to the public or a section thereof'. Thus Labour Club and Institute v RRB (1976) AC 285 demonstrates the broad exclusions created by a restrictive interpretation of this provision. Clubs are now specifically provided for by Section 25 which, generally, makes discrimination by clubs and associations (this would include housing associations) in respect of membership, etc., unlawful.

Section 20 applies only to discrimination by persons who are 'concerned with' the provision of goods, facilities or services to the public or a section of the public. Thus the section does not apply where the transaction is one of a purely private and personal nature. As with the 1968 and 1975 (Sex Discrimination) Acts, this Act does not provide any definition of 'goods, facilities or services'. Accordingly they should be given their ordinary and natural meaning; 'goods' are any movable property, including merchandise or wares, 'facilities' include any opportunity for obtaining some benefit or for doing something; 'services' refer to any conduct tending to the welfare or advantage of other people, especially conduct which supplies their needs. Each of the terms is deliberately vague and general and, together, they cover a wide range of human activities (Lester and Bindman, 1972: 260).

Two Court of Appeal decisions, Gill and Coote v El Vino Co. (1983) 2 WLR 155 and Jones v Royal Liver Friendly Society (1982) The Times 2 December, CA, which related to the 1975 SDA equivalent provisions, suggested that the court will not be impressed by attempts to imply exceptions and limitations into the statute to excuse latent (sex) discrimination. Nonetheless the expression 'public or section of the public' clearly implies some limitation and three cases under the Race Relations Act 1968 (Charter, Applin and Dockers Labour Club and Institute v RRB (1974) 3 All

ER 592) attempted to define the application of this concept. The Law Lords decided that 'private' rather than 'domestic' is the antithesis of 'public'; and a club which selects its members according to their personal qualities does not cater for a section of the public (but see particular provisions of the 1976 Act relating to clubs); Lord Wilberforce observed (Applin: 278):

> the area in which discrimination is forbidden is that in which a person is concerned to provide something which in its nature is generally offered to and needed by the public at large, or a section of it, which is offered impersonally to all who choose to go through the doors or approach the counter ...

In relation to housing, Section 20 would therefore appear to cover such activities as: the provision of advice and guidance at a housing advice centre; access to local authority grants and loans; the provision of boarding or hotel accommodation not otherwise covered by Section 21; and probably the provision of facilities or services associated with the declaration and implementation of Housing Action Areas and other area treatment as well as the issuance of repair notices; and activities associated with the declaration and implementation of individual housing treatment such as closing orders and demolition orders.

Provision of housing accommodation

Section 21 of the 1976 Act, replacing a similar provision (Section 5) in the 1968 Act to which reference has already been made, states that discrimination on racial ground in the disposal or management or premises is unlawful. A person is exempt from this section where he owns an interest or estate in the premises and wholly occupies the premises provided he neither uses an estate agent nor advertises in connection with the disposal. The provisions apply to companies as well as local authorities and individuals. 'Premises' include land, houses, flats and business premises but not accommodation in a hotel or similar establishment which, as we have noted, is covered by section 20(2)(g) above.

Discrimination by a tenant in respect of the assignment or sub-letting of his tenancy to another person is unlawful under Section 21. Where the tenant requires the consent or licence of the landlord to assignment or sub-letting to another person, it is unlawful for consent or licence to be granted or withheld by the landlord in a way which is discriminatory against the prospective assignee or sub-lessee.

Exceptions from Sections 20 and 21

Section 20 does not apply to acts done by a person as a participant in arrangements under which he takes into his home, and treats as if they were members of his family, children, elderly people or people in need of a special degree of care and attention. To that extent these provisions exclude fostering and boarding-out arrangements entered into by a local authority, so far as they are activated by the recipient families.

It is not unlawful to discriminate in the provision or disposal of accommodation or premises or by withholding consent or licence to assignment or sub-letting where the person providing or disposing of or withholding consent or licence, or a near relative of his, lives and intends to continue to live on the premises and shares accommodation (other than storage accommodation or means of access) with other persons living there, who are not members of his household, and where the premises are small. For the purposes of the Act, a person is a near relative of another if that person is the wife, or husband, parent or child, grandparent or grandchild or brother or sister (including, for example, half-sister, step-sister and sister-in-law). Small premises are, in the case of premises comprising residential accommodation for one or more households, those in which there is not normally residential accommodation for more than three households (including that of the person providing the accommodation or of a near relative of his); in any other case, small premises are those in which there is not normally residential accommodation for more than six persons (not counting the members of the household of the person providing the accommodation or of a near relative of his).

Local authorities

Section 71 of the Race Relations Act 1976 as amended by Section 56 of the Housing Act 1988 is addressed to local authorities generally and to the Housing Corporation (including Scottish Homes), Housing for Wales, and Housing Action Trusts. It has not been extended to all public bodies as the CRE advocates (CRE, 1985b: 35). It requires that every local authority, etc., should make appropriate arrangements with a view to securing that their various functions are carried out with due regard to the need: (a) to eliminate unlawful discrimination; and (b) to promote equality of opportunity, and good race relations, between persons of different racial groups. Although these provisions may be seen as largely cosmetic, and so far as they are enforceable may only be instituted, implicitly, by the Secretary of State, they do at least facilitate action by housing authorities to secure compliance with the Act (MacEwen, 1985). It should also be recognised, however, that the provisions of Section 71 make no allowance for local authority expenditure although no doubt such expenditure may be sanctioned by Section 71 so far as they are within the general housing authority budgetary provisions. The power of local authorities to use Section 71 to promote good race relations is limited: in Wheeler v Leicester City Council (1985) CLY 17 it was held unlawful for the local authority to withhold the use of a rugby ground because of the connections of members of the club concerned with South Africa; in R v Lewisham London Borough Council ex p. Shell (UK) (1988) 1 All ER 938 the council's boycott of Shell (UK) because of their commercial links with South Africa was also held to be unlawful. In both cases the authority was held to be acting ultra vires. In addition Section 18 of the Local Government Act 1988 states that Section 71 of the Race Relations Act 'shall not require or authorise a local authority to exercise any function regulated by Section 17 [relating to local and other public authority contracts] by reference to a non-commercial matter'. The exception to this exclusion relates to asking 'approved questions' (see Annex B to Department of Environment Circular 8/88 dated 6 April 1988) and asking for, and considering, responses to terms or provisions relating to workforce matters in draft contract or tender documents. In effect Section 18 emasculates any stringency relating to local authority contract compliance to secure equal opportunity monitoring of contractors: the 'approved' questions are totally inadequate, as presently framed, for this purpose (but see CRE, 1989g).

Remedies in respect of individual complaints

No anti-discrimination law will operate successfully unless violations are costly and appeal to the law is worthwhile to the victim. Most complainants in housing matters, as in other areas, will not merely be seeking a vindication of their claim to fair treatment, but some form of tangible restitution or damages in respect of the loss or injury suffered by the discriminatory act, or omission. The remedies provided by the Act (Section 57) include monetary damages and compensation for hurt feelings (Section 27(4)). In respect of indirect discrimination (1(1)(b)), where the respondent proves that the offending condition or requirement was not applied with the intention of treating the claimant unfavourably on racial grounds, damages are not payable. While proceedings in respect of housing services are within the exclusive original jurisdiction of a designated County Court or Sheriff Court, the remedies available are those which would have been obtainable in the High Court or the Court of Session, as the case may be (Section 57(2)), including orders of mandamus, certiorari and prohibition and equivalent orders in Scotland.

Guidance on damages is provided by Alexander v Home Office (1988) IRLR 190 CA. In respect of injury to feelings, while the Court of Appeal states that awards should not be minimal, because that would tend to trivialise or diminish respect for the public policy to which the Act gives effect, it opined that to award damages 'which are generally felt to be excessive' does almost as much harm: injury to feelings, moreover, is likely to be of relatively short duration and is less serious, the court advised, than physical injury. Damages should in some instances include aggravated damages where the defendant behaved in a high-handed, malicious, insulting or oppressive manner and exemplary damages may be made in appropriate circumstances. In practice although awards for injury to feelings have ranged from £30 to £5,000 (CRE, 1985b: 27) typically awards are £300 or less. For example, estate agents who admitted racial discrimination in refusing to sell to two Asian brothers were required to pay £100 damages to each (Bradford Telegraph and Argus, 25 November 1982). Such

129

minimal awards provide little financial incentive for complainants. Given the reluctance of the courts to use interim injunctions or interdicts to secure the potential of reinstatement by court order after the successful pursuit of claims (for example, the allocation of a specific house by the recalcitrant housing authority) other remedies in restitution have not generally been available. Moreover the absence of class actions in both jurisdictions provides limited opportunity for collaborative pursuit of claims against major estate agents or housing authorities, rendering such ventures as highly speculative.

Statutory exemptions

Section 41 of the 1976 Act provides a blanket exemption of any act of discrimination, first, in pursuance of 'any enactment or Order in Council' or 'any instruction made under any enactment by a Minister of the Crown' and, second, in order to comply with any condition or requirement imposed by a Minister by virtue of any enactment. The 1976 Act, then, by its own terms, is subordinate to any rule or order, past, present or future, with which it conflicts and a Minister, for example, may by statutory instrument, circular or otherwise - so far as done by virtue of any enactment - extend direct and indirect racial discrimination with impunity. Not surprisingly this has attracted severe criticism from the CRE (CRE, 1983c: 23; 1985b: 9).

Formal investigations

The CRE is given power to conduct formal investigations under Sections 48 to 52 of the Act. The purpose of such investigations is to allow the CRE to inquire into the policies and practices of a named respondent, or an area of activity, to determine whether discrimination has occurred, to make recommendations for improvement and, ultimately, to issue a Non-discrimination Notice. Such a notice is to ensure that the named person, or a person identified and notified during the investigation, no longer commits his unlawful discriminatory act and that his practices are changed in order to prevent repetition of like acts. The power to conduct the formal investigation is one element within the general duty placed on the CRE under Section 43(1), which comprises three elements, as follows:

1. To work towards the elimination of discrimination.
2. To promote equality of opportunity and good relations between persons of different racial groups generally.
3. To keep the working of the Act under review.

Under the terms of Section 48 any formal investigation must be connected with the carrying out of these above-mentioned general duties but need not be for the purpose of disclosing unlawful discriminatory acts. Thus the fact that the immigration service of the Home Office was outwith the purview of the legislation has been held insufficient, in so far as its duties relating to immigration control affect individuals or groups on a discriminatory basis, to prevent an investigation into that service (R v Immigration Appeal Tribunal, ex p. Kassam (1980) 1 WLR 1037 and Report of CRE Formal Investigation into the Immigration Service (CRE, 1983a)).

The Home Secretary has the power, which has yet to be exercised, to require the CRE to conduct a particular investigation but otherwise a formal investigation can only be undertaken if the CRE thinks fit. In conducting an investigation the CRE can delegate its functions under the Act to one or more Commissioners to act on its behalf. In addition Section 48 enables the CRE to appoint further Commissioners to conduct the investigation. Moreover the CRE may also delegate its functions under the Act to its servants to act on behalf of the Commissioners (R v CRE ex p. Cottrell and Rothon (1980) 1 WLR 1580).

The CRE is obliged to draw up its terms of reference for the investigation as the first step in the procedure. Where the investigation will examine the activities of a named person, following on the CRE's belief that there has been a breach of the Act by such a person, it must inform that person of its belief and of its intention to conduct an investigation. In such circumstances the CRE is obliged by the terms of Section 49(4) to provide the named person with the opportunity of making oral and/or written representations with regard to the terms of reference of the inquiry. A person who wishes to make such representations can be represented by counsel or a solicitor or by some other person of his choice. It is a condition precedent to embarking on a formal investigation in respect of a named person that there should be some particularisation of the kinds of acts of which the respondent was suspected by the CRE (R v CRE ex p. Hillingdon London Borough Council

(1982) IRLR 424 per Lord Diplock). Further Lord Diplock's views (later confirmed in R v CRE ex p. Prestige Group (1983) IRLR 408) that the CRE must have a reasonable belief about the existence of discriminatory conduct which can be inferred from the specified acts, have imposed a further constraint on the ambit of formal investigations.

The CRE has the power, under Section 50, to obtain documentary and oral evidence in its conduct of a formal investigation and the person who fails to provide such evidence may be the subject of an order of compliance issued by a County Court or the Sheriff Court. Anyone who wilfully conceals or destroys or alters a document which has been the subject or an order of compliance or who knowingly or recklessly makes a statement which is false in complying with such an order is guilty of a summary offence and liable to a fine.

On completion of a formal investigation the CRE is obliged to prepare a report of its findings, which must be published or made available for inspection. At any time during the course of an investigation the CRE may make recommendations, by dint of Section 51, to the respondent on changes which he can make to his practices to promote equality of opportunity. Thus where the respondent, at some stage in the investigation, accepts the validity in whole or in part of the terms of reference of the inquiry a settlement may be negotiated which may preclude the issuance of a Non-discrimination Notice. Recommendations might also be made in the situation where the CRE comes to the conclusion that the respondent has discriminated but the nature of the discrimination is not such as would permit the Commission to issue a Non-discrimination Notice (Home Office v CRE (1982 1 QB 385). In such cases the CRE's power would be restricted to making recommendations to the respondent about changing his procedures and practices but, in the absence of the sanction of issuing a Non-discrimination Notice, the recommendations of the CRE may be ignored without further action in terms of the Act.

Non-discrimination Notice

Following a formal investigation which concludes that an act (or acts) of unlawful discrimination has been committed, the CRE may issue a Non-discrimination Notice, after the grounds on which it is to be based have been made known to the party/parties on whom it is to be served. Thereafter the respondent is entitled to make oral and/or written representations which must be considered before a Non-discrimination Notice is issued. In R v CRE, ex p. Cottrell and Rothon (as above) the CRE, after conducting a formal investigation, concluded that a Non-discrimination Notice should be issued against a firm of estate agents. The firm was given the opportunity to make representations as required by Section 58(5) and both written and oral representations were made. In respect of the latter, however, the firm was not given an opportunity to cross-examine certain witnesses upon whom the CRE was relying and in these circumstances the firm made application for judicial review before the Divisional Court of the Queen's Bench. The estate agents contended that there had been a denial of natural justice on the grounds, first, that the Commission was relying on hearsay evidence and, second, that it had refused the firm the right to cross-examine witnesses. Lord Justice Laing held that the rules of natural justice were flexible so that 'what may be the rules of natural justice in one case may very well not be the rules of natural justice in another'. In these particular circumstances Lord Laing decided that all the safeguards that the rules of natural justice demanded were contained in the provisions of Section 58 and, since no right of cross-examination was referred to therein, the respondents had not been prejudiced unfairly by the procedure adopted and the application was therefore rejected.

A specimen copy of the notice is set out in Regulation 6 of Schedule 2 of the Race Relations (Formal Investigations) Regulations 1977. Where the notice requires the respondent to change his practices he must inform the Commission that he has carried out the necessary changes and, so far as is reasonably required by the notice, he must provide this information to other persons concerned. Effectively a Non-discrimination Notice enables the CRE to monitor compliance up to five years (Section 58(4)) after the notice has become final. The notice becomes final and operative either when six weeks have elapsed from the date of its service or, if an appeal against it has been made, when that appeal is either lost or discontinued. This six-week period is the time within which an appeal may be lodged either to an industrial tribunal in respect of notices relating to employment matters or, in other instances, to a County Court or Sheriff Court. In CRE v Amari Plastics (1982) IRLR 252 the respondent company sought to challenge, by way of such appeal, the facts upon which the CRE had based its decision to issue a notice. It was submitted by the CRE that an appeal under Section 59 was intended only to concentrate upon the reasonableness of the requirements laid down by the Commission to ensure that the respondent complied with the terms of the notice. This view found favour with neither the EAT nor the Court of Appeal, which concluded that a statement of facts on which a finding of unlawful discriminatory conduct was based should be included in the notice and could be subject to challenge by way of appeal. Where the notice becomes final it will be entered in a CRE Register of Non-discrimination Notices, which is open to the general public for inspection. No other sanction attaches to the respondent by receipt of the notice alone. Other enforcement procedures are competent only where the CRE becomes convinced that the respondent is breaching the terms of the notice and is a persistent discriminator. In such a case the CRE can raise proceedings in the County Court for an injunction or in the Sheriff Court for an interdict restraining the respondent from continuing to breach the terms of the notice. These proceedings are competent at any time within five years of the date of the original Non-discrimination Notice.

PROMISES AND MISGIVINGS

The potential strengths of the 1976 Act may be summarised as follows:

First, in extending the definition of unlawful discrimination to include indirect discrimination it provided the potential both for the CRE and for the individual complainant to pursue an attack on institutional practices, including those in the public and private sector in housing which had the effect of disadvantaging individuals and groups by reference to their racial origin.

Second, it strengthened significantly the enforcement agency, the Commission for Racial Equality, not only by combining the previous responsibilities of the Race Relations Board and the Commission for Racial Equality but also by extending the power of conducting formal investigations. These may now be instituted by reference to matters or strategic importance, including public and private housing, even when the evidence available does not point to unlawful discrimination. Moreover in combining the enforcement and promotional activities the CRE was given greater potential than that afforded the Community Relations Commission for targeting the promotional work to reflect successes in enforcement and thereby achieve a more profound impact.

Third, the newly formed CRE now had the potential for utilising the existing local networks of Community Relations Councils to act as local policemen not only by referring complaints to the CRE for assistance but also by providing an opportunity for the local CRC to plug into the expertise of the CRE in directly assisting complainants in pursuit of individual claims. The new structure within which the local network was located had the potential, therefore, to be mutually reinforcing in respect of the range and efficiency with which complaints might be pursued. In addition the potential access to local information might prove of critical importance at a strategic level, both in respect of enforcement and in respect of promotional activities.

Fourth, now that individual complainants were given direct access to the courts and tribunals the charge, under the 1965 and 1968 Acts, that the enforcement agency was a bureaucratic and unsympathetic body and was occasionally gutless in the pursuit of complaints would be put to the test. Consequently the credibility of the CRE with ethnic minority groups might be revived as the CRE would shed the filtering responsibility relating to conciliation and pursuit of complaints in the courts previously placed on the Race Relations Board. Individual access might have two benefits: the new legal framework might be tested more probingly by individual complainants with direct access to the courts than was possible by one enforcement agency and, second, in removing the absolute discretion of the enforcement agency as to whether individual complaints should proceed to litigation, the success or failure of individual enforcement would not necessarily impinge on the credibility of the CRE.

Fifth, with the passage of the 1975 Sex Discrimination Act and the creation of a sister commission, the Equal Opportunities Commission, operating within a similar legal framework, a novel opportunity arose for a concerted attack on unlawful discrimination with the potential for each agency to benefit from strategic casework and investigations undertaken by the other, whether as a consequence of a concerted strategy or unplanned, but nonetheless complementary, activity.

Lastly, the obligations placed on local authorities by

Section 71 had the potential for beneficial monitoring and promotional work by housing authorities who might now more readily seek the co-operation and advice of the CRE in securing compliance with the letter and, more importantly, the spirit of the Act.

Against these potential advantages of the new Act could be set a number of weaknesses of both an extrinsic and an intrinsic nature.

In respect of the former, first, the law itself, as Government recognised in the 1975 White Paper, merely provided a framework of operation: without Central and Local Government to secure active promotion of equality of opportunity in housing and other spheres this framework was bound to prove inadequate. Moreover, while Central Government, in particular, was prone to somewhat vacuous declarations of good intent, the record of action from 1965 was not promising. There was a distinct lack of a strategic plan backed up by political commitment to ensure that Central Government, both in respect of employment and service delivery, not to mention its supervisory functions in respect of Local Government, public bodies and quangos, would demonstrate how good intentions were to be translated into good practice. At the time of the passage of the Act Central Government could not be held out as a model employer or service agency. As there was no intention to change this position at the time the CRE was created the potential for conflict between the CRE and Central Government was clearly present.

Second, and of particular relevance to housing, Local Government was not operating from a clean sheet: existing policies which had led to structural disadvantage might be altered but there was no provision for positive discrimination to counteract patterns of disadvantage which had emerged. This is illustrated by housing developments in Birmingham's inner-city wards from the 1960s. John Rex (Benyon and Solomos, 1987: 103) has referred to the Labour-led Birmingham City Council deciding to appoint a liaison officer for coloured people, a former colonial policeman, which in itself symbolised the need for policing an alien element in the population. Newcomers were not allowed to go on the council housing list for five years and even when they did get on the list they were not given points for length of residence. At first those excluded from council housing were left to solve their housing problems themselves, a number buying large old terraced houses and financing the purchase by letting rooms. As things developed all those who had difficulty in getting council houses turned to the immigrant lodging house proprietors and the result was a concentration of all the problems of social pathology and of racial conflict in one area. This was the situation in Sparkbrook which was described by Rex and Moore (1967). By the early 70s Birmingham City Council had set about providing an alternative housing system, including a policy of giving council mortgages to immigrants in areas where the building societies would not lend and also encouraging the housing associations to convert properties for letting to categories of tenant excluded from the council housing system. Together these two policies provided housing for immigrants and especially West Indian immigrants in segregated areas. Thus Sparkbrook became largely Pakistani, Handsworth Caribbean and Soho Indian. The successive censuses of 1961, 1971 and 1981 showed an increase in the proportions of such ethnic minorities in these areas and Rex concluded:

> It was now clear that Birmingham's policy for Black and Asian immigrants was to ghettoise them, when in due course it came to have race relations equal opportunity policies there were also inner city policies for the inner city internal colonies where the colonial immigrants lived. Birmingham was, in fact, an apartheid city.

Clearly then the de facto situation in respect of certain local authority housing had already crystallised and, however successful strategic formal investigations and individual complaints might be in terms of the revitalised legislation, the impact of the new law was bound to prove problematic and had substantial hurdles to overcome.

Third, the experience of the Race Relations Board in respect of judicial interpretation of the 1965 and 1968 Acts was similarly unpromising. In describing the various provisions of the Act in outline and their interpretation by the courts, this chapter has largely confirmed what might readily have been anticipated at the time of the passage of the 1976 Act. As there was no tradition in either jurisdiction of interpretation of civil rights legislation to promote the interests of disadvantaged groups, a fact to which more extensive reference has been made in the opening chapter, there was a need first for guidance and direction on the approach to be adopted to statutory interpretation by the judiciary, second for a significant input of training in respect of judges and tribunal members and third for a commitment to changing the white face of the legal establishment: these weaknesses the Act did not address.

Fourth, there was general acknowledgement that the new provisions were a significant improvement on the 1965 and 1968 Acts but there was no expectation of an enthusiastic pursuit of individual complaints. While the PEP reports in the mid 1970s confirmed the continuance of widespread discrimination, with the exception of discriminatory advertisements, it was apparent that the 1976 Act would have to prove itself before those discriminated against would seek to take advantage of the potential remedies. Moreover it was also apparent from previous court decisions that monetary damages awarded by the courts were not going to prove an incentive to litigation even if marginally improved by the prospect of a compensatory award in respect of hurt feelings.

Fifth, among the intrinsic shortcomings of the Act it was apparent that indirect discrimination would prove a novel concept to courts and tribunals. The extent to which the CRE and individual complainants could rely on disclosing a pattern of events prior to the Act complained of as demonstrating the need for a conclusion to be drawn that discrimination had occurred and, thereby, shifting the onus of proof, was likely to be problematic. Moreover the background information and analysis required for such an approach might well be beyond the individual litigant, particularly in respect of industrial tribunal proceedings where legal aid was not available. The complexity in proving discrimination is illustrated in the diagrammatic maze comprising Fig. 5.1. It became apparent therefore that the CRE would play a critical role, not only in respect of formal investigations but also in respect of its ability to assist a wide range of complainants in terms of section 44 of the Act: even where such assistance had been made available and the requisite analysis had been undertaken the willingness of the courts to accept an innocent but unverified explanation of events remained to be tested. The lack of resources available to the CRE as the enforcement agency together with the lack of structures within Central and Local Government to complement its work was bound to inhibit the implementation of an effective strategy.

Sixth, the idea of the Non-discrimination Notice was

Figure 5.1: Unlawful discrimination: Race Relations Act 1976

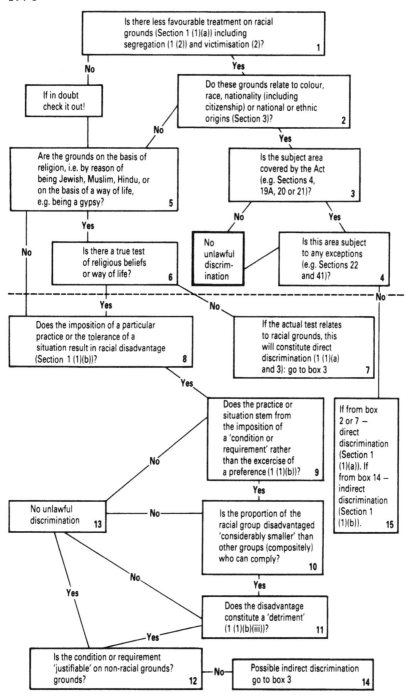

untested: the extent to which compliance was necessary to avoid sanctions and the mechanism for enforcement were issues given cursory consideration.

Seventh, the Act was largely reactive rather than proactive. The opportunity for positive discrimination, for ensuring adoption and compliance with the Codes of Practice and for ensuring that Government consulted the CRE in the adoption and review of all policy was minimal or non-existent.

Lastly, as with the 1965 and 1968 Acts, the 1976 Act was based essentially on individual redress: despite the fact that the concept of indirect discrimination, by its essence, addressed policies, procedures and practices which disadvantaged a whole section of the community, that section had no access to class action. With few exceptions both remedies and sanctions would be piecemeal. The salutary lesson of a multi-million-pound award of damages, as experienced in the US, was not available and, to that extent, financial prudence alone was not going to demand a review of existing practice whether in the public or in the private sector. It seemed more likely that those concerned with housing provision, as with other services, who discriminated indirectly on racial grounds would not be caught and, even if caught, would be released unchastened and, given the climate of public opinion, probably unrepentant.

Part II

THE LAW IN PRACTICE

CHAPTER 6

URBAN PLANNING

PLANNING AND THE HOUSING MARKET

One of the most fundamental changes in the structure of our cities since the war can be found by looking at the housing market. In 1947 only 26 per cent of households in Britain owned their own home, only 13 per cent occupied council housing and a massive 61 per cent rented from private landlords. By 1972, half the households in Great Britain were living in owner-occupied houses, the local authorities and new town corporations provided accommodation for one-third and the private sector had been reduced to a mere 18 per cent (Abrams and Brown, 1984: 31). There were three principal reasons for this restructuring of the housing market. First, when the Tories came to power in 1951 they were committed to a property-owning democracy and were able to extend private ownership in a favourable economic climate: economic and social conditions facilitated the widespread formation of small households able to acquire a mortgage and the benefits of owner-occupation, accelerating the decline of the private landlord. Second, just as Labour had come to accept, at least in part, the Tory commitment to private housing, so the Tories when in government accepted the need for local authority housing as a necessary provision for a sizeable number of the population and this recognition was reflected in many of the renewal and urban development programmes of the 50s and 60s, resulting in local authorities supplanting many private landlords and where need be many owner-occupiers. Third, the economics of private landlordism had become increasingly unattractive. David Eversley commented in 1975 that the rising price of even old property, the improvement grant obtainable on a very large scale after 1969 with no ties as to the future of the premises and the accelerated processes which created more small households with relatively high purchasing power all played a role in the demise of the landlord (New Society, 12 January 1975). Evidence in the first report of the Environmental Committee of the House of Commons in 1982 (House of Commons, 1982: Vol. 3, Appendices) demonstrated that rents between 1970 and 1980 in the private sector failed to keep pace with inflation. Even in London, where the figures for unregistered rents were some 70 per cent above those charged elsewhere, the returns on private rental were deemed by most landlords to be grossly inadequate and in consequence many were anxious to sell (Elliot, in Abrams and Brown, 1984: 31).

From 1979 and the election of the Thatcher Government, important changes in the housing market have been effected. Through the housing legislation of 1980 enabling the sale of local authority housing, which increased from 85,000 in 1980 to 205,000 in 1982, together with the severe restrictions on local authority spending, particularly in respect of new public housing, for the first time in post-war years there has been a diminution in public-sector housing. As a proportion of all housing stock it diminished from 29.6 per cent in 1977 to 26.9 per cent in 1982 in England, with a marginally greater diminution in Wales and a lesser diminution in Scotland over the same period (Department of Environment, 1983a). The GLC survey of private tenants in London (GLC, 1986b) estimated that the decline in the private rented sector was continuing at 7.5 per cent per annum and, in London, was attributable to three main mechanisms: first, slum clearance, second, the transfer of existing properties to other types of tenure, principally to owner-occupation, either through sale with vacant possession or to a sitting tenant, and third, through the concentration of new housing investment almost exclusively in other tenures. With reference to the first the Department of the Environment estimates that of post-war demolitions nationally, 80 per cent have been of private rented dwellings and that these 'slum clearance' losses account for around a quarter of the sector's post-war decline in England. A GLC survey of demolitions in the early 1970s found that private rented dwellings accounted for around 56 per cent of London's demolitions in the post-war period. With reference to investment, the Government's public expenditure figures show that by 1984/85 the annual cost of subsidy to owner-occupation by way of mortgage tax relief was £3.5 billion and this figure contrasted sharply with the Government's subsidy to the local authority housing sector of £1 billion for that year: although the private rented sector received no comparable subsidy, indirect subsidy through housing benefit expenditure was estimated at £500 million for that year. The Government Expenditure Plans 1986-87 to 1988-89 (Treasury, 1986: 20) demonstrate a sharp diminution in the proportion Central Government allocates to local authority housing, decreasing from 7.6 per cent of public expenditure in 1978-79 to 4 per cent in 1985-86 with continuing diminution to 3.7 per cent in 1988-89.

While fiscal policy, in relation to physical land use planning, may not have directly affected the assembly and release of land for both public and private housing, it has certainly affected the radical change in tenure patterns which have been effected post-war. It is indicative of the low priority given by Central Government to planning as an element in social engineering in order to create the desired mix of housing tenure and a redistribution of resources to enable those on low incomes to have access to decent housing of whatever tenure. Nonetheless, physical land use planning is far from irrelevant even in the eyes of the Thatcher Government, whose views, expressed in the White Paper Lifting the Burden (HMSO, 1985: see DoE Circular 14/85), clearly characterise both development plans and the planning control system as, at least on occasion, an unnecessary incursion into free market forces inhibiting the release of private capital into desirable development. In Government's view there should always be a presumption in favour of development unless it would 'cause demonstrable harm to interests of acknowledged importance'. Reasons for refusing planning consent must always be 'precise, specific and relevant' and planning authorities should avoid unnecessarily onerous and complex controls. Inherent in the concept of lifting the burden of planning controls was the 1980 Local Government Planning and Land Act establishing Enterprise Zones and the Housing and Planning Act 1986 introducing the concept of Simplified Planning Zones, both of which are designed to accelerate development and to free planning permission from unnecessarily bureaucratic control.

In addition the proposed abandonment of structure plans and the introduction of district-wide development plans in England and Wales will remove a strategic element in local authority planning which will accentuate piecemeal planning for local housing need. The 1986 consultation paper The Future of Development Plans intimated the death sentence on twenty years of county structure planning in England and Wales and this was confirmed by Government in its White Paper issued in 1989. Many of the consultees shared the view of the Association of County Councils that the proposed reform was throwing out the baby of coherent planning with the bureaucratic bathwater (Planning, 1989). Without the legal status of structure plans, the replacement - strategic county planning statements - may prove an ineffective device for guiding District Plans and development control. There is existing evidence that the interpretation and revision of plans in respect of land supply for housing are private-sector demand-led (Bate and Burton, 1989) and that the building lobby, particularly the House Builders Federation, has been influential in this process: the new planning regime will accelerate this trend. It must also be recognised that the Government's promotion of Urban Development Corporations effectively bypasses local democracy and facilitates the dominance of commercial interests in housing development proposals. This is in keeping with the Government's promotion of tenure change from the public sector and the diminution of local authority influence over housing provision. In the metropolitan districts and London boroughs Unitary Development Plans (UDPs) are replacing the two-tier development plan (see Local Government Act 1985, Section 4 and Schedule 10, and DoE Circular 33: 1988).

Perhaps the greatest concern about strategic planning being capable of reflecting a cross-section of need as opposed to commercial demand has emerged in London. The London Planning Advisory Committee (LPAC) commissioned a report on housing and planning by Coopers and Lybrand which forms part of the LPAC response to the DoE's strategic planning guidance issued in March 1989 (Inside Housing, 1989, Vol. 6, No. 20: 8). This report concluded that access to housing in London would only be secured for all those who legitimately require it if action was taken on a scale 'not yet seen or even contemplated'. The report suggested the need for a housing agency capable of overall planning and market intervention. The zoning of land for rented accommodation and extended use of Section 52 agreements between planning authorities and developers to increase the provision of low-cost housing was also advocated.

The LPAC was not convinced that the provision of 200,000 extra houses in the capital from the private sector in the period 1991-2001 (as proposed in the DoE draft) would be sufficiently geared to meet the need for low-cost housing, of whatever tenure. In the draft the only mention of low-cost housing and ways of achieving it was through conversions. Land zoning and the use of special planning controls, as adopted in rural areas, provide evidence of the legitimacy of planning in securing housing provision to meet special needs, but, given the Government's ambivalence towards planning as market intervention, the reconstructed development plan process is unlikely to be used to pursue such objectives on a broader scale.

The Nuffield Committee of Inquiry into Planning asked the question 'Why have planning at all?' (Nuffield, 1986: 55) and in answer suggested that the fundamental principle of planning control appeared to be based on two propositions: a common recognition of land as a unique resource and the central importance of environmental controls in creating and conserving acceptable living standards. There is, it stated, an inherent public interest in all decisions about development and it is necessary that the 'social costs' of any development should be taken into account before it goes ahead. Such a restricted interpretation might be characterised as 'damage limitation'. But many go beyond this and see the need for planning as a mechanism for conserving and creating communities with the quality of life sustained by access to a full and balanced range of services, facilities and opportunities for work and leisure. A view was also expressed that planning may be seen as an instrument for the redistribution of wealth so that in the interests of social justice land is allocated according to the needs rather than the means of the people. The Thatcher Government clearly sees planning as being confined to an essentially regulatory role, standing back from the market and judging the initiatives coming from that market by reference to whether or not the 'public interest' is served. On such a view the essential role of the planning system is to minimise the disharmony or the social costs that arise from individual decisions about land use and development. The difficulty with such a view is that market forces and private investment may well not respond to the directives inherent in the development plan system in the way anticipated and consequently development in certain areas anticipated by the plan may well not take place because of the lack of private interest and conversely development not anticipated in the plan may be effected because of the continuing pressure from the private sector and its refusal to be channelled into the preferred options. Moreover, the absence of substantial public investment, whether in infrastructure, public housing or community and recreational facilities, may well not only inhibit private investment but also make that private development which has taken place unattractive or socially undesirable because of the lack of a coherent context.

In such a public investment vacuum in development planning it is the most disadvantaged who will suffer because private investment will not make a sufficient return on investments designed for their benefit. Accordingly the planning system so far as it fails to be involved in the redistribution of wealth does not merely perpetuate the status quo in relation to disadvantage but becomes, albeit involuntarily, an agent of its extension. As Elliot has observed (Abrams and Brown, 1984: 23) the 1947 town planning legislation saw the convergence of two very different views of planning. One saw it as the underpinning of private enterprise by the public provision of infrastructure and incentive and the other saw it as an instrument for a more equitable distribution of resources: because of the fear of urban sprawl and the need to preserve the countryside the principal aim of post-war planning was deeply conservative, seeking to prevent change and seeking to prevent the continuation of haphazard suburban sprawl. But the alliance made between these two essentially conflicting views was largely removed in the 1950s under a Conservative Government when planning in Britain was developed as a regulatory mechanism and the concept of the creation of balanced communities was largely based on an assumption of leadership by the middle class, as evidenced by the creation and development of new towns and the creaming off in their expanding populations of the skilled and semi-skilled manual workers from the declining core of inner cities.

While many saw this as the provision of a broad social mix in a movement towards the destruction of class barriers others appreciated that the worst-off and least powerful political groups, including ethnic minorities, were constant losers in the planning initiatives that took place post-war. The cracks in the fragile alliance of divergent interests exposed by Tory policies in the 1950s were translated into social conflict in relation to the effects of urban renewal programmes, the boom in office building and city-centre development in the 1960s with the resultant recognition that planning was not the rational non-partisan business which its proponents had frequently professed (Elliot, op. cit.). The 400,000 housing units a year promised by the Tories in the 1960s and matched by Labour from 1964 were translated, by inaccessible and insensitive planning machinery, into ill-conceived, high-rise accommodation, frequently dependent on unproved system-build techniques and located in vast suburban estates far from areas of work, lacking social infrastructure and of a size which defied any regeneration of a sense of community enjoyed by the inhabitants previously located in inner-city slum clearance areas. Such developments exposed the planning apparatus in its then form as being ill-equipped to deal with complex economic and social problems in the cities.

The introduction of the Urban Programme in 1968 marked a temporal acceptance that the mainstream town planning legislation would not prove a useful vehicle for focusing Central Government policy on urban regeneration. Thereafter the traditional Urban Programme was adumbrated by the creation of partnership and programme authorities by the Inner Urban Areas Act 1978. Together with the shift to rehabilitation through Housing Action Areas created by the 1974 Housing Acts and more specific initiatives such as the Priority Estates Project established in 1979, these illustrated a move towards 'additionality': inner-city problems were localised and particularised and solutions became tailored additions to mainstream planning. In essence, these processes crystallised land use planning within a straitjacket of the physical environment and, by segregating issues of community development into 'additional' programmes, structured an explanation of deprivation and racial disadvantage into a local social pathology: in doing so the State transferred responsibility to the victim and denied the causal relationship between land use planning and inner-city deprivation.

Nuffield observed that the most general complaint is of the irrelevance of land use planning as presently practised to the main contemporary trends which will determine the problems with which planning will have to deal: the growth of giant corporations, the shift of manufacturing industries to the countries of the Third World, the growth of the nuclear industry and of other technologies deemed capable by their opponents of destroying the environment, population growth, changes in demographic structure and the centralisation of government. There was a considerable body of opinion that current planning policies failed to give enough weight to problems of the distribution of wealth and opportunity, particularly to the problems of regional inequalities and the fate of declining areas. Particular stress was laid on the lack of serious response to the problems of inner-city deprivation and the problems of creating robust economies in the poorer regions because of changes in regional policy, Local Government failure and the inability to channel development to needy areas. However, Nuffield, almost inevitably, referred to another strong body of opinion that current planning practices gave insufficient weight to the needs of development and did not allow in structure and local plans for the release of enough land for commercial and housing needs. Failure to release enough land for housing produces high prices, creates scarcity and inhibits the mobility of labour, especially in rural communities. Similarly, refusal to use the market or to release public land for development in inner-city areas has produced dereliction rather than development, confused planning intentions with market inclination and led to false notions of land values and land market glut (Nuffield, 1986: 71).

THE SUPPLY OF LAND FOR HOUSING

Both the conflict over the aims of the planning process and the conflict over its impact have been reflected in relation to the supply and allocation of land for housing. In some countries physical land use planning has involved mandatory zoning and has permitted the identification of low-cost public housing, which in turn may result in a dispro-portionate allocation of public tenancies to minority groups. While in the UK the allocation of land for housing purposes both in the development plan and in the exercise of development control functions through the grant of planning permission, has attracted litigation over the designation of land for public housing purposes (Lowe and Sons v Burgh of Musselburgh (1974) SLT 5 and R v London Borough of Hillingdon, ex p. Royco Homes (1974) 2 All ER 643), the dominant view is that planning legislation should not be used to regulate matters which can be dealt with under other statutes, including those relating to housing (Young and Rowan-Robinson, 1985: 227).
    With regard to slum clearance and the rehabilitation of

housing, the deployment of Housing Action Areas and Environmental Improvement Areas has, in general, not depended on the statutory planning process. This is principally because demolition does not of itself constitute development and because repair and improvement work will generally fall outwith the definition of development. Generally speaking a material change of use as defined in the planning legislation will not be involved and Central Government approval of a scheme such as the declaration of a Housing Action Area may often constitute deemed planning approval so far as this is required. In addition, much of the development carried out by a local authority, or a housing authority, in the normal course of its duties constitutes permitted development in terms of the relevant town and country planning General Development Order and will not require express planning permission. Where express planning permission was required, until fairly recently, there was not requirement to disseminate a notice of intention to develop. Not infrequently, therefore, in such circumstances, the planning authority might grant planning permission to itself as a housing authority without widespread knowledge of what was occurring.

In summary, then, while the public might have an opportunity of expressing a view in respect of the designation of land for housing development in the local plan, that plan would not allocate such land specifically for public or alternatively for private housing purposes. Although specific approval would be required in order to develop land so allocated for private housing, the fact that it had been so designated in the local plan would be a powerful and relevant consideration for the local planning authority in determining whether to approve or refuse the individual application. Where such land was being developed for public housing the planning authority, most frequently being the housing authority, could grant itself permission to develop, with restricted opportunity for public challenge. In respect of housing improvement and rehabilitation the need for express planning permission was frequently avoided. Strategic decisions on the allocation of land for housing purposes in the development plan will be located in the structure plan, which is a written document. These strategic decisions contained in the structure plan have attracted the concern of private developers, as evidenced by their attendance and submission of evidence at Examinations in Public, the vehicle by which strategic issues identified by the Secretary of State are exposed to public debate, but they have seldom attracted either broad public concern or the involvement of local community groups: a partial explanation for this differing interest is the perception of relevance.

The major developer who has interests throughout the region or county covered by the structure plan will see his potential for land assembly and development being enhanced or retarded by strategic decisions within the planning process at structure plan level but this will not be true of the local amenity group, a local tenants' association or a black residents' group. In contrast Central Government will frequently see the structure plan as an important process in securing what it considers to be the appropriate degree of land allocation, particularly for private development. Thus, while some authorities, such as Strathclyde Regional Council, have attempted to use the structure plan to ensure infill housing development by restricting the release of housing land on green-field sites, Government may revise such plans, virtually by dictate, to meet the preference of developers in the supply of housing of previously undeveloped or derelict sites (see Strathclyde Regional Council v Secretary of State for Scotland (1989) 27 SPLP 47).

THE DEVELOPMENT PLAN AND RACE

While some discretion is given in the regulations as to what matters may be covered by the structure plan, generally the legislative reference to 'development and other uses of land' has been interpreted in a restrictive manner, forcing structure plan policies to be essentially physical and land use-based. This approach has been endorsed by the Secretary of State for the Environment (see Circular No. 23/81). However, each of the Town and Country Planning (Structure and Local Plans) Regulations operating in England and Wales and Scotland requires structure plans to indicate the regard the planning authority has given to social considerations. Government circulars, whether issued by the DoE, the Welsh Office or the Scottish Development Department, have failed to identify the racial dimension in the performance of the planning function. Accordingly the failure of planning authorities to include in their surveys of their areas an assessment of the needs of different ethnic minority groups has been ignored by Central Government. As a result it is no surprise that a survey of all Scottish planning authorities in 1980 demonstrated that there was no reference to ethnic minorities in any structure or local plan (MacEwen, 1980a). Cowen (1983: 17) recounts that at the Gloucester Plan public inquiry the City Planning Officer contended that the District Plan was not the place to make any statements about the black community's needs. He contended, however, that they would benefit from the plan's housing proposals, particularly for designated improvement areas. But since ethnic monitoring was not undertaken, such a contention seemed speculative if not spurious. In this respect the requirement of Section 71 of the Race Relations Act 1976 for local authorities to promote equality of opportunity in their various functions has been a dead letter. Both the RTPI and the CRE have recognised this and, as a consequence of representations made to the DoE, the revised memorandum on structure and local plans includes the CRE and where appropriate local Community Relations Councils and racial minority groups as appropriate consultees (see DoE Circular 22/84 dated 8 September 1984). That memorandum by paragraph 2.25 (dealing with publicity and public participations) advises planning authorities:

> To bring policies and general proposals to the attention of identifiable groups of people who will be particularly affected by them.

And they may make a particular provision for sections of the community which, for example, because of language problems, may not be reached by the authority's normal publicity measures.

This advice, which relates to structure plans, is repeated (para. 3.37) in respect of local plans. In both instances social considerations are recognised as relevant to policies and proposals. Paragraph 4.10 states that policies should be related 'to social needs and problems, including their likely impact on different groups in the population such as racial minorities and the elderly'. Annex C to this memorandum includes the CRE as one of the twenty-three public bodies to be consulted in England and one of the eighteen to be consulted in Wales regarding the preparation of development plans. No such Central Government advice is proffered in Scotland, where general guidance on the development plan process is provided in Planning Advice

Note (PAN) 30 (local) and PAN 28 (structure). Moreover the reference to ethnic minorities in the DoE/Welsh Office memorandum is likely to prove inadequate. Gillon (1981: 162) observed that even where a local authority had policies on racial discrimination the results in terms of planning were usually disappointing: in submitting comments to the Lambeth Draft Development Plan, an Asian community action group observed that it was totally lacking in explicit policies to deal with problems experienced by Asians in employment, education and housing. In recording that black groups had found planners' and councillors' knowledge of the community structure to be very limited, Gillon observed that there was an urgent need for plans to adopt a new perspective. Brown (1981) argues that planners generally venture no further than charting the spatial distribution of minorities, based on census data which is invariably out of date: 'the analysis stops there because other facets of race are not understood or are considered the province of housing or social service departments, and traditional Local Government antagonism prevents realistic dialogue between them'. Cowen (1983) emphasises this point. A team of sociologists from Liverpool University, in evidence on racial disadvantage submitted to the Home Affairs Committee, noted that in neither of Liverpool's two major planning documents - the 1979-82 Partnership Programme (1979) and the Merseyside County Structure Plan (1979) - was there one explicit reference to the needs and problems 'of what is in fact one of the most long-standing black communities in Britain, whose disadvantages have by now received sufficient documentation to merit specific attention' (Ben-Tovim, 1980). Consequently a reference to consultation and social needs will be meaningless without the radical change advocated by Gillon and ignored by the DoE.

Clearly, in law, the household structure and socio-economic and religious/cultural characteristics of particular ethnic minority groups are relevant to the planning process and the traditional 'colour-blind' approaches, in so far as they ignore these factors, may result in indirect discrimination. Such discrimination in the planning process relating to development control was found unlawful in the case of CRE v Riley (Manchester County Court, July 1982, unreported). However, given the decision of the House of Lords in the case of R v Entry Clearance Officer, Bombay ex p. Amin (1983) 2 AC 818, where it was held that the expression 'provision of goods, facilities and services' in the anti-discrimination statutes applied only to activities or matters analogous to those provided by private undertakers, some doubt was cast on the likelihood of the Riley decision being upheld on an appeal on similar facts to the House of Lords. As a result of lobbying by the CRE and RTPI the Housing and Planning Act 1986 effected amendments to secure that the planning functions of planning authorities fell within the scope of a new section (19A) of the Race Relations Act 1976.

A survey of planning authorities conducted under the auspices of the National Development Control Forum (NDCF), while confirming the persistence of colour-blind attitudes, also provided examples of good practice in local plans (Owen, 1988).

Southampton (Policy 13 of draft City Plan) Specialist Housing: 'To encourage the provision of specialist housing to meet the needs of the elderly and minority groups disadvantaged in the housing market by permitting schemes for change of use of dwellings or new development provided that these are similarly located for the purpose intended and would not adversely affect neighbouring residential areas.'

Brighton In monitoring the effect of the policies in the plan and in collecting further information necessary for the effective planning of the borough, the Council will pay particular regard to the current lack of comprehensive information on the composition and needs of ethnic minorities in the town.

Nottingham 'By recognising that different communities have different cultures, values and aspirations this Plan can help to reflect these needs and by implementing its policies and proposals in a positive way the City Council can contribute to the elimination of discrimination, the promotion of equality of opportunity and the improvement of race relations to the benefit of all.'

Leicester Some local plans reflect the needs of ethnic minorities by specifically relating to:
1. Need to retain larger dwellings.
2. Provision of community facilities.
3. Places of worship.
4. Shopping policies geared to specific needs.

Consequently, plan-making should accommodate two areas of concern: first, in so far as development plans are a vehicle for meeting community needs, the needs of ethnic minority groups should be assessed, like those of other groups, and reflected in policies and proposals put forward for consideration. Second, where proposals, such as major shopping outlets or the designation of Action Areas - may have a disproportionately adverse impact on ethnic minority groups such impact should be made known and alternatives explored. However, it must be recognised that with the exception of the special provisions to meet education, training and welfare needs (sections 35 and 37 of the Act) the Race Relations Act does not generally provide for positive discrimination in favour of any particular group and, consequently, while it may be unlawful, by way of indirect discrimination, to ignore the needs of a particular group in relation to planning, it would also appear to be unlawful to advantage one group at the expense of another. A prerequisite to achieving such objectives is an assessment of the characteristics and needs of various groups as provided, for example, by the Lothian/Edinburgh Ethnic Minority Profile (SEMRU, 1987) at the survey stage of the plan-making process and the establishment of appropriate consultative arrangements. Although few planning authorities have, as yet, developed such assessments and arrangements, no court action has been instituted, nor formal investigations initiated by the CRE in respect of those authorities who have ignored these requirements. One consequence of assessing needs will be the emergence of unsuspected disadvantage. A study of black financial businesses in Brixton (including banks and building societies) found that they were being prevented from establishing a foothold in the retail core of Brixton town centre, in accordance with local planning policy. White building societies and banks were established before the policy had been adopted. To redress the imbalance, the town centre plan now includes a policy in favour of 'non-retail' in respect of black business directly serving the ethnic minority community as an exception to the 'retail only' policy in the core area. The former policy was clearly discriminatory indirectly while the latter is directly discriminatory, equating with 'affirmative action' to redress historical racial disadvantage (Duffield, 1986). Because the Race Relations Act 1976, as we have noted, does not accommodate such exemption, the latter policy is likely to breach section 19A -

even when it merely compensates previous disadvantage - while the former policy - which created the racial imbalance - may not be unlawful because it is indirect and potentially justifiable on non-racial grounds. In these circumstances, the Act appears to be powerless to prevent racial discrimination and yet outlaws positive measures to promote an equitable resolution. Although this dilemma is not addressed generally, advice on what should be done is contained in 'Planning for a Multi-racial Britain', CRE/RTPI (1983) and 'Planning in a Multi-racial Scotland', SCRE/RTPI (Scottish Branch) (1983). Prior to its demise the GLC also produced guidelines in this area (GLC, 1986c).

Lambeth, which, as noted above, ignored ethnic minority needs in its Draft Development Plan, following a change of administration took heed of its own communities and of the RTPI/CRE advice: a planning resource centre was set up - financed by but independent of the Council; a standing black community working party was established; local plan documents, from the District Plan to Brixton Town Centre Action Area Plan, all have specific chapters on black needs and each policy is assessed regarding its impact on the black community and race relations (Duffield, 1986).

In England and Wales, as noted previously, the introduction of Unitary Development Plans for metropolitan areas and District-wide Development Plans elsewhere together with the substitution of county planning strategic policy statements for structure plans will provide a new development plan regime for the 1990s. The impact of this regime on strategic planning for ethnic minority needs at the county level is difficult to predict: it may be argued that the vagueness of structure plan policies in this area, when indeed it is addressed, leaves much to be desired, enabling district planning authorities such breadth of discretion that the extent and method of implementation depend on the political commitment of the local planning authority. More frequently, as demonstrated in the relationship between the Leicester County Structure Plan and the East Leicester and North East Local Plans (Farnsworth, 1989) beneficial provisions in local plans are a response to local surveys and initiatives concerning which the structure plan policies are largely irrelevant. On balance, therefore, the demise of structure planning in England and Wales is unlikely to have a significant effect on planning for ethnic minority needs. This will remain the discretionary, and largely ignored, responsibility of the district authorities.

In October 1988 Planning Practice and Research (PPR) in conjunction with the Association of Black People and Planning (AB Plan) held a conference on Town Planning and Racism from which ten strategic recommendations emerged (Griffith and Amooquaye, 1989). One related to structure planning expressly: planning gain - the public benefits which a planning authority may secure by the discharge of its responsibilities - it was argued, should be seen as one part of an overall anti-racist strategy of local authorities and planning departments in particular. Having assessed local black needs, planning gain should be incorporated into planning policies such as Unitary Development Plans, Structure Plans and planning briefs. It was further suggested that detailed research was needed on the legal basis of 'social gain' in order to establish how far it is permitted by law.

DEVELOPMENT CONTROL AND RACE

A number of studies of development control, for example, those in Leicester (Farnsworth, 1989) and Glasgow (Richards, 1981), have shown a disproportionately high refusal rate in respect of planning applications from ethnic minority applicants: it has also been demonstrated that such refusals, while frequently relating to applications for change of use to hot food shops and minor operations affecting housing in particular in conservation areas, have not been confined to any particular category of development proposal. Both advisory documents, referred to above, endorsed the concept of monitoring applications and revising development control policies in the light of appropriate analysis. An RTPI (Scottish Branch) survey of Scottish planning authorities was conducted in 1987: this concluded that a negligible number of authorities had taken any of the advice. The NDCF Survey (Owen, 1988) showed that, of the District Authorities in England and Wales, only Leicester monitored refusal rates in planning applications and it was the only authority to embark on major research projects or to use Section 38 of the Race Relations Act to employ minority graduate trainees. Rochdale introduced a scheme in 1988 to provide ethnic minority places. Clearly the majority of authorities see the ethnic minority dimension of planning as being largely peripheral. To some, therefore, development control is concerned only with land use issues, something which should lead to little racial controversy. Town planners, as with other Local Government officials, may have to face issues in respect of applications for the conversion or extension of dwellings for culturally defined modes of life such as the conversion of two adjacent dwellings into one to accommodate a large extended family or an extension to a dwelling to provide a private prayer room or the provision of hostels for special cases such as refugees, contract or guest workers. What the studies undertaken clearly indicate is that in considering the exercise of discretion in respect of such applications many authorities have little familiarity with, or understanding of, local community practices and attitudes to environmental change and, moreover, have very little understanding of the concept of indirect discrimination contained in the race relations legislation.

PLANNING CONSIDERATIONS

While recognising the importance of such institutional disadvantage in respect of planning applications both to the individual applicant and cumulatively to the ethnic groups affected, of equal importance must be the ability of the development control process to accommodate broader social and economic considerations. In R v Hillingdon London Borough Council ex p. Royco Homes (1974) QB 720, the Queen's Bench Division quashed planning permission for the development of land for residential purposes which was subject to conditions restricting the first occupiers of the proposed houses to persons on the council's waiting list and according them security of tenure for ten years. The court took the view that the conditions effectively required the applicants to take on a significant part of the local authority's housing duties and as such were unreasonable. Similarly in Lowe and Sons v Musselburgh Town Council (1973) SC 130, the First Division of the Court of Session quashed planning permission for a residential development which was subject to a condition allocating four out of every five of the houses to meet local authority housing need. In the view of both the Lord President (Lord Emsley) and Lord Cameron the condition went far beyond matters relating to the development of land (Young and Rowan-Robinson, 1985: 223). In relation to an application for housing, it would be difficult for the planning authority to attach conditions to the grant of planning permission which, in effect, would secure that such need would be met by the provision, for example, of sheltered housing: because building regulations may prescribe minimum standards in relation, for example, to space and ventilation, the courts have decided that the internal arrangements of a building for which planning permission is sought cannot generally be a proper planning consideration (see Sutton London Borough Council v Secretary of State for the Environment (1975) 29 P&CR 350). Because the commercial viability of a particular undertaking is the risk of the developer and is not a relevant concern of the planning authority, in housing matters it would appear that the presumption in favour of the grant of planning permission will enable a developer to obtain planning permission for private housing development provided that is not incompatible with adjacent existing uses and does not conflict with proposals in the development plan irrespective of whether or not the housing provided meets the priority needs of the community in a particular area. Self-evidently the system, in this respect, fails to protect the interests of disadvantaged groups, including ethnic minorities. The physical land use planning system and the decision of the courts which regulates its application will frequently appear powerless in pursuit of socially desirable goals.

PLANNING, RACE AND THE DOCTOR'S PRESCRIPTION

Dr Bersano (1970) suggested that the fractures of American society were a reflection of unfulfilled political promises and a rejection of national ideals on a local plane which only gained broad acceptance in the 60s. The fractures are too deep to be cured through a bettering of physical environmental conditions - thus urban planning can do little. He observed that, if the underlying reasons for grievances were deeply rooted in slavery, segregation and discrimination, going back to the tradition of hundreds of years, additional provision of open space, more efficient transport, etc. can introduce only insubstantial modifications in a situation of social tensions where the basic demands are for employment, economic opportunities, equity of rights and the abolition of discriminatory practices. He suggests that physical planning can only focus upon, through analysing and advising, the physical goals that must be achieved to provide a decent environment and indicate the political course to achieve that end: it cannot guarantee political decisions, nor does it have real instruments to deal with socially and politically excluded groups or modify the class structure. Yet projects in health, education, law enforcement, employment, recreation, welfare and housing have proliferated. Their relative failure, he opines, is due to the fact that such effort has been short of real necessity and has not coincided with improvement in the conditions of society or with decreasing social tension. The blacks are more or less in a colonial situation of dependence and distinct from the whites, who enjoy economic privilege. Black workers are still concentrated in the least skilled and lowest-paid jobs, and as unskilled jobs decrease there is increased competition from unskilled whites and the ability of blacks to hold on to such jobs becomes doubtful. Owing to city in-migration in the 1950s and 1960s, blacks are more urbanised and metropolitan than whites: the pattern has been black concentration in the core of cities with a white flight to the suburbs, aided and abetted by Government investment in urban motorways, increasing commuter accessibility. Urban planning, in his view, is likely merely to tackle the symptoms without touching the economic and social problems that are the core of a complicated social process. Given these constraints, however, physical planners, in his view, can still aim at limited goals. Better housing and facilities, speedy rehabilitation of city centres and environmental improvement are possible with the commitment of planners amongst others. From this analysis Bersano identified three principal factors in the ability of planning to combat ethnic minority disadvantage:

1. Strong overall metropolitan Governments.
2. Public ownership of development land.
3. Sufficient tax revenues and Central Government aid for comprehensive housing projects.

An unqualified transposition of the American experience into the problems of the inner cities of England and Wales as well as in Scotland would be inappropriate but many of the structural problems of black disadvantage are not dissimilar. As a consequence it may not be purely academic to gauge to what extent his structural remedies are reflected in Britain today.

Previous attempts to secure the public ownership of development land such as the Land Commission Act 1967 and the Community Land Act 1975 were short-lived. The Thatcher Government is ideologically opposed to any further attempts at such control. Moreover policies regarding physical land use controls and Government aid for comprehensive housing projects have had the effect of denying further public investment in the provision of housing for local authority tenants on a relatively low income: the abolition of the Fair Rent system will surely result in a failure of the private rented sector to meet such need unmet by local authority housing, even where subsidised by Housing Benefit and Income Support effective from April 1988. With reference to strong overall metropolitan governments, the abolition of metropolitan counties and the GLC anticipated by the Government White Paper Stream-lining the Cities (HMSO, 1983) not only demonstrates the Government's unwillingness to brook opposition from political opponents at the Local Government level but also its lack of commitment to the concept of an independent Local Government with sufficient resources and independence to secure revenue required to give effect to policies devised at the local level. It is also noteworthy that the Government White Paper, in describing how the functions of the GLC would be carried out after abolition, made no mention of those aspects of its work which had been developed in a novel and comprehensive fashion, at least in comparison with other local authorities, to combat racism and attack racial disadvantage which included the funding of a significant number of ethnic minority projects.

Cumulatively these developments constitute a fundamental attack on the ability of the planning system to be involved in the redistribution of resources to the less well-off.

But the conservative approach to town planning dates back to the period of post-war reconstruction. By the late 1960s it was evident that, whatever the benefits of a narrowly focused physical land use planning system, it required both reconstruction to meet community needs, as advocated by the Planning Advisory Group (PAG, 1965), and supplementary measures as a response to inner-city deprivation and the needs of ethnic minority communities. The latter need was to be addressed by the Urban Programme.

THE URBAN PROGRAMME

The traditional Urban Programme was launched by Harold Wilson in May 1968 in response to the growing recognition that poverty and disadvantage had not disappeared. The work of Abel-Smith and Townsend (1965) along with Government-sponsored investigations such as Plowden (1967), Milner-Holland (1965) and Seebohm (1968) had shown, in contrast, that the gap between those in poverty, including significant numbers of black people, was if anything getting wider. The solution was seen as an area-based approach, attacking pockets of relative deprivation and disadvantage. The idea of positive discrimination at an area level was initially developed in the United States and was adopted in the Plowden Report which advocated the establishment of educational priority areas.

The concept of spatial disadvantage implies that disadvantage is either closely tied to the family and the individual's behaviour or, so far as it is related to structural processes operating within society, these processes are of a localised nature and could be effectively combated by enhancing public service provision in such localities up to the norms experienced by the rest of society (Edwards and Batley, 1978). The Local Government (Social Needs) Act 1969, which authorised the Urban Programme, enabled local authorities to obtain 75 per cent Central Government grants for various schemes in recognised areas of urban deprivation. Although Central Government had a clear intention of assisting ethnic minorities, the racial dimension to the programme was largely implicit in that the emphasis was on an area-based approach: the reason for such a strategy was seen as twofold. First, the political sensitivity of discriminating in favour of black communities was a nettle Government was not prepared to grasp and, second, it was also clear that there were significant pockets of deprivation which were either exclusively or predominantly white. This failure to recognise, or at least to tackle, racial disadvantage as a separate component in the experience of urban deprivation has been criticised as one of the main failings of the Urban Programme approach (Edwards and Batley, 1978, Stewart and Whitting, 1983). In July 1974, Roy Jenkins, the Home Secretary, asked the Community Relations Commission (CRC):

to consider the extent to which the needs of ethnic minority communities differ from those of the rest of the population in areas of urban deprivation; to report; and to advise on the implications for community relations policy.

The CRC report was published in November 1976. This report (Home Office, 1976) suggested the racial dimension had three components: first, a greater degree of deprivation, irrespective of socio-economic grouping; second, special clusters of deprivation not experienced, together, by the white population; and third, distinctive causes of deprivation, racial discrimination being a major element which was frequently overlooked or by-passed. The CRC report did not refer to the fact that racial discrimination had been omitted from its terms of reference. It made a number of recommendations relating to monitoring, funding, self-help and community involvement. In addition it recommended that the new race body, the CRE, should work in co-operation with officers in each Central Government department specifically designated and empowered to develop policies to eliminate racial inequality in housing, employment, social services and education and to issue guidelines to local authorities on meeting their duties under the Race Relations Act. A 'Programme for Racial Equality' was advocated to concentrate resources in multi-racial areas and to aid local authorities to examine the effectiveness of their current services in meeting the needs of racial minorities; to adapt services to allow equality of access to minorities (with emphasis on the in-service training of staff and the appointment of professionals from minority groups); and to provide for special needs. These recommendations were largely ignored.

The loose criteria applied to the programme, its speed of construction and the political desire to see it implemented quickly precluded any real analysis of deprivation and disadvantage. Politicians and officials tended to group all problems associated with deprivation, both structural and pathological, together, including racial disadvantage, and label them 'multiple deprivation'. As a result of the aims and objectives of the Urban Programme being confused and ambiguous its implementation became ad hoc and expedient.

Through the absence of any centrally coordinated strategy, funding was dependent on how well organised and experienced local authorities were rather than on how they fared in the table of deprivation. Even those authorities that were able to plug into the financial resources available and establish innovatory projects saw the programme as a small extra resource rather than a coherent structural mechanism for tackling broadly based and deep-rooted problems (Edwards and Batley, 1978; Jacobs, 1986).

The Urban Programme was radically altered by way of the Inner Urban Areas Act 1978, which aimed at tackling economic and physical decay as well as social disadvantage. This Act created a limited number of partnership, programme and 'other designated' authorities with increased resources. Despite the greater structural emphasis and the White Paper's more explicit recognition of ethnic need, the basic perception of deprivation and disadvantage within the Act had not changed significantly. While the traditional Urban Programme remained available to local authorities throughout Britain, all the targeted authorities which have been designated are in England, where the inner-city problems are perceived to be more acute. The areas falling under the responsibility of partnership, programme and 'other designated' authorities contain some 60 per cent of all UK black residents and the racial dimension to the Urban Programme in England and Wales is clearly recognised in the various circulars and ministerial guidance issued by the DoE. Revised ministerial guidelines were issued in May 1985 and stated Government's objectives for urban policy as:

1. To improve employment prospects in the inner cities by increasing both job opportunities and the ability of those who live there to compete for them.
2. To reduce the number of derelict sites and vacant buildings.
3. To strengthen the social fabric of the inner city and encourage self-help.
4. To reduce the number of people in acute housing stress.

The DoE employs eight indicators of deprivation which refer to (1) employment, (2) overcrowding, (3) single-parent households, (4) pensioners living alone, (5) basic amenities in housing, (6) ethnic origin, (7) population change and (8) standardised mortality rate (National Audit Office, 1985: 26). With reference to ethnic origin the relevant indicator is the percentage of residents in households where the head of household was born in the New Commonwealth or Pakistan.

Figure 6.1 shows the increase in the overall allocation to the Urban Programme between 1980 and 1986 in England and Wales, and Figure 6.2 demonstrates a substantial proportional increase to projects with an explicit ethnic minority focus from just over 2 per cent in 1980-81 to approximately 12 per cent in 1985/86. Figure 6.3 gives the distribution of Urban Programme resources in 1985-86, including allocation for the urban development grant introduced from 1983/84. Both programme and partnership authorities are required to produce inner area programmes (IAPs) for the purpose of monitoring but in 1985 the National Audit Commission concluded that there was considerable scope for improvement and, from an appraisal of DoE records, it was not satisfied that the appraisal and monitoring carried out by the department was adequate. In May 1985 the DoE announced new proposals under the Urban Programme Management initiatives (UPM), which addressed a number of the issues raised in the report. However, while the Urban Programme may be designed to assist in the creation of inner cities as places where people want to live, work and invest the Government considers that the achievement of the objectives of the Urban Programme would be difficult and long-term: no grand strategy was appropriate (National Audit Office, 1985: 2). The absence of any grand strategy is fairly clear by reference to Figures 6.2 and 6.3, particularly in the impact on ethnic minorities. Thus while partnership authorities received the largest slice of the Urban Programme cake in 1984/85, comprising 40 per cent of the total allocation for both 1985/86 and 1986/87, only 18 per cent of the total ethnic population are resident in partnership authority areas.

While both programme and partnership authorities allocated 10 per cent of Urban Programme resources to housing projects (7 per cent and 9 per cent respectively in 1986/87) only 5 per cent of the Traditional Urban Programme (TUP) was allocated to housing. In 1986/87 of the £35 million allocated to ethnic minority projects 11 per cent concerned housing. Among all authorities the largest share of Urban Programme funds goes to culture, recreation, health and welfare projects. Whether this allocation reflects the needs and demands of the ethnic minority community throughout Britain can be questioned, as no real mechanism for measuring ethnic need exists, owing to the lack of basic information on socio-economic characteristics at both national and local level.

Figure 6.1: Expenditure on the Urban Programme and on ethnic minority projects (England and Wales) 1980/81 to 1986/87 (£ million cash)

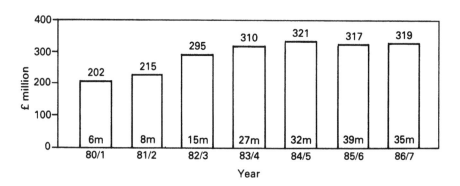

Source: Department of the Environment 1985, 1987

Figure 6.2: Expenditure on Urban Programme ethnic minority projects as a percentage of total Urban Programme (%)

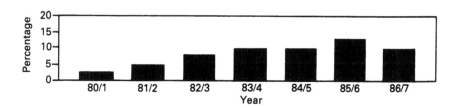

Source: Department of the Environment 1985, 1987

Figure 6.3: Ethnic minority projects 1985/86

(a) Classification of approved total expenditure on Urban Programme (England and Wales)

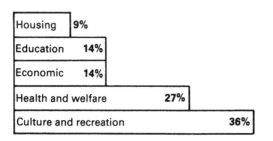

Capital £13.6 million
Revenue £25.2 million
Total £38.8 million
Projects 1,378

Housing	9%
Education	14%
Economic	14%
Health and welfare	27%
Culture and recreation	36%

(b) Partnership authorities

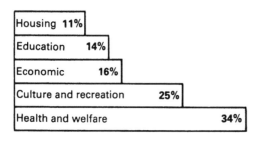

Capital £6.1 million
Revenue £12.3 million
Total £18.4 million
Projects 589

Housing	11%
Education	14%
Economic	16%
Culture and recreation	25%
Health and welfare	34%

(c) Programme authorities

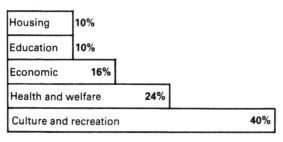

Capital £13.6 million
Revenue £25.2 million
Total £38.8 million
Projects 1,378

Housing	10%
Education	10%
Economic	16%
Health and welfare	24%
Culture and recreation	40%

(d) Traditional Urban Programme

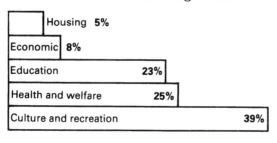

Capital £1.1 million
Revenue £6.2 million
Total £7.3 million
Projects 319

Housing	5%
Economic	8%
Education	23%
Health and welfare	25%
Culture and recreation	39%

Table 6.1: Distribution of Urban Programme resources in 1985/86 and 1986/87

| | 1985/86 | | 1986/87 | |
	£ m.	%	£ m.	%
Partnerships	129	40	127	40
Programme authorities	96	30	102	32
Other designated districts	10	3	9	3
Traditional Urban Programme	44	14	37	12
Subtotal	275	87	275	86
Urban development grants	22	7	25	8
Other UP authorities (including Merseyside Task Force)	20	6	19	6
Total	317	100	319	100

Source: Department of the Environment, October 1987

The most recent ministerial guidelines on the Urban Programme phased out the Traditional Urban Programme and the 'other designated districts' categories - 1986/87 being the last year of operation - and placed emphasis on targeting UP resources 'within a clear overall strategy for tackling the problems of a particular urban area': all 'other Designated Districts' (except Ealing) and ten others were invited to submit Inner Area Programmes which enable local projects to be pursued in terms of the approved programmes but without individual scrutiny by the DoE. Given the diminishing resources made available to the UP, in real terms, such targeting was inevitable and the IAP system, which highlights a priority for meeting ethnic minority need and creating local employment opportunities, has a logic and coherence which the TUP demonstrably lacked. However, the emphasis on local business and revitalising local economic activity is a double-edged sword. Previous projects may have stimulated some economic activity but the benefits to local residents have frequently been marginal.

The Inner Area Studies, initiated by Peter Walker in 1972, had provided an analysis of the Urban Programme in Lambeth, leading to its partnership status being announced in

April 1977. This analysis brought out the association between lack of skills and lack of employment opportunities and the effects of disadvantage produced by racial discrimination. The study's findings also identified two distinct fears of poverty - income poverty and housing poverty - and, somewhat controversially, suggested that the two were to a degree independent phenomena (Deakin et al., 1983: 137). Although meeting ethnic minority needs was, and continues to be, identified as an objective of the programmes, racial discrimination generally or its effect on housing opportunities in particular is not addressed directly: in the course of implementing the programmes the issue of race receded to a position of secondary importance or disappeared altogether (Higgins et al., 1983: 190).

Three explanations have been suggested. First, compensation for disadvantage had legitimacy in relation to place, in keeping with an area-based approach, but not for people. Second, there is perceived political disadvantage in positive action for ethnic minorities because of the electoral white backlash. Third, although the Home Office had initial responsibility for race relations it failed to inject any positive identification of issues into the programme and, when responsibility was assumed by the DoE, the DoE refused to ensure that other departments did so: without any lead the CRE and Community Relations Councils had little leverage to influence strategic policy although, at a local level, race might be given a high profile (Higgins et al., 1983: 192). Despite the IAP approach, the disparity in resourcing ethnic minority projects, let alone in tackling racial discrimination, illustrated by the 1987/88 allocations (DoE, 1987b) clearly indicates that the programme, on the issue of race, is ad hoc and continues to reflect a lack of Central Government planning.

Lawless (1988: 271) in a critique of British inner urban policy post-1979 concluded that of the four objectives in the 1985 guidelines (supra) two, alleviating unemployment and reducing housing stress, did not appear to have been met. Unemployment doubled within the central local authorities of the seven major conurbations between 1979 and 1987 (Department of Employment, 1979, 1987). In housing the £75 million allocated to the Government's Urban Housing Renewal Unit in 1987/88 has to be contrasted with the halving of local authority housing allocations as a whole, £4 billion in 1979/80 and less than £2 billion in 1985/86 in real terms. Moreover, as Lawless also observed in respect of a third objective - strengthening the social fabric - it would be 'problematic' to sustain the position that those ills identified by Scarman (1981) and the Gifford Report (1986) on the Broadwater Farm disturbances have been or are likely to be moderated to any substantial degree by inner urban policies. In this light the fact that the last objective of reclaiming derelict land has been at least partially successful has failed to qualify a view that the policy has been 'totally inadequate' (AMA, 1986: 4) and 'a dismal failure' (Town and Country Planning Association, 1986: 8). In respect of ethnic minorities a recent study of the Partnership Programme in Birmingham (Ball, 1988: 7-22) shows that while the percentage of funding in the core area to ethnic minorities had increased from 10 per cent to 13.8 per cent between 1985/86 and 1986/87, the fact that the core area is nearly 50 per cent black and Asian is a less reassuring statistic: moreover white voluntary groups applied for 12 per cent of the 'Capital Sum' in 1987/88 and were granted 61 per cent: Afro-Caribbeans applied for 14 per cent and received 2 per cent while the Asian groups applied for 22 per cent and received 14 per cent. The same rank order applied to Revenue Grants.

In March 1988 the Government announced its 'Action for Cities' programme for the regeneration of inner cities (Cabinet Office, 1988). The key elements in the programme, while not too distant from the priorities outlined in the ministerial guidelines of May 1985, were presented by Government as a striking new initiative. They were:

1. To encourage enterprise and new business and help existing business grow stronger.
2. To improve people's job prospects and provide training to develop skills and motivation.
3. To make the inner-city environment more attractive to residents and business by tackling dereliction, preparing sites and encouraging development, bringing buildings into use and improving the quality of housing.
4. To ensure that inner city areas are safe and attractive places in which to live and work.

Although little reference was made to ethnic minorities in this programme, Government Ministers and departments referred to ethnic minorities' experience and needs in some statements and accompanying documents (Ollerearnshaw, 1988), the emphasis being to ensure that local communities,

171

including ethnic minorities, benefit from the initiatives. While £250 million of Government investment in twelve projects was specifically detailed, it was made clear that the initiative was not about money but was characterised by two themes: a focus on private sector led initiatives and the exclusion of local authorities, who were identified as obstacles to action through, for example, hoarding developed land required by developers (Mullins, 1988). Certainly there will be opportunities for revitalising inner area jobs through the private sector (Banham, 1988), and the creation of the CBI Task Force, whose primary objectives are to ensure that business optimises its contribution to this process, illustrates the willingness of the private sector to be identified with this programme. In addition despite the poor record of the Urban Programme in creating long-term employment with a significant ethnic minority element, the new initiative together with stringent goal setting and project monitoring by Central Government does provide potential for action on equal opportunities through strategic policy formulation, targeting and monitoring (Ollerearnshaw, 1988).

What remains at the heart of these initiatives, however, is that they are significantly under-resourced and fail to harness or build on local authority networks and experience. The experience of the London Docklands demonstrates that bypassing Local Government and attracting private investment may result in significant improvements to the physical environment but the processes of gentrification which result neither generate local employment nor benefit local residents. While this may not be a necessary consequence, it seems probable that 'Action for Cities' is insufficiently geared up to securing a concerted attack on structural disadvantage in employment or housing: even if it is more successful than commentators anticipate, it is clear that, in comparative terms, it is a mere palliative when contrasted with a decade of disinvestment reflecting macro-economic policy.

It will be apparent from the above brief description of the Urban Programme that the legal framework provided by the 1969 Act is of a loose and open texture, allowing for significant Central Government discretion in the setting and variation of objectives as well as in the criteria for assessing Local Government and private sector project evaluation. It must also be recognised that in terms of the Race Relations Act 1976 any ministerial guidance given under any other legislation escapes the requirements to be non-discriminatory and accordingly even if the policy and/or practice relating to the operation of the Urban Programme in respect of housing, or indeed any other activity, is shown to be discriminatory, recourse to law would be to no avail. As a consequence the failure of the Scottish Office even to mention ethnic minorities in any of its guidance relating to the Urban Programme and the fact that the £341,810 allocated to Urban Aid projects in Scotland in 1985/86 with a specific ethnic minority targeting represented a mere 1 per cent of the total Urban Aid Programme in Scotland for that year has no legal significance.

Although it is recognised that other Urban Programme funded projects, while not specifically geared to ethnic minority concerns, may undoubtedly produce benefits (DoE, 1986b) there is the danger that these generalised community-wide projects will develop into 'colour-blind' initiatives and end up discriminating unintentionally against minority groups (Higgins et al., 1983: 190).

PLANNING LAW AND REDISTRIBUTITVE JUSTICE

The 1980s witnessed a significant retreat of activity in the public sector and the resurgence of the belief that progress is best left to market forces. Not only is public investment, whether by way of public housing or job creation and support, seen as a distorting and self-defeating intrusion, but development plans and development control are viewed as necessary evils which must, however, be streamlined and geared more sensitively to the needs of market economics. Thus Enterprise Zones, Simplified Planning Zones, joint venture schemes and inner-city partnerships and programmes are designed to stimulate private-sector investment and growth while Central Government's view of development control is that is should first presume that applications should be granted unless there are sound reasons to the contrary and, second, be streamlined so that delays in the system are minimised to avoid financial loss in the private sector because of unnecessary bureaucracy.

The post-war consensus regarding the advantages of a mixed economy have largely evaporated under the Thatcher Government and this shift of view is reflected in planning as in other sectors affecting economic development. Put crudely, Friedman economics presumes that private wealth generation will have a multiplier effect on the less well-off, who, while not enjoying an equal distribution of wealth and resources, will benefit from general economic expansion and profits. Because ethnic minorities have largely operated as replacement labour in respect of the least attractive and worst paid jobs and suffer from the cyclical effects of deprivation and poverty, they, along with other economically disadvantaged groups, will benefit least from market-led growth where the redistributive effects of State planning and intervention are marginalised. The present trend in tax liability has, from 1980, witnessed a lowering of the highest rates of tax, and amendments to social security legislation, housing benefits, national insurance criteria and income support individually and cumulatively adversely affect a majority of claimants. Moreover the reduction in tax revenue, although largely offset by oil and privatisation revenues, has, in the long term, to be met by expenditure cuts. With continuing high rates of unemployment significant cuts in social security provision cannot be expected but savings on public expenditure have been made elsewhere, including those on local authority rate support grant, which now represents some 45 per cent as opposed to 65 per cent of expenditure some ten years ago. Loney and Allan (1979) have shown that the needs element in the Rate Support Grant (RSG) has provided much higher subsidies to the wealthier suburbs than the inner cities. Moreover the fall in RSG has not in any way been mediated by expenditure through the Urban Programme and partnership schemes (Johnson, 1987). This has affected local authorities, budgets both in relation to infrastructure provision and in respect of significant aspects of welfare 'redistribution' such as education and housing. Not only have local authorities got to resource a greater element of such budgets but, through Central Government fiscal controls, they must not over-spend restrictive target figures if further penalties are not to be imposed. While domestic rating, as the Layfield Report on Local Government Finance (Layfield, 1976) demonstrated, has been open to much criticism, it has been a less regressive form of local taxation than the community charge or poll tax. It has been noted that ethnic minority groups have experienced relative disadvantage by the processes described because of their disproportionate representation in lower income groups but this is compounded by two additional factors.

The electoral support for Friedman economics cannot

be explained solely in terms of its attraction for the better-off. Lower income groups, it must be assumed, see the potential for upward mobility and their current status ascription is not, generally, perceived as an insuperable barrier to progression. Consequently, it is only those groups who are severely handicapped in economic terms who will not identify with the potential within the present economic system of achieving upward mobility by breaking the cycle of deprivation. Social status, however, while frequently a reflection of economic status is not absolutely correlated with it. The status of being a single parent or being black will adversely affect 'life chances' as well as present income and other features of deprivation such as job status, living in an inner city and housing tenure to the extent that while economic upward mobility is possible such potential is further constrained by social status. In terms of jobs this may be illustrated by the fact that in the early 1980s 'the faster growth of unemployment among blacks than whites has opened up an area of massive racial disadvantage that was absent in 1974' (Brown, 1984: 174). An analysis of trends between 1963 and 1981 showed that when unemployment rises nationally, it rises much faster for blacks, and the gap between white and blacks widens. But Brown has shown that there has been upward mobility amongst black employees between 1974 and 1981 and this is largely explained by improved job levels (by socio-economic group) for young black (Asian and West Indian) workers, where the job levels are closer to those of young whites than for other age groups. While there is evidence of convergence of job levels amongst those employed the opportunity for employment remains vastly different. Amongst 16 to 25 year olds the proportion of men in employment was 61 per cent for whites, 42 per cent for West Indians and 48 per cent for Asians. For women the figures were 56 per cent, 38 per cent and 25 per cent respectively. Brown (1984: 293) concluded that the tenacity of the pattern of poor jobs among black workers was due, <u>inter alia</u>, to the following factors:

1. Different educational backgrounds.
2. Lack of fluency in English amongst Asians.
3. Different residential locations.
4. A distinctive pattern to the ethnic minority labour market.
5. Direct and indirect racial discrimination.

Although these factors are interrelated it is the third aspect, that of residential location, in which planning may have the most significant input. The 1982 survey (Brown, 1984) showed that at a local level white people living in those census enumeration districts identified as areas of ethnic minority settlement have lower-paid work than whites elsewhere. Thus it can be argued that black people tend to be found in localities where people - black and white - have jobs at a generally lower level. The jobs in which blacks are well represented are those which they filled when they first moved to this country and amongst Asians those jobs related to Asian businesses. Crudely, some jobs are labelled 'multi-racial' or for 'blacks and poor whites' while others remain largely the preserve of white workers. Although the job market has changed little between 1974 and 1982 in its hostility to black workers, the few changes that have occurred for the better have been from a consolidation of gain in the public sector such as the health and transport services and from the development of black businesses. Moreover some economic gains are at a heavy cost. Thus the overall wage differential has been reduced by a greater proportion of blacks working shifts, which is work at the foot of the occupational ladder, with little opportunity for upward mobility, particularly for those less fluent in English.

The original pattern of owner-occupation amongst black people in Britain was associated neither with affluence nor with good property, unlike the pattern for whites. Blacks took replacement labour and occupied replacement housing when whites moved out of the inner cities. This is illustrated by the fact that nearly 50 per cent of Asian families housed by councils up to 1974 were allocated council housing because their previous dwelling was to be demolished. Between 1971 and 1982 the black population, by natural increase and immigration, grew by 60 per cent but there has been little change in geographical distribution. Given the high rate of mobility amongst blacks - 46 per cent moved house within the period 1977-82 - and their relative length of residence in Britain, it is remarkable how their position, both in the labour market and in residential location, has changed so little. Both factors have experienced further concentration; the areas of high density tending to have the worst housing and employment problems. While this period has witnessed a relative concentration of black households in existing locations, such locations are still predominantly 'white' in proportion to the total population except at street level. Although the significance of geographical location is a complex one it is clear that while one reason for black preference for existing locations is community support and access to available housing, another is the lack of opportunity to locate elsewhere because of restricted access to housing and job opportunities. Moreover the PSI study demonstrates that existing residential patterns restrict upward mobility and reinforce patterns of disadvantage. Despite the increased access to public housing in the late 1970s, particularly in respect of the West Indian community, those sections of the ethnic minority population who have changed tenure to the public sector have not affected the overall pattern of residential location nor the incidence of relative disadvantage.

In these processes the planning legislation, principally the consolidating codes of 1971 and 1972 and the Urban Programme introduced in 1968, provided fundamental opportunities for improved housing in either the public or the private sector. The reasons for their lack of impact have already been alluded to. The Urban Programme has affected some aspects of disadvantage associated with residential location in deprived areas but it is evident that the law relating to planning is not only irrelevant to the process of combating racial disadvantage but may well accentuate the gap between the rich and the poor and the divided community which is emerging.

Planning, however, was but one vehicle for tackling poor housing. The local authorities in Britain, through housing, public health and environmental legislation, have a substantial battery of powers to effect demolition, closure, rehabilitation and repair both in respect of individual dwellings and at area level.

DISREPAIR AND IMPROVEMENT

Investment and maintenance

Thomas (1986) in his examination of housing and urban renewal has shown that while investment in housing as a capital asset has increased post-war in respect of the owner-occupier sector, a picture of economic dependent and independent variables shows a relative disinvestment in housing repair and maintenance: while public and private investment has resulted in a diminution in houses lacking standard amenities, the provision of repair grants at 50 per cent in 1969 in respect of approved costs, increasing to 70 per cent following the housing Acts of 1980 and rising to 90 per cent during the period 1982 to 1984, has not stemmed the relative deterioration of the housing stock. The process of disinvestment has been aggravated by the abandonment of slum clearance as a central policy and the very substantial diminution in public housing new build.

Between 1976 and 1980 slum clearance accounted for 45,000 dwellings per annum. Each subsequent year the figure has fallen until 1986, when 10,000 dwellings were cleared. Between 1976 and 1980 the public sector built an average of 139,000 dwellings per annum, marginally fewer than the private sector average of 146,000. By 1986 the former figure was 36,000 and the latter 169,000 (HMSO, Social Trends 18, 1988: Table 8.2). Ball (1983: 5) has suggested that the rate of new housing provision is a very good barometer, with housing conditions for most people tending to get better during periods when house building is expanding. An alternative indicator is the volume of housing investment within the overall distribution of national resources. From the 1970s the total capital investment has fallen, with increased expenditure on repair failing to offset the losses in new housing output (Thomas, 1986: 31). At 1975 constant prices the total investment in the UK as a percentage of GDP relating to investment in new housing repair and maintenance was 5.9 per cent in 1970 and 3.6 per cent in 1981 (Thomas, 1986: 33). According to the English Housing Conditions Survey of 1981 the outstanding bill on disrepair was in the region of £35 billion, an increase of £3.3 billion on 1976. This total represented 17 per cent of total GDP. The 1981 survey estimated that £6.1 billion was spent, in that year, by all households on repairs, improvement, maintenance and decoration of the housing stock, of which £160 million came from Government grants (DoE, 1983: 14). But a high proportion of this expenditure on the owner-occupied stock had nothing to do with remedial work, being related to comfort and convenience through the provision of central heating, replacement bathroom and kitchen fitments, external windows (mainly double glazing), and electrical rewiring (DoE, 1983a: 15).

Between 1975 and 1981, on average, 100,000 renovation grants were given each year, but this figure jumped to 340,000 in 1984 and fell to 177,000 in 1986 (HMSO, 1988).

While there is little doubt that the housing stock and living conditions have improved as a consequence of grant aid to the private sector, the outstanding needs are exponential. The survey Greater London Housing Statistics 1984 (GLC, 1985a) found that levels of renovation work would require to be topped up, in London alone, to 4,000 extra dwellings per annum with a further rehabilitation of 70,000 extra dwelling units per annum for ten years to eliminate serious disrepair and unfitness in London's housing stock. Public expenditure in this area would have to increase by £860 million per annum to meet a narrowly defined programme of essential work.

In this light, the disadvantage of grants is that they represent a major but infrequent injection of capital without being able to influence the future pattern of maintenance activity. As fire-fighting measures they fail to tackle the reasons why owners in aggregate continue to invest in property repair at a level which is insufficient to prevent its deterioration (Thomas, 1986: 36).

Ethnic minorities and improvement grant take-up

That evidence which is available on the take-up of improvement grants (e.g. Simpson, 1981; MacEwen, 1976; Brown, 1984; Dalton and Daghlian, 1989) suggests that there is regional variation in the proportional take-up of the ethnic minority groups and a lack of distinct trends. Brown (1984: Table 81) shows a higher percentage of West Indian (35 per cent) and Asian (28 per cent) applicant owner-occupiers than whites (22 per cent) but a less significant difference in those receiving grants, 15 per cent and 17 per cent as opposed to 14 per cent in respect of the respective groups surveyed. When expressed as a percentage 'success rate' of applications the West Indians at 45 per cent were appreciably less successful than both Asians and whites at 64 per cent. Given the nature of residential property in which the black community lives, a higher take-up rate by both ethnic minority groups might have been expected. Brown (1984: 80) concluded that, despite the lack of direct evidence, the scheme was not being used by those most in need, a possible explanation being found in inadequate targeting of ethnic minorities regarding information on the terms of eligibility, the application procedures and the type of work covered. Although some surveys (Field and Hedges, 1979; Johnson and Cross, 1984) show a proportionately higher take-up of grants, there is some question as to whether such improvements translated into ethnic gain equating with the white experience (Johnson, 1987). There is no evidence that the Race Relations Act 1976 (Section 20) has been invoked successfully in respect of improvement grant practices and the general lack of ethnic record keeping renders legal action an improbable source of redress unless evidence of direct discrimination is adduced. Currently Brent BC is investigating the consequences of its improvement grant and enforcement of standards policies for black and ethnic minority tenants in the private sector with a view to reviewing its policy (LRHRU, 1987). While this project may provide lessons beyond Brent, until the DoE, Welsh Office and Scottish Office undertake their own analysis and offer guidance to local authorities, such initiatives will have little impact on the national scene: indirect racial discrimination, an inevitable but implicit conclusion from the PSI survey, will remain untouched by the legislation in this area.

Ethnic minorities and area treatment

National figures showing a breakdown by ethnic group of households resident in housing treatment areas, of various descriptions, are not available, but given the profile of ethnic minority locations and housing conditions, the expectation would be that such minorities would be disproportionately affected. This explanation is generally borne out by local studies (Jacobs, 1986: 112; Ouseley, 1981). Until the late 1960s Government policy focused on slum clearance (DoE, 1968). So far as black concentrations were seen as undesirable ghettoisation, clearance areas provided an opportunity for local authorities, which were legally bound to provide appropriate alternative accommodation for displaced residents, to introduce policies of dispersal and both Lewisham and Birmingham did so (McKay, 1977: 166). To the extent that such policies applied quotas in respect of allocation by race to housing estates and were involuntary they clearly breached the provisions of the Race Relations Act 1968 - a fact referred to expressly in a special memorandum published in the Race Relations Board's 1974 Annual Report. While the Cullingworth Report (1969) called for dispersal and this was echoed in the recommendations of the Select Committee on Race Relations and Immigration (1971) Government's response was more cautious:

> If, for instance, a Housing Action Area, General
> Improvement Area or Priority Neighbourhood Area
> is declared ... [the] authority ... will want to ensure
> that the declaration is not seen as an attempt to
> enforce 'dispersal' against the wishes of those
> concerned. (HMSO, 1975a)

In addition to less favourable treatment through dispersal,
clearance areas might adversely affect ethnic minorities by
reason either of their owner-occupied status disqualifying
them from mainstream council housing or through rules
relating to longevity of residence in an area having like
consequences (Henderson and Karn, 1987). A potential
consequence was the hazardous 'homeless' route to council
housing frequently resulting in allocation to less popular
estates and, on occasion, to housing purchased in advance of
area housing programmes: by definition such housing was on
the margins of habitability.

Conversely the decision to declare a Housing Action
Area, or its associates, entails potential benefits to
residents whether through subsidised improvements to the
housing affected or through preferential access to council
housing through displacement, where residents meet the
local criteria. The housing authority's decision is seldom
based solely on objective criteria. Value judgements may
include the political repercussions of advantaging ethnic
minority groups. Such considerations may not be explicit but
nonetheless, as Pareto (1963) has suggested, constitute the
subliminal backdrop to the decision-making process. Clearly
Government sees the need for local housing authorities to
address the racial dimension in the declaration and
implementation of housing area treatment but it avoids
prescriptions, expecting local authorities to 'take a balanced
view' regarding the 'pattern of settlement of coloured
people ... as it changes over time', 'cultural, social and
family patterns' and regarding the wishes of ethnic minority
groups (HMSO, 1975a): no advice is proffered as to what a
'balanced view' is or even what factors are to be weighed in
the balance. In this vacuum housing authorities may consider
this dimension explicitly, as recommended, may ignore it - in
a colour-blind approach - or may, whether consciously or
otherwise, permit it to remain implicit in the decision-
making process. Indeed, given the diversity of considerations

relevant to different ethnic minority groups, elements of each approach may simultaneously influence the one decision.

It must also be acknowledged that the extent to which the racial dimension is not perceived as a substantial issue in declarations not only encourages the 'colour-blind' approach but also creates the opportunity for subliminal racial accountancy. In the comprehensive slum clearance policies in Nottingham and Moss Side described by Simpson (1981) and Ward (1975) respectively, the extent of the areas included and the multi-racial nature of the residents preclude the conclusion that there was racial targeting - that black concentrations were transmogrified into poor housing in the eyes of the authority - irrespective of the physical housing conditions. Nonetheless the most significant issue in the context of renewal policies is the extent to which comprehensive redevelopment largely missed areas of substantial black concentration (Paris and Blackaby, 1979: 20). Referring to a national sample, English et al., (1976) concluded that slum clearance procedure does not deal with transitional areas of multi-occupation and immigrant housing.

In part the explanation may be found in settlement patterns at the time of Government subsidy to local authority clearance programmes and in part by the logistics of clearing high-density housing but the factor of white hostility to rehousing substantial numbers of black residents cannot be discounted (Paris and Blackaby, 1979: 21), even if not of necessity an explicit policy (as argued by Rex and Moore, 1967: 35). In some instances, such as Birmingham in the mid-60s, the political assumptions regarding multi-occupation, unfit housing, black landlords and tenants and problems of racism on rehousing were quite explicit (Paris and Blackaby, 1979: 42) and the shift from clearance to rehabilitation created an important escape route.

As with all other areas of the housing authority's responsibility, ethnic record keeping was the exception in respect of housing treatment. Indeed some local authorities decided not to keep records to avoid charges of discrimination (Cowen, 1983: 6). Despite the serious implications for residents and the racial dimension to inner-city programmes, the declaration of Housing Area Treatment has attracted little litigation generally, and there has been no successful action in respect of indirect unlawful discrimination in terms of the Race Relations Act 1976

(Section 20). Perhaps the most substantial explanation for this may be the lack of information on a policy's general impact on particular ethnic groups and the access of such groups to legal services but the law itself requires examination.

Legislation and housing treatment

The present legislation relating to housing treatment is contained in the Housing Act 1985 for England and Wales, parts VI to IX dealing successively with repair notices, Improvement Notices, area improvement and slum clearance. Similar provisions for Scotland are contained in the Housing (Scotland) Act 1987, part IV addressing sub-standard housing - including improvement orders and Housing Action Area treatment, part V addressing repairs and part VI closing and demolition orders. The general legal implications of such provision have been discussed by others (Hughes, 1987; Himsworth, 1986) and need not be reiterated here. In addition to the opportunities for intervention by the housing authority to secure habitability or the closure of unfit or unsafe dwellings a plethora of legislation enables intervention by local authorities. Thus the Public Health Act 1936 for England and Wales and the Public Health (Scotland) Act 1897 allow the control of statutory nuisances (given a wide definition by statute and the courts), enabling intervention where disrepair affects health or safety. Similarly the building authority may intervene to secure structural safety and compliance with the Building Standard Regulations. Local legislation, such as the Birmingham City Act 1965, or the Edinburgh Corporation (Provisional Order) Confirmation Act 1967, has been extensively deployed in respect of housing clearance and repair notices in the past but public general Acts such as the Civic Government (Scotland) Act 1981 have largely replaced it. The array of powers available is therefore extensive. Most of these legislative provisions allow for appeal against the service of a notice relating to repair, which, at least in theory, provides a method of judicial challenge either on the merits -as to whether or not the dwelling meets the criteria, so far as provided, of the notice - or as to the procedural requirements being met. To that extent any arbitrariness in the selection of dwellings, so far as it breaches the legal requirements, may be challenged. Where such arbitrariness in selection nevertheless meets the criteria but is expressed

183

in the targeting of specific houses, perhaps defined by the 'character' of the householder, challenge would be much more difficult. The individual affected might seek redress by reason of maladministration through the Local Government Ombudsman and, where racial discrimination was alleged, challenge the relevant notice as a breach of Section 20 of the Race Relations Act 1976 relating to the provision of goods, services or facilities.

However, judicial pronouncements leave this area in doubt (Gardner, 1987). In Hasnain Tejani v Sup. Registrar for the District of Peterborough (Times Law Reports, 10 June 1986) the opinion was expressed, obiter, that the services of the District Registrar in providing marriage licences fell within Section 20. But in Sanjani v IRC (1981) QB 459 a distinction was made between 'primary' and 'secondary' functions, with the implication that the specific service offered must be examined rather than the totality of functions provided by an agency or particular official. In this instance the collecting of tax revenue fell outside Section 20 while repayment of tax would not. In R v Immigration Appeal Tribunal ex p. Kassam (1980) 1 WLR 1037 per Stephenson LJ, in determining that granting leave to remain was not a service within Section 20 the court suggested that only 'market place activities' fell within its ambit. Given the examples listed in Section 20, including facilities for grants (20(2)(c)) and the services of 'any local or public body' (20(2)(g)) this is clearly wrong. In R v Entry Clearance Officer, Bombay ex p. Amin (1983) 2 AC 818 Lord Fraser, giving judgment in the House of Lords regarding an application for a voucher (facilitating leave to enter) determined that the issuance of vouchers was not the provision of a service to would be immigrants but rather the performance of a duty 'of controlling them'. He introduced three restrictions to the ambit of Section 20:

1. 'Facilities' do not include permission to use a facility.
2. 'Services' do not include acts done in the course of performing a duty of control, and
3. Section 20 does not apply to actions of Crown officers which are dissimilar to actions which could be performed by private individuals (Gardner, 1987).

The last restriction follows the cul-de-sac signposted by Kassim but marked 'no entry' by Section 20(2)(g), referred to above, and the first two are clearly contrary to the intentions of the 1976 Act.

It is perhaps more remarkable that, prior to the decisions of the Court of Appeal and the House of Lords in Kassam and Amin respectively, there was no substantial testing of the application of Section 20 in relation to the local authority housing function generally or specifically in relation to orders and notices in respect of closure, demolition, repair or improvement, including the declaration of Housing Action Areas and General Improvement Areas.

CONCLUSIONS

The potential of urban renewal policies to improve the living conditions of those resident in the inner cities including ethnic minorities is largely threefold. First, the physical improvement of the housing stock is, self-evidently, designed to improve the living standards of residents. Second, there has always been a dramatic relationship within older urban areas between poor housing conditions and unemployment (Townsend, 1979: 528). Housing problems are really problems of unemployment, poverty and inequality: a concern for housing is inseparable from the need to provide work opportunities. Building work then promises an important opportunity to generate local employment. Third, the economic impact of renovation activity in the local community has a potentially significant multiplier effect. However, Government policy has tended to use urban renewal as an economic regulator without reference to the need for continuity, assuming that if capital investment is made available the building industry will respond by doing the work (Thomas, 1986: 190). The ability to gear up to the programmes of rehabilitation is dependent on continuity (RTPI, 1981: 23). However, Central Government has demonstrated a singular lack of ability to combine funding arrangements and reinvestment strategies which would offer the potential for sustained work opportunities, using local labour for projects aimed at stimulating employment growth. As a consequence, while the overall lack of public investment in rehabilitation is open to criticism, that investment which has taken place has seldom optimised the potential benefits to local residents. The stop-go measures of Government investment in urban renewal by way of improvement grants have not infrequently resulted in higher costs and poorer performance in renovation because of the reluctance of the building industry to commit sufficient capital investment in expansion in view of the longer-term prospects of a downturn in investment. This creates a vacuum filled by the building 'cowboys'. Moreover there is some evidence that the encouragement of lower-income groups into owner-occupation has resulted in a relatively higher proportion of the income of such groups being devoted to mortgage repayment with little or no flexibility to invest in repairs and maintenance. Such over-commitment to repayment of mortgages in low-income groups is demonstrated by figures released by the Building Societies Association which show that across the UK the number of homes repossessed when their owners defaulted on repayments went up by almost 650 per cent between 1979 and 1987 (Scotsman, 10 August 1988). In Scotland 1,471 houses were repossessed in 1987 in comparison with 200 in 1979 and blame has been attached to readily available credit encouraging people to shoulder commitments they could not sustain and at the same time a decrease in private rented accommodation pushing people to buy property for the first time in their lives. If rehabilitation and repair costs cannot be met by present residents because of their low income, the process of gentrification is encouraged.

In a summary of the discrict councils in England responsible for administering the Housing (Homeless Persons) provisions, the incidence of homelessness attributable to building society repossessions increased from 218 to 748. Half of the repossessions involved homes acquired from public authorities under the 'right to buy' provisions introduced by the Housing Act 1980 (Hawes, 1985). Given that council house 'privatised' stock represents about 6 per cent of the total owner-occupied sector, this survey indicates a substantial extent of financial difficulty experienced by such purchasers with repercussions not only in respect of homelessness or repossession but also in respect of an owner's ability to effect running repairs, let alone maintain the property in good condition.

In view of the foregoing it will be seen that the housing legislation establishing the parameters of general improvement areas, Housing Action Areas, individual house improvements by way of improvement and repair grants and the provision of standard amenities is of secondary importance in relation to its impact on low-income groups and ethnic minorities to the overall strategies for public investment and at the micro level the processes adopted by the local authority to secure optimum advantage to local communities within the constraints of existing resources and legislation. Whether or not, therefore, racial discrimination occurs directly or indirectly in the declaration of Housing Action Areas and the deployment of grant aid for housing rehabilitation (and the evidence suggests that it does), this is likely to be a less important factor in respect of racial disadvantage than public investment strategies.

In the public sector reductions in Central Government subsidies for council housing have led to reduced expenditure on repairs and maintenance and increases in rents in real terms. In Scotland more than half of Scottish housing authorities are no longer in receipt of housing support grant and these developments restrict the ability of housing authorities to discharge their statutory functions and to meet the requirements of the regional strategic schemes, particularly for those areas identified as in need of priority treatment (NFHA, 1985: 70). The English Housing Conditions Survey showed that over 20 per cent of pre-war housing in the public sector was unfit or otherwise in a condition to merit clearance or compulsory renovation. In addition local authorities face major problems as owners of dwellings that are fundamentally defective because of design, construction or material faults. The legacy of system-build housing promoted by Central Government in local authorities' post-war public housing development programmes has imposed an additional burden on local authorities in respect of maintenance and repair which is by no means confined to older housing stock. Today the majority of investment in the renovation of local authority stock relates to purpose-built council housing. The demand for investment has mushroomed not only as a result of Government initiatives such as the Housing Defects Act, the Priority Estates Projects, and the · Urban Housing Renewal Unit (renamed Estate Action in 1986), but also as awareness of the scale of problems in the traditionally built stock, the 'non-traditional' stock of the 1950s and 1960s and the industrial and system buildings of the 1960s and 1970s has grown (Malpass and Murie, 1987: 121). However, in contrast with the expansion of improvement grants to private owners, which peaked in the year 1983/84, and represented a sevenfold increase in four years when spending on enveloping schemes and environmental improvements in declared improvement areas is included, the capital expenditure on renovation of local authorities' own stock remained constant or declined.

That local authority policy concerning a designation of areas for clearance and improvement may potentially amount to indirect discrimination has been demonstrated in relation to Lambeth (Rex and Tomlinson, 1979) and Liverpool (CRE, 1984a), to provide but two examples. This may also extend to the allocation of housing investment programmes (CRE, 1989e). However, the law has had no impact in respect of the designation and implementation of housing action or improvement areas, in relation to the allocation of discretionary improvement grants or mandatory repair grants. This is as much a reflection of the inadequacy of the analysis of local authority policy in respect of its impact on ethnic minorities, aided and abetted by Central Government's failure to pay more than lip service to the need for local authorities to take on board the ethnic dimension in the development of such policies, as it is of the inadequacy of information available to individual applicants and ethnic minority groups which might indicate to what extent they might be beneficiaries or otherwise in the policies adopted at a local level. Accordingly until ethnic record keeping by local authorities is a statutory requirement with regard to the implementation of their various housing functions the capacity of the law to secure equality of opportunity in these processes will remain a dead letter. Having said this, it would be simplistic to suggest that record keeping and data analysis together with easier recourse to the courts to secure formal equality of opportunity would result in a radical change in housing practice. Ethnic preferences in the housing process may well expose inconsistencies in relation to tenure and location. Formal equality of housing destination may obscure or restrain opportunities for cultural and political cohesion as well as constrain opportunities in other aspects of life, including employment and educational opportunity. Accordingly, while record keeping may be a prerequisite of analysis it is only part of the equation: information dissemination, consultation and explanation of housing options must be part of the machinery of housing policy formulation and this 'open' approach is particularly important in a period of radical change imposed by Central Government, including its present proposals to introduce means testing in respect of housing grants.

The failure of strategic Central Government planning, whether through the physical land use legislation, the Urban Programme or the development and support of housing clearance and improvement, to secure adequate housing in the private sector to meet ethnic minority needs provides a key structural explanation of increasing homelessness, and it is this concern that is now addressed.

CHAPTER SEVEN

HOMELESSNESS

INTRODUCTION

The problems of homelessness (IoH, 1988), of race and residence (Smith, 1989b), of race and law (Lustgarten, 1980) and of homelessness and law (Watchman and Robson, 1989) are well documented. But the combination of all three, homelessness, race and law, has lacked commentary. This chapter does not attempt to fill the void systematically. It merely attempts to map out a topography deserving further exploration and in doing so to provide a key to notable landmarks which future travellers may wish to visit. The starting point for this exercise is the intersection of the grid references - the homeless persons legislation which provides the horizontal line along which the reader is directed and the Race Relations Act 1976, through which the reader is presumed to have passed. The examination of the subject is fourfold: a description of homelessness, including the ethnic minority dimensions as a context for the statutory definition; an examination intrinsic to the homeless provisions - how they are interpreted and applied; a look at the interrelationship between the homeless provisions and essentially extrinsic considerations; and lastly a concluding overview.

DEFINING HOMELESSNESS

As the authors of the Greve Report Homelessness in London (1971) noted, there is no universally accepted definition of homelessness: the narrowest definition is one which equates with 'rooflessness' those sleeping rough in derelict buildings, barns, hedgerows, under the sky, wrapped in newspapers, old sacks and old clothes (Sandford, 1971: 13). Until 1977 and the passage of the Housing (Homeless Persons) Act the

Government did not recognise even this category as homeless unless the individual concerned had applied to the welfare department of the local authority for assistance and had been admitted to temporary accommodation (David Ennals, press statement, 29 April 1969). In considering homelessness, however, it is more realistic to take into account those who are houseless: that is, those who are either roofless or who are in long-term institutions or have bed-and-breakfast provision, lack secure accommodation or are living in intolerable housing conditions. Voluntary housing bodies have suggested the use of the last criterion, i.e. intolerable housing conditions. Given that, at the time of the passage of the Act, almost three million households in the UK would be classified as living in overcrowded accommodation or accommodation which lacked the exclusive use of a hot water supply, fixed bath or an inside water closet, it is not surprising that the Government accepted a more restrictive view of homelessness in the statutory provisions of that Act. These are now reflected in the provisions of the 1985 Housing Act and, for Scotland, the Housing (Scotland) Act 1987.

Essentially, statutory homelessness, which includes those immediately threatened with homelessness, depends on the following tests being applied by the recipient authority:

1. Accommodation: the applicant must lack access to 'accommodation' of a kind now partially defined by the legislation and partially dependent on the courts' interpretation of this term.
2. Priority need: the applicant must be in priority need as statutorily defined. But while certain categories such as those who are pregnant or have dependent children are clear-cut, other categories of vulnerability depend on the authorities' judgement as aided by the relevant statutory Code of Guidance and by court precedent.
3. Intentionality: the applicant must not be intentionally homeless. Intention may be inferred from a positive act, such as vacating available accommodation, or an omission, such as failure to pay the rent leading to eviction. The question of intention lacks any clear-cut statutory definition. This must be provided by the recipient authority and court precedent.

Once an applicant is found to lack accommodation, to be in

priority need and not to be intentionally homeless, he/she is entitled to permanent accommodation. But if the recipient authority demonstrates a closer association between the applicant and another housing authority it is the latter which must secure such provision (the local connection test).

THE HOMELESS CRISIS AND STATUTORY DISCRETION

Clearly the ability of a local authority to meet these statutory requirements to house the homeless is critically dependent on two factors: the nature and extent of available accommodation in its area and the number of homeless who make application and meet the statutory requirements. Where there is low-cost housing scarcity in both the private and public sectors, coupled with high rates of unemployment and a high incidence of workers on part-time work or low pay, not only will there be an increase in homelessness but also in the difficulties experienced by local authorities in meeting their statutory obligations.

The Government's statistical service housing survey (DoE, 1987a) shows that the number of homeless rose by one third in 1986 and continues to be a significant and growing problem throughout the UK. Homelessness acceptances by local authorities have more than doubled from 53,100 in 1978 to 112,500 in 1987 (IoH, 1988: 2). It should also be noted that acceptances run at less than half the number of applicants (1986, 120,000 and 244,000 respectively) and that many single people and childless couples, knowing that they are unlikely to be considered in priority need, do not even apply.

Information on the comparative homelessness of members of ethnic minority groups is not well documented (AMA, 1988a) but, evidently, settlement patterns, the nature of housing tenure, the experience of employment/ unemployment, the nature of community support, and the relative incidence of vulnerable groups will affect the significance of homelessness amongst ethnic minorities in the UK (CRE, 1974, Bonnerjea and Lawton, 1987).

It is estimated that black families represent 56 per cent of the homeless in Haringey, 46 per cent in Lambeth and Wandsworth and over 95 per cent in Tower Hamlets (FBHO, 1987). In a study of homelessness in Brent (Bonnerjea and Lawton, 1987), 70 per cent of the sample survey (in comparison with 50 per cent of the population) were from ethnic minorities: city-wide ethnic minorities constitute 40 per cent of homeless acceptances (London Housing Unit, 1989).

But while the position of the homeless, and the black homeless, in London is particularly acute, recent surveys of homelessness demonstrate that it now constitutes a national crisis (IoH, 1988).

A London Research Centre survey in addition to suggesting that black households were three or four times as likely to become homeless as white households demonstrated, in an analysis of metropolitan and sixty non-metropolitan authorities, that a high concentration of black households was a feature of many areas with acute housing stress (London Research Centre, 1987). Advice and referral agencies also report a high incidence of young black people amongst their clients (GLC, 1986d).

In situations of housing stress, local authorities are tempted to avoid their statutory obligations by a narrow construction of key terms in the Act. This is illustrated by their interpretation of accommodation, priority need and intentionality and, when this fails, by their interpretation of local connection to offload responsibilities on to other authorities. The last is exemplified by the fact that although London contributes disproportionately to the national total of households living in temporary accommodation pending enquiries or awaiting permanent accommodation, 40 per cent of London's total in 1987 had been placed there by non-London authorities (AMA, 1988a).

The AMA observed (AMA, 1985) that whilst the Homeless Persons Act itself has shortcomings, many black people experience further and unnecessary problems in relation to homelessness. Some local authorities exercise their not inconsiderable discretion to evade the spirit of the Act and the Department of the Environment Code of Guidance. The Greve Inquiry (1985) stated that the operation of the Act was characterised by 'ethnocentricism and racially discriminatory practices'. Because of the pressure on housing from the general waiting list, applicants from homeless families are stigmatised, generally, as queue jumpers and frequently, as a matter of policy, offered the least desirable housing. To the extent that disadvantage in employment renders black families less able to seek alternative accommodation in the private sector, their over-representation in the homeless categories illustrates the association of unemployment with the low-paid and of subemployment with poverty (CRE, 1974, Townsend, 1979).

The homeless are often portrayed as part of the undeserving poor, the 'dangerous classes' for whom the workhouse rather than welfare is the appropriate response (Watchman and Robson, 1983: 9). Such attitudes, so far as they affect ethnic minorities, are no doubt reinforced by Central Government's stance that immigrants should not become 'a burden on the state' (Waddington, in Stearn, 1987: 12). Black people may therefore find themselves categorised as less deserving than the undeserving.

The legislative provision on homelessness noted above provides ample scope for the exercise of discretion by the housing authority and such discretion is not subject to any statutory means of appeal on the merits of a case to any tribunal or court. As a result there is every opportunity for housing authorities, in fulfilling their legal obligations under the Act, to exercise subjective and partial judgements which will not always be open to challenge by way of judicial review in the High Court or, in Scotland, the Court of Session. Nonetheless it has been one of the facets of the homeless provisions that the exercise of local discretion by housing authorities has been subject to considerably judicial supervision. It is that supervision which provides an opportunity for commentary on how the Act has been interpreted in its affect on ethnic minorities. It must be remembered, however, that this supervisory function of the courts is very restricted and does not afford an appeal on the facts or merits of an individual application against the decision of a housing authority. As a consequence there will be many decisions which may be directly or indirectly discriminatory on the grounds of race or colour which have adversely affected ethnic minority families but which have not been subject to any challenge either because of the lack of knowledge as to the reasons on which the decision was based or because of the lack of opportunity to introduce legal challenge.

ACCOMMODATION

Although the Act defines when accommodation is available for occupation, it does not define 'accommodation' itself. Consequently no minimum standards of existing accommodation - whether by reference to provision of amenities, overcrowding, structural stability or general fitness for habitation - have been set by which local authorities might judge whether an applicant who was not roofless was nevertheless homeless. Those occupying hostels, common lodging houses, resettlement units or squatters or those in accommodation lacking privacy, security of tenure or access to cooking facilities, or sharing basic sanitary arrangements or living in damp, cold or otherwise barely tolerable circumstances would not of necessity be considered by the relevant authority as homeless in terms of the Act. It is this and like avenues of discretion which have given rise to such a plethora of judicial law-making by way of application for judicial review.

A temporary refuge (R v Ealing London Borough Council ex p. Sidhu (1982) 80 LGR 534), a night shelter (R v Waverley District Council ex p. Bowers (1983) QB 238) or uninhabitable accommodation (R v South Herefordshire Borough Council ex p. Miles (1984) 17 HLR 82) may qualify the occupant as homeless. But the decision in Puhlhofer v Hillingdon London Borough Council (1986) AC 484 established that accommodation that was neither large enough to accommodate people 'expected to reside' with the applicant nor providing the basic amenities of family life was, nonetheless, still 'accommodation' sufficient to disqualify the occupants from claiming homelessness. In the view of the House of Lords, accommodation does not become 'no accommodation' simply because it is statutorily unfit or overcrowded. Thus Part III of the Housing Act 1985, which replaces the Housing (Homeless Persons) Act 1977 in England and Wales, does not entitle the homeless to short-circuit the normal local authority selection and allocation procedures automatically - it must now be seen as an emergency provision.

In Scotland the rigours of the Puhlhofer decision have been attenuated by a subsequent amendment now contained in the Housing (Scotland) Act 1987 to the extent that statutory overcrowding or a failure to reach the tolerable standard would render occupants 'homeless'. For England and Wales, Section 14 of the Housing and Planning Act 1986 amends Section 58 of the Housing Act 1985: a person shall not be treated as having accommodation unless 'it would be reasonable' for him to continue to occupy it. In determining 'reasonable' regard may be had to the general housing circumstances pertaining to the area. While the latter test of accommodation is harmonised with the provision on intentional homelessness, the reference to the housing circumstances in the area suggests the opportunity for special pleading to minimise obligations in areas suffering from poor housing which are likely to coincide with areas of proportionally high rates of black residents, reinforcing structural disadvantage for black families (Brown, 1984).

PRIORITY NEED

The Acts specify four categories of priority need: applicants (1) with dependent children; (2) who are pregnant or who reside (or who might reasonably be expected to reside) with a pregnant woman; (3) who become homeless because of an emergency such as flood, fire or other disaster; or (4) who are vulnerable. It is this fourth category of vulnerability which has, generally, proved most problematic.

The Act specifies that vulnerability may stem from old age, mental illness, handicap, physical disability 'or any other special reason'. The DoE and Scottish Development Department Codes of Guidance refer to those above retirement age who are frail or in poor health, the blind, deaf or dumb or persons otherwise substantially disabled. They also refer to those threatened with violence and the young homeless at risk of sexual or financial exploitation (see Kelly and Mallon v Monklands District Council (1985) SLT 169). The Government survey of 1981, Single and Homeless dispelled the popular image of the single homeless as drunken dossers, identifying the major causes of homelessness amongst that group as being the lack of adequate housing and people moving between institutions, charities, hostels and sleeping rough. Although proportionately there are more white than black single homeless (Bonnerjea and Lawton, 1987), there are more black (predominantly Afro-Caribbean) than white single parents falling into the priority need category. Those black homeless surveyed in the study of Brent (Bonnerjea and Lawton, 1987), did not, generally, identify direct racial discrimination as a factor in the local authority's treatment of the homeless, including those in priority need, but there was a feeling that the system was unfair and that blacks suffered disproportionately. For example, nearly half the survey sample had family abroad with whom their housing histories were interwoven: the housing system failed to recognise the complexity of housing needs in such situations. In the absence of explicit recognition of such factors, including racial violence (AMA, 1988a) and the need for extended family provision (Henderson and Karn, 1987), leading to vulnerability (and priority need classification), local authorities exercise a discretion towards a narrow, entrenched view of the legal requirements, with a disproportionately adverse impact on black homeless.

INTENTIONALITY

The concept of 'intentional homelessness', which was described by Lord Soper as 'gobbledegook' (Hansard, H.L. Debates, 15 July 1977, Col. 1157) was a major concession to the local authority lobby during the passage of the Bill and effectively reinstituted the discredited notions of the deserving and undeserving poor and foreseeable homelessness (Watchman and Robson, 1983). If the applicant to the housing authority has left 'available accommodation' elsewhere then the authority may determine that such departure has rendered him or her intentionally homeless and therefore outwith the scope of the Act. In R v Wandsworth London Borough Council ex p. Nimako Boateng (1984) HLR 192, an applicant was found by the housing authority to be intentionally homeless in leaving accommodation in Ghana without having previously ensured that suitable accommodation was available in this country, a decision upheld by the courts. Self-evidently an individual may feel that the availability of accommodation in another country is essentially a fiction because the potentiality of continued residence in such accommodation is untenable. But this may not, of itself, render the person homeless in terms of the statutory provisions. Thus in R v Crawley Borough Council ex p. Mayor and Burgesses of the London Borough of Brent (Queen's Bench), 24 March 1983, it was held that a person fleeing from Uganda, although having a reasonable apprehension as to his safety in Uganda, was not entitled to be considered unintentionally homeless, as the individual should have an official refugee status in this country and therefore recognition as such by the Home Office. In contrast in R v Hillingdon London Borough Council ex p. Wilson (1983) 12 HLR 61 it was held that a pregnant woman who had no right to remain in Australia on a permanent basis because of the expected expiry of her leave to remain should not be considered intentionally homeless in this country; 'this lady is an English national, who had no right to remain in Australia and was coming back to this country'. Although the law does not give any recognition to the status of 'English national' it would appear the courts have taken into account the English origin of an applicant in determining the issue of intentionality and how it must be interpreted in terms of the Act. A further contrast is provided in R v Tower Hamlets London Borough Council ex p. Monaf, Ali, etc. (1988) 20 HLR 29. Mr Ali had come to the UK in 1963 when he was 23 years old and had brought his wife and six children in August 1986. Because he had entered the country before 1973, that is, before the provisions of the 1971 Immigration Act and the Immigration Rules thereunder had come into effect, he had no obligation to secure the availability of accommodation for his family 'without recourse to public funds' in respect of their entitlement to settle with him. It should be noted, here, that one of the central concepts of the Act is that of family unity, the objective being not only to keep families together but also to bring them together when they have been kept apart for various reasons (Din v Wandsworth London Borough Council (1981) 1 AC 688). Ali had visited Pakistan fairly frequently in the intervening years and there had occupied with his wife and family the property belonging to his mother. Tower Hamlets, who averred, in error, that the house was his, rather than his mother's, had argued that he was intentionally homeless by deserting available accommodation in Bangladesh. They argued that as he did not have 'a settled placed of residence' in the UK he and his family were not entitled to housing provided by Tower Hamlets under the provisions of the Act. The court held that although Tower Hamlets had been mistaken with regard to his ownership of the house and that the reference to 'settled place of residence' in the UK was misplaced, nonetheless the fact that there was a continuing availability of a residence in Bangladesh entitled the local authority to conclude that he and his family were intentionally homeless and therefore had no entitlement to housing in terms of the Act. The Court of Appeal upheld the appeal and remitted the decision to Tower Hamlets solely on the ground that the decision letter failed to disclose that the Council had considered the factors involved in deciding whether or not it was reasonable for the appellants to continue to occupy settled accommodation available to them in Bangladesh. In June 1988 Tower Hamlets evicted ten Bangladeshi families (sixty-six people), having 'reviewed' their previous decisions to evict in the light of the Court of Appeal decision and the eviction of a further thirty families was expected (Guardian, 7 June 1988: 3; see also Boparai, 1987).

THE CHILD CARE ACT 1980

The Act imposes a duty on local authorities in England and Wales to make available such advice, guidance and assistance as may promote the welfare of children by diminishing the need to receive them into care. In the Monaf case (above) the court decided that these provisions were discretionary. Once a family were determined to be intentionally homeless because of leaving 'secure' accommodation in Bangladesh, the housing authority was not obliged to secure accommodation to prevent families breaking up, beyond the requirements of the homeless persons provisions. It would appear, therefore, that neither the Child Care Act 1980, nor the Social Work (Scotland) Act 1968 in Scotland, will prove of any benefit in helping the homeless.

DEPENDENCE ON PUBLIC FUNDS AND FAMILY UNITY

Leon Brittan, then Home Secretary, announced in July 1985 changes in the immigration rules affecting those settled in this country after 1 January 1983: in future spouses and children would only be entitled to join the 'head of household' from abroad if, on doing so, they would not then have 'recourse to public funds'. Public funds were defined as supplementary benefit, housing benefit, family income supplement and housing under the homeless persons legislation. Allocation of local authority housing through the waiting list, other forms of housing subsidy like mortgage and tax relief, in reality a greater form of public subsidy than subsidised council house renting, were not included in that definition. In 1982 an immigration adjudicator ruled that

> public housing does not enjoy a sufficient degree of public subsidy to constitute recourse to public funds. If an immigrant were to be denied access to public housing on grounds of public subsidy, logically they must be denied the enjoyment of services such as public transport or refuse disposal - all of which the person concerned themselves pay for through their rent and rates.

However, the Government's concern was expressed with alarming candour by the Home Office Minister, David Waddington, during the 1985 rules debate when he said:

> I have looked recently at the situation in Tower Hamlets. Men living in single persons' accommodation marry women overseas, bring them to this country and expect the council to rehouse them. The councils cannot, and the couple then have to be put up in hotels. In April 1985, 418 families - virtually all with one parent an immigrant - were being maintained in bed and breakfast accommodation in Tower Hamlets at an annual cost of £5.5 million. I do not think that any sensible person would fail to realise that there is a real problem.

The Tower Hamlets Homeless Families Campaign, however, suggested, in evidence to the Home Affairs Sub-committee on Race Relations and Immigration, that, rather than immigration, the principal reason for large numbers of families being put in hotels was council policy and poor housing management practice. The group said that although the council at that time had more families in hotels than any inner London borough it had the second largest public housing stock: yet it allocated only 20 per cent of housing to the homeless - about half the allocations made by neighbouring councils. Although most heads of Bangladeshi households had been settled in England for more than twenty years council policy meant that they could only register their wives and children on the council waiting list when they obtained entry clearance and arrived in the borough. As a result they were placed in hotels even though the spouse might have been registered on the waiting list for a substantial period. Jonathan Stearn (1987) demonstrated that there was a clear connection between information supplied by one housing department, i.e. Tower Hamlets, and subsequent changes in national immigration policies. His conclusion concurred with that of the Homeless Families Campaign

> it is clear that the policies and practices of Central Government coincide in this case with those of the local authority and adversely affect a particular section of the community making it more difficult for people of Bangladeshi origin to get a home, or even live together in this country.

In April 1986 the DoE's Priority Estate Project in Tower Hamlets issued a report which concluded that the inefficiencies and under-resourcing of the authority had the affect of depleting the housing stock and impoverishing the service to tenants and would-be tenants.

> The intricacies of the interacting allocation categories, priorities and procedures make it difficult to draw firm conclusions. What has emerged very clearly from this study is the relative disadvantage of the homeless applicants, and, through their over-representation in this housing category, the Asian minority.

The relationship between the immigration rules and access to public housing is becoming more patent. While access to local authority housing through the general waiting list does not, as yet, affect the entitlement to entry clearance of dependants of applicants with settlement status, such applicants may be denied points allocation reflecting the needs of the legitimate family unit which would secure appropriate council housing on the grounds that all members of the family unit must be UK resident before such needs are to be taken into account. Consequently the actual housing needs of the family will only be assessed on arrival in the UK and when it is in immediate need of housing. At this juncture the waiting time points, frequently of critical importance to the chances of speedy allocation, which the applicant would otherwise have secured by reason of early planning for family reunion, will be discounted. However, where the family does secure sufficient points to warrant immediate allocation from the general waiting list the choice of accommodation is likely to be more limited than that offered to other applicants on the general waiting list because the circumstances will oblige acceptance of the first offer of accommodation. The condition or requirement imposed by some housing authorities for an applicant's family to be resident in the UK before their housing needs are taken into account constitutes indirect discrimination in terms of Sections 1 and 21 of the 1976 Act. Such rules have only recently, however, been challenged in the UK, perhaps because of the view that the courts would find them otherwise 'justifiable' on non-racial grounds. But such rules will now, not infrequently, foreclose access to all housing except through the homeless channel, the effect of which will be to render the family dependent on 'public funds' and ineligible for entry clearance to the UK.

The only people not affected are EC nationals, who, under Community law, have a right to bring in their spouses, children under 21, their parents and their grandchildren from anywhere abroad.

Local authority housing departments will in future be in the situation of making decisions that affect immigration: spouses and children wanting to secure entry to the UK will be refused if housing under the homeless persons legislation is offered and yet those authorities like Tower Hamlets who refuse to use the alternative of giving waiting list points to spouses and children who are still living abroad effectively foreclose any reasonable alternative. For many practical purposes therefore the housing authority has become the immigration authority.

THE IMMIGRATION ACT 1988

The Government described the Immigration Bill 1987, then in Standing Committee at the end of that year, as 'loophole repairing' but there remains some ambiguity as to the Government's motivation for bringing the Bill forward. Primary immigration from outside the Common Market effectively stopped with the Immigration Act 1971, as we have already noted. Between 1979 and 1985 the numbers settling annually in Britain fell from 70,000 to 55,000.

In the opinion of the Joint Council for the Welfare of Immigrants the Government had three motives in introducing the Bill. First, it found it politically expedient to promise further restrictions during the election; second, the Civil Service wanted to overturn some legal reverses; and third, successful experiments on DNA testing meant it would become easier for immigrants to prove right of entry. The provisions outlined above repeal a guarantee in the 1971 Act to all male Commonwealth citizens settled before 1973 that they could bring their wives and children to this country. In future these men will have to pass the same tests as those settled after 1973, that they will not be dependent on the public purse (supplementary benefit, family income supplement and housing benefit) and that the primary purpose of their marriage was not to circumvent immigration law. The apparent motive for the change was the European Court ruling that the existing law was discriminatory in giving a right to males that was not available to women. Mr Renton, then Minister responsible, accepted that he could have extended the right to women but argued:

> We believe that those who come to this country and want to bring their spouses and dependants here should show that they can maintain and accommodate them without immediate recourse to public funds.

As we have noted the change will affect most severely the Bangladeshis, as the most recent immigrant community, in their desire to unite with their families: in 1986 the provisions now repealed protected 780 Bangladeshi women and 2,400 Bangladeshi children. Section 1, then, by repealing the protection to the families of those settled prior to 1 January 1973 previously afforded by Section 1(5) of the Immigration Act 1971, places a further obstacle in the way of divided families. For many years large numbers of mainly Bangladeshi families have remained divided because they were unable to prove to the satisfaction of British entry clearance officers that they were related as claimed to their husbands/fathers in the UK (UKIAS, 1987). When DNA blood tests became commercially available in 1987, for the first time they had a means of proving the claimed relationship and securing entry to the UK as dependants, albeit after frequently excruciatingly long waiting periods - a ploy deliberately used by the Immigration Department to reduce entry numbers, whether by the death, despair or division (CRE, 1983a). The further requirement - not to be dependent on public funds - was introduced, ostensibly to comply with decisions of the European Court of Human Rights finding that Section 1(5) in conferring benefits on dependants of husbands and not wives was discriminatory. The obvious remedy, which would have honoured the commitment made by Government when the clause was debated, would have been to extend the benefit to dependants of wives. Given the small numbers involved, such a concession would not only have been logical and honourable but would not have constituted a departure from the principle of 'firm' immigration controls. Moreover the Act does not amend any other discriminatory provisions in the 1971 Act, for example Section 5(4), which provides for the deportation of the wife as the member of the family of a man subject to deportation (ILPA, 1987).

However, in addition to affecting housing directly, it is worth mentioning the other aspects of the Act which will affect black immigrants most adversely. The Act puts new restrictions on the right of appeal against deportation. Previously overstayers who were threatened with deportation had the right of appeal on compassionate grounds. Now anyone who was given permission to enter Britain less than seven years after a deportation decision was made will no longer be able to argue compassionate circumstances at an appeal tribunal. More importantly the Act overturns a court ruling and makes it possible for the Home Office to institute criminal proceedings against overstayers, however long they have been in the UK. Previously such cases could only be brought by the Home Office within three years of the date when the overstaying began. One by-product of the change may well be a return to stop and search. It may not be coincidental that on 17 December 1987, while the Bill was in Standing Committee, Tim Renton in a written answer to a planted question announced that passport frauds were increasing. In 1986, the only complete year for which figures were available, the passport department of the Home Office recorded 173 cases where standard British passports had been fraudulently altered and nine cases involving counterfeit documents. In the period to 11 December 1987 there were 183 such alterations and fifteen counterfeits. In truth these were far from significant increases but the time may have proved opportune to reinforce the 'closed texture' of the Government's debate on immigration.

REFUGEES

On the same day, i.e. 17 December 1987, the Government must also have been applauding the decision of the House of Lords in R v Home Secretary ex p. Sivakumaran and Others: United Nations High Commissioner for Refugees as Intervenor (1987) Guardian Law Reports, 17 December 1987. This case overturned the decision of the Court of Appeal that in order to qualify for refugee status a claimant need merely assert his genuine apprehension or fear of prosecution in his own country. The House of Lords determined then that in order to qualify for refugee status a claimant must have a well founded fear of persecution in his own country and for that purpose he must demonstrate a reasonable degree of likelihood that he would be persecuted if he returned there. The necessary degree of likelihood may be described as 'a reasonable chance', 'substantial grounds for thinking', or a 'serious possibility'. In this instance each of the six applicants was a Tamil who arrived in this country on various dates between 13 February and 31 May 1987 and upon or shortly after arrival applied for asylum in the United Kingdom, claiming to be a refugee from Sri Lanka.

The Home Secretary refused their applications and Mr Justice MacCowan dismissed their applications for judicial review. The Court of Appeal reversed that decision, quashed the Home Secretary's decision and held that he should consider the applications anew. The United Nations Convention on the Status of Refugees dated July 1951 (Cmnd 9171), as amended by the Protocol dated December 1966 (Cmnd 3096), provides by Article 1A(2) that the term 'refugee' applies to any person who

> owing to well founded fear of being persecuted for reasons of race, religion, nationality, membership of a particular social group or political opinion, is outside the country of his nationality and is unable or, owing to such fear, is unwilling to avail himself of the protection of that country; or who, not having a nationality and being outside the country of his former habitual residence ... is unable or, owing to such fear, is unwilling to return to it.

Lord Keith said that the critical words were 'well founded fear' of being persecuted for what might compendiously be called a Convention reason. The Home Secretary had expressed the view that army activities aimed at discovering and dealing with Tamil extremists did not constitute evidence of persecution of Tamils. It appeared that the Home Secretary also considered whether any individual applicant had been subjected to persecution for Convention reasons and decided none of them had been. In this present case, an examination of the decision-making process did not disclose, in the court's view, any error on the Home Secretary's part, or justify the court in contradicting his view that the applicants would not be in danger if they were returned to Sri Lanka. The appeal by the Home Secretary, therefore, was unanimously allowed. Clearly then the Home Secretary's subjective judgment of whether persecution is a substantial possibility has been substituted for that of the person seeking asylum. As a result it would appear that so long as the Home Secretary goes through the motions of serious deliberation, his ultimate decision is unlikely to be successfully challenged.

All this may appear a diversion from the issue of homelessness and ethnic minorities but it serves to illustrate a mealy-mouthed and widely held chauvinism towards ethnic minority immigrants of whatever status. It is also a reminder of the Government policy of dispersal towards Vietnamese refugees, effectively offloading a responsibility of Central Government to a diversity of local housing authorities without providing either the refugees or the authorities with guidance or resources to secure effective support. Many such refugees, without relevant employment skills or opportunities, gravitated back to major conurbations, principally London, not infrequently adding further stress to Homeless Persons Units in the boroughs affected. Some families would then find themselves 'intentionally' homeless and, if anything, provided with short-term temporary accommodation.

LOCAL AUTHORITY ALLOCATION POLICIES

The particular allocation policy of a local authority will determine not only the priority afforded to different categories of applicant within the general waiting list but also the relative proportion of available lets assigned to the homeless. CRE investigations demonstrate the variety of ways in which the former may discriminate against ethnic minority applicants (CRE, 1984a, 1984b). Policies such as those employed by Tower Hamlets in Bethnal Green (Runnymede Trust, 1989a) to favour 'sons and daughters' of existing tenants inevitably steer ineligible applicants, frequently comprising a disproportionate number of black applicants, into the homeless category when their existing housing provision becomes intolerable and access to general waiting list lets is inhibited. Similarly, where a relatively small proportion of lets, often from the less desirable housing stock, is earmarked for the homeless further disadvantage ensues: not only may this result in unsatisfactory hostel and boarding accommodation being provided but in addition many London boroughs are placing the homeless outside their own boundaries. This may result in poor or impossible access to employment opportunities and a lack of community support (CRE, 1988c).

The CRE draft Code of Practice in the Field of Rented Housing, enabled by the Housing Act 1988 (CRE, 1989b), clearly advises against such policies, warning that they may constitute unlawful discrimination in terms of the Race Relations Act 1976. However, even when the draft is approved by the Home Secretary (after a period of consultation) it will not have statutory force in the sense of obliging compliance. Just as the Code of Practice on Employment has been ignored with impunity - not infrequently by major employers - it is unlikely that the Housing Code will strengthen the resolve of many housing authorities to take their statutory responsibilities (under Section 71 of the 1976 Act) more seriously than before. This duty is too vague to be enforceable (MacEwen, 1985).

Following the decision in R v Tower Hamlets London Borough Council ex p. Camden London Borough Council (1989) 21 HLR it is possible for the homeless to embark on shopping expeditions to different local authorities: provided one of these, in applying all the correct criteria - including consultation with the 'responsible' authority (i.e. that with which the applicant has a close connection and which he has previously consulted) concludes that the applicant meets the necessary requirements then the responsible authority will be required to find housing. In truth, however, while this technique may be used to attenuate the harshest decisions of some authorities regarding homelessness, it constitutes a mere palliative. The lack of Central Government guidance to and monitoring of local authority allocation policies has been the subject of trenchant criticism generally (MacEwen, 1987b) of some longevity (e.g. Cullingworth, 1969) but, despite the problems recently exposed (SCPR, 1988), there are no current governmental proposals to strengthen the statutory provisions regarding allocation. Consequently applicants will remain subject to the vagaries of local rules and policies and discretion. This is not to deny the need for some discretion at the local level but merely to question, in the light of experience, the absence of a more clearly defined statutory framework and opportunities for administrative review. Moreover while in proportional terms the most serious incidence of racial disadvantage in respect of homelessness may be located in England, there is growing evidence that ethnic minorities in Scotland experience significant disadvantage in access to public and housing association lets (Sim et al., 1989; Dalton and Daghlian, 1989; SEMRU, 1987; Hancock and MacEwen, 1989).

THE JUDICIAL APPROACH

The Puhlhofer case was important not only because of the interpretation of the 1977 Act which it provided but also for Lord Brightman's statement that he was troubled 'at the prolific use of judicial review' by those challenging local authority performance. He advised that 'great restraint' should be exercised in giving leave to proceed by judicial review, which should be reserved for exceptional cases. Given the extent to which the High Court has made reference to that opinion it may be argued that such restraint on challenge will be more influential in permitting local authority discretion than the subsequent legislative amendments, which were confined to the problem of defining 'accommodation' in particular circumstances (Mullins, 1988). While little solace is found in the court's view that a pigsty (obiter in Brown v Hamilton District Council (1983) SLT 397) would not constitute 'accommodation' or that an infested hut (R v South Herefordshire District Council ex p. Miles (1984) 17 HLR 82) was below the accepted borderline in respect of a family of five but not of a family of four, the willingness of the courts to invoke a duty to ratepayers to prevent queue-jumping (Din, above), and to conclude that premises offered to an Asian family then occupied by squatters (R v Westminster City Council ex p. Wahab (1983) QB) was sufficient to discharge the authority's duty on the grounds that it could be made available within a few days, illustrate a judicial approach to the Act which, at least on occasion, is more sympathetic to local authority views than the homeless. The fact that judicial distinction has been made between immigrants and 'English' nationals evokes little confidence in the objectivity of judicial interpretation.

INVESTIGATIONS BY THE CRE

IN R v CRE ex p. Hillingdon London Borough Council (1982) WLR 520 the CRE investigated Hillingdon's handling of applications from homeless immigrants on the basis of two cases: in the first case the Janmohamed family (an East African family with four children aged between 12 and 21) arrived at Heathrow airport from Kenya on 5 November 1978. Hillingdon London Borough Council provided them with temporary accommodation and on completion of its inquiries determined that although they were homeless and in priority need its duty was confined to providing temporary accommodation for a period sufficient for them to be afforded a reasonable opportunity of securing accommodation elsewhere. In the event, however, even this limited obligation was not met in that the chairman of the council's housing committee, Mr Terry Dicks, decided to have the Janmohameds taken by taxi to the Foreign Office and dumped on the pavement in a gesture later described by Lord Justice Griffiths as in the worst possible taste and with inhuman disregard for the feelings of the unfortunate family. In contrast a Mr Turvey, an Englishman returning to England from Rhodesia, with his children aged between 18 months and 13 years, was found by the same authority to be homeless and in priority need but despite the fact that he had been employed as a school-teacher and had voluntarily relinquished his accommodation in Rhodesia, not to have become homeless intentionally. He was therefore secured permanent accommodation by the housing authority.

Hillingdon was advised that the CRE was considering conducting a formal investigation under Section 48(1) of the Race Relations Act 1976 on the basis that the treatment of these two families constituted a <u>prima facie</u> case of unlawful racial discrimination. Hillingdon challenged this investigation on two grounds: first that the CRE did not have reasonable grounds for believing that the council had unlawfully discriminated against the Janmohamed family on the grounds of race and second that the CRE's frame of reference was not specific enough and this resulted in the Commission exceeding its powers. Lord Denning, with reference to the former ground, found that it would be unfair to subject Hillingdon to an inquisitorial investigation on such an insubstantial foundation as the disparity of treatment between the Janmohamed and Turvey families, as this disparity could, he believed, be explained 'perfectly well by the honest decision of the council's housing officer' in that the former family was intentionally homeless and the latter unintentionally homeless. Hillingdon was similarly successful in respect of the second ground. On appeal to the House of Lords the decision of the Court of Appeal was confirmed. Lord Diplock advised that the CRE could not embark on a formal investigation without being specific about the types of unalwful acts which it considered had been committed by Hillingdon. It could neither 'throw the book' at Hillingdon nor tell the borough that it might have committed or be committing some acts capable of

amounting to unlawful discrimination. There was a requirement for 'a particularisation of the kinds of acts of which
the borough was suspected'. Section 49 of the 1976 Act
permits three kinds of investigation by the CRE: general
investigations into a particular activity such as mortgage
allocation or immigration; investigations into named persons
where an unlawful act was suspected - as in the Hillingdon
case, or investigations into named persons where no
unlawful act was suspected. Only in the second of these was
the CRE given power to compel the production of evidence
without the Secretary of State's consent. The House of
Lords, in holding that the terms of reference should not go
beyond the suspicions held by the CRE, per Lord Diplock in
Hillingdon, cast doubt on the validity of any investigations
where no unlawful act was suspected - doubts confirmed by
the decision in CRE v Prestige Group (1984) 1 WLR 355. The
Hackney investigation (CRE, 1984b) clearly demonstrated,
however, that the CRE may be fully justified in
investigating named persons where no act of unlawful
discrimination is suspected because of the covert nature of
indirect discrimination generally and specifically in relation
to housing allocations, including, by inference, those
affecting homeless persons. Given the improbability of any
individual having either access to the records or sufficient
time and resources to mount a thorough examination of
procedures relating to a specific housing authority, the
Hillingdon and Prestige decisions limit the use of the last
resort - that of a CRE formal investigation.

However, one of the most recent CRE reports of formal
investigations concerned homelessness in Tower Hamlets
(CRE, 1988c). The investigation was based on the belief that
Tower Hamlets might have discriminated in four areas:

1. Treatment of council tenants forced out through
   emergencies.
2. Treatment of homeless families, particularly in respect
   of poor quality accommodation for Bangladeshi homeless in Southend-on-Sea.
3. Treatment of applicants, particularly those with
   families overseas.
4. Allocation of families to worse quality housing.

Although the size of the Bangladeshi population, its
dominance in the public sector and public profile of the
authority's policies facilitated a broad structuring of the
CRE's terms of reference, enabling, in turn, a wide-ranging

inquiry, it is evident that such preconditions are less likely to be present in other local authority areas, particularly outside London. Consequently the limits imposed by <u>Prestige</u> remain a limitation on the CRE's investigative powers. Nonetheless the conclusions reached by the investigation are noteworthy. In respect of discriminatory treatment the investigation found:

1. Direct discrimination against Bangladeshi homeless families in respect of waiting time for accommodation.
2. Indirect discrimination in respect of the treatment of separated families where part of the family was resident outside the UK.
3. Indirect discrimination in the allocation of a dispropor-tionate number of ethnic minority families to John Scurr House, a poor estate.

The housing department was described as 'a shambles undergoing improvement'. The Non-discrimination Notice (CRE, 1988c: Appendix D) provided detailed requirements for compliance, including ethnic monitoring and reporting progress to the CRE. The extent to which this investigation will inform the practice of other housing authorities is, as yet, too early to gauge but, as the CRE chair observed (CRE, 1988c: 7) 'although there are no easy answers ... no improvement can be made without firm and determined Government intervention'.

CONCLUSIONS

In summary the law and its interpretation on homelessness are critical for ethnic minorities in the following areas:

1. The definition of key terms - 'homeless' (and available 'accommodation') 'priority need', 'intentionality', 'local connections' etc. - by which the homeless are given or denied access to local authority provision.
2. The quality and quantity of local authority housing stock made available for letting to the homeless.
3. The use of hostels and boarding houses as temporary and even permanent accommodation for those deemed homeless.
4. The extent to which local authorities refuse to permit general waiting list applicants to register needs by reference to dependants currently overseas.

5. The inclusion of the allocation of council housing through homelessness within the scope of dependence on public funds and the consequences for immigration status.
6. The attitude of the courts to the needs of the homeless.

The above considerations are, self-evidently, additional to those relating to council house allocations, which have been demonstrated to impose a particular burden on black applicants. In the light of the restrictions imposed by the courts on the effectiveness of the CRE as a policing agency by way of future formal investigations, the impact of the Race Relations Act 1976 on the duties of local authorities to the homeless is likely to be limited. Where discrimination is covert - the real nub of the problem - the Act remains essentially of declaratory effect. This, in itself, is unlikely to influence the vast number of authorities who continue to turn a colour-blind eye to the red warning signs displayed from existing investigations.

The general situation regarding homelessness, which has doubled since 1979, is now critical and in London it is desperate, 51 per cent of all allocations being earmarked for the homeless, rising in some London boroughs to 90 per cent. The obligations which are to be imposed on housing authorities to sell off large estates to Housing Action Trusts (HATs) and private landlords will result in fewer properties being earmarked for higher proportions of homeless families. Although HATs and other housing landlords are obliged to give councils 'such assistance as is reasonable' in housing the homeless - equating with the obligations placed on New Town Development Corporations - these have been deployed in a piecement and unconvincing fashion (Greenwood, 1988). Black families, being disproportionately affected by homelessness, would be significantly disadvantaged by the provisions of the Housing Acts 1988. Although such disadvantage would otherwise constitute unlawful discrimination, unless it might be considered justifiable on non-racial grounds, the 1976 Act does not safeguard against discrimination in primary or secondary legislation. Miles (1989) suggested that the DoE had considered new rules which, in cutting thousands from the official homeless statistics, would:

1. Require three year's local residence (currently six months) before eligibility as homeless.

2. Restrict eligibility for those rendered homeless following disputes with their family or friends with whom they had been staying.
3. Restrict the definition of priority need, making it harder for single mothers to meet the legal requirements.

Such proposals would have had a disproportionately adverse impact on certain ethnic minority groups. Because the changes would be effected by statute, Section 41 of the Act exempts the proposals, and any act done to secure compliance with them, from falling within the definition of unlawful discrimination. In the event the DoE review concluded that changes in the homelessness provisions were not necessary, that £250 million should be allocated to London and the South East and that a revised Code of Practice should be issued after consultation (IoH, 1989).

But it is likely that much greater scrutiny will be given to council house access channels, and the homelessness route, in a period of sharply diminishing stock and resources; this will attract even more criticism than at present, as it will be blamed for precluding allocations from the general waiting list.

The Institute of Housing, in response to such concern, published a report in 1988, Who will House the Homeless? which identified the weaknesses in the current legislation, the Code of Guidance, and practices by local authorities (IoH, 1988). Its main conclusions were, first, that Government should not now start to dismantle the legislation; second that it must accept responsibility for the needs of the homeless instead of expecting local authorities, with their diminishing resources, to bear the brunt; third that it should ensure that implementation of the present legislation is 'strengthened and tightened'; and that, in using information available to measure need and adjust policy, it should require housing associations, the Housing Corporation, building societies and private landlords to contribute to solving the problems in a more active fashion. No mention was made of the particular needs of or problems faced by ethnic minority families.

In 1989 the National Federation of Housing Associations (NFHA), the Association of Metropolitan Authorities (AMA) and the Association of District Councils (ADA) issued a joint statement stressing the responsibilities of housing associations in respect of the homeless. In part this was a response to the Housing Corporation advising housing associations of their objective to provide 'reasonable assistance' under Section 72 of the 1985 Housing Act (Stearn, 1989). In addition NFHA has produced guidelines, 'Tackling Homelessness'; the London Housing Association Council and the Association of London Authorities/London Boroughs Association have produced guidelines on local authority nominations, 'Partners in Meeting Housing Need'. Historically a very small proportion of the 50 per cent local authority nomination rights are allocated to the homeless. In 1987/88 only 13 per cent (London 34 per cent) of local authority nominations fell into this category (Hansard, 15 February 1989), while a number of boroughs failed to nominate any homeless applicants to housing association tenancies (Stearn, 1989). 'Partners' shows that 25 per cent of London Association nominations fail, the major reason being a failure of the prospective tenant to show up or to accept the tenancy. But for large associations, 21 per cent of refusals were because the nominees' income was too high! Clearly local authorities fail to use existing nomination rights to cater for the needs of the homeless and pre-empt housing association discretion. The Liverpool investigation (CRE, 1989e), in demonstrating racial discrimination in local authority referrals, gives cause for concern that ethnic minority homeless may suffer at both local authority and housing association access points for housing association tenancies.

Whatever emerges from this debate the pervasive stigmatisation of the homeless as less deserving than general waiting list applicants has been apparent not only in housing authority allocation and Governmental responses to housing shortage but also in the attitude of the courts. The categorisation of 'homeless' as those in priority housing need as the rationale for obtaining direct access to council housing in preference to applicants on the general waiting list is testimony to the fact that the majority of allocation rules give greater weight, through a points or merit system, to time on the list as opposed to need itself: if this were not the case the need for special provision would not be manifest. But as those with waiting time points are squeezed out of allocations through the processes of council house residualisation, the contrary view that housing need is less important than longevity of application is being revitalised and the concept of belonging to the community is a factor inherent in this realignment. To the extent that ethnic minorities are seen as immigrants and newcomers and are disproportionately represented in homeless categories, particularly in the inner London boroughs, their lack of belonging, obstensibly because of their more recent presence but in reality because they are black, has facilitated this shift of opinion in London and the Home Counties and, by a process of lobbying and osmosis, in the Conservative Party. In such a climate of opinion even if the Race Relations Act had been effective in securing equal access to council housing for the black homeless - which it has demonstrably failed to do - it seems unlikely that this would have stemmed the demands for homelessness to be redefined to exclude a high proportion of black families. Indeed the Tower Hamlets predicament suggests that an effective Race Relations Act in the area of homelessness might well have accelerated the process.

CHAPTER EIGHT

PUBLIC HOUSING: ADMISSION AND ALLOCATION

INTRODUCTION

Collectively local authorities are the greatest landlords in the country and at the time of post-Second World War immigration provided the most coherent form of access to decent housing to a wide social and economic spectrum of society. Certainly owner-occupation then matched and now exceeds council house tenure in becoming the dominant form of residential holding and this essential duality in the housing market formalised and sustained social divisions. But the stigma of welfare provision, currently reinforced by the residualisation of council housing, was less prominent in the 1960s and early 1970s. Despite the increase in the bridging tenure of housing associations, the privatisation of 10 per cent of council housing stock since 1980 and the introduction of other initiatives such as homesteading and staircasing, council tenancies remain accessible and desirable to many applicants with more limited housing options.

However, for a complexity of reasons including indirect discrimination ethnic minorities have only fairly recently become council tenants in significant numbers (see Table 2.1 and Smith, 1989b). West Indians and Bangladeshis, within the ethnic minority groups, are, in comparison with the rest of the population, now disproportionately within this tenure, while Indians and Pakistanis remain predominantly in owner-occupation, irrespective of socio-economic class. Before examining the Race Relations Act 1976 respecting its impact on council housing admissions and allocations a general outline of the legal framework is provided as an essential background for considering the racial dimension.

THE GENERAL LEGAL FRAMEWORK

Rules, discretion and information

After the obligation to provide housing for the 'working classes' had been removed by the legislation of the late 1940s and before the passage of the Housing Act 1980 and, for Scotland, the Tenants Rights etc. (Scotland) Act 1980, public housing authorities enjoyed virtually untrammelled discretion regarding the choice of tenant. That statutory provision which existed tended to be couched in vague and general terms. Thus the Housing Act 1957, and the Housing (Scotland) Act 1966, obliged local authorities, in their selection of tenants, to secure that a reasonable preference was given to persons who were occupying insanitary houses or overcrowded houses, had large families or were living under unsatisfactory housing conditions. These Acts also obliged local authorities to reserve houses for the 'agricultural population' and rural workers so far as the relevant Secretary of State had effected payments towards such provision in terms of various enactments and so far as demand from that section of the population required such reservation.

Additional obligations were imposed by the Land Compensation Acts of 1973 and the Housing Acts of 1974 to rehouse persons displaced by an authority's own demolition and improvement schemes within a reasonable distance of their previous accommodation. Discrimination on the grounds of sex and race - whether direct or indirect in respect of allocations - was made unlawful by the Sex Discrimination Act 1975, Section 30, and, as we have noted, the Race Relations Act 1976, Section 21, and the provisions of the Housing (Homeless Persons) Act 1977 required local authorities to give preference to the homeless in their allocations.

Nonetheless in aggregate these provisions made few inroads, whether procedural or substantive, into the broad array of allocation systems and policies adopted by local authorities throughout England, Wales and Scotland; no machinery for selection was prescribed, rules were not required to be published and there was no opportunity for scrutiny, challenge or review. Indeed such discretion had been viewed by government as advantageous in allowing maximum freedom to respond to local circumstances, although it was recognised that there was still scope for

217

considerable improvement in this field (Scottish Development Department, 1977a).

In Scotland this need for improvement was emphasised by a report, Allocation and Transfer of Council Housing (SDD, 1980), issued by a sub-committee of the Scottish Housing Advisory Committee before its demise. The report argued that allocation and transfer policies should be seen to be fair, meet housing needs, maintain balanced communities, be flexible, promote mobility and make best use of the housing stock. Authorities were criticised for some rules relating to eligibility for the waiting list, such as age and income criteria, allocation methods - including those enabling councillors to assess the merit of applicants - and the selection of priority groups within the general waiting list. Similarly waiting list restrictions were criticised by the Central Housing Advisory Committee in 1949, 1953 and 1955 and in the 1969 report Council Housing Purposes, Procedures and Priorities and in the 1978 Housing Services Advisory Group's report Housing for People (Hughes, 1987). Some authorities operated income bars, others applied very strict residence qualifications, coupled with income criteria, some excluded single people from their normal allocation procedure, some discriminated against owner-occupiers and some went so far as to operate dual lists, only one of which was active so far as considering allocations was concerned.

Although Central Government advice on admissions and allocations was generally couched in sympathetic and liberal terms, the nature of existing council stock and the severe financial restrictions on local authorities in respect of housing improvement had resulted in a vast diversity of the quality of stock in relation to location, type and standard of repair. Consequently many local authorities were faced with the need, first, to exercise some selection criteria because demand greatly outstripped supply and, second, in determining allocations, to devise some system which would secure that the worst quality stock was tenanted in order to secure necessary income from rent. Inevitably the quality of property was matched or attempted to be matched with the quality or status of the tenant. Unsuitable tenants would be turned away because they represented a potentially considerable drain on management resources and 'more suitable' tenants would represent minimising the likely rent arrears and disturbances, etc. Newcomers and latecomers might be rejected for reasons of 'fairness' while for other minority groups a classification of deserving or undeserving might apply as a means of filtering. Frequently the decision as to who should be excluded was made by housing investigators or visitors who interviewed the household concerned while they were still living in the dwelling to be vacated. The visit was a means of assessing the family, the dwelling, the condition of the furniture, household cleanliness, rent arrears and so on and a not infrequently subjective assessment of the 'type of applicant' was made and degrees of suitability and eligibility reflecting both group and individual prejudices and beliefs were determined (Merrett, 1979: 216).

The current law with regard to the allocation machinery - the procedural aspects - is found in the Housing Act 1985 and the Housing (Scotland) Act 1987. In these measures the concern is restricted essentially to openness. Thus housing authorities, including New Town Development Corporations, Scottish Homes and housing associations are obliged to publish any rules in existence governing admission to the waiting list, allocations, transfers and exchanges and to make them available ad longum for a charge (if imposed) and in summary form without charge. Applicants also have a right of access to any information which they have provided to the housing authority. The Local Authorities (Access to Information) Act 1985, which deals with public access to local authority documents and meetings, enables applicants, amongst others, to inspect and obtain copies of minutes, reports and documents relevant to committee or sub-committee deliberations so far as they are neither confidential nor exempt. These may relate to allocation policies and procedures which have not been incorporated in the published rules as amended from time to time.

Despite these provisions an applicant may remain ill informed as to his prospects regarding allocations. If there are no rules, a situation possible in Scotland, there is no obligation to draw them up and publish them. If the rules enable official discretion, so far as such discretion has not been considered by a committee or sub-committee, its exercise confers no right of public scrutiny. While an applicant has access to information provided by him or her, there is no right of access to information provided by others even where it may critically affect the chances of allocation. Where a points system is employed the applicant has no right to know how many points he had got or how many he needs to secure accommodation either generally or specifically in relation to housing type (cottage/tenement as opposed to high-rise or maisonette) or location. Consequently the right to know is fettered and does not give rise to a right to question. However, following on the Access to Personal Files Act 1987 and the publication of the rules thereunder, an applicant should have access to all information relating to his application and consequently be provided with an opportunity to challenge factual inaccuracies not only in respect of information provided by him or herself but from whatever source. Until now the housing authority would only be obliged to release factual information about the applicant to the applicant when it was not supplied by the applicant when such information was held on computer by reason of the Data Protection Act 1984.

Admission to the waiting list

In England and Wales the Housing Act 1985, Section 22, requires the housing authority to give a 'reasonable preference' in allocation to those occupying insanitary or overcrowded houses, those living under unsatisfactory housing conditions or those having large families. There is no equivalent restriction to that now imposed by the Housing (Scotland) Act 1987 (Sections 19 and 20).

Thus local authorities (islands and district councils) in Scotland in considering admission to the waiting list must now disregard the following:

1. The age of the applicant (if over 16 years).
2. The income of the applicant and family.
3. The ownership of (or value attached to) heritage or movables past or present of the applicant or his family.
4. Rent arrears for which the applicant was not liable as tenant.
5. Whether the applicant is living with his spouse or another as husband and wife.
6. Non-residence within the area by the applicant where:
   (a) He is employed or has been offered employment in the area of the council.
   (b) He wishes to seek employment in the area - this being the purpose of his move to the area in the eyes of the council.
   (c) He is 60 years old and wishes to move to be near a younger relative.

(d) There are special social or medical reasons for requiring to be rehoused in the area.

The requirement in Scotland to disregard non-residence in the area in the above circumstances will not only secure admission to the waiting list but also ensure that where allocation rules apply such applicants should not be disadvantaged in comparison with tenants applying for transfers in similar housing need.

In Scotland there are also additional provisions relating to allocation where the following considerations must be disregarded:

1. The length of the applicant's residence in the area.
2. The considerations listed above in respect of age, income, ownership and arrears not attributable to the applicant as a tenant.

In addition the housing authority should not impose a requirement that:

1. The application should have remained in force for a minimum period.
2. A divorce or judicial separation be obtained.
3. The applicant should no longer be living with some other person.

Despite the additional provisions applying in Scotland, additions which were proposed unsuccessfully to be applied to England and Wales by a Private Member's Bill in 1988, it is evident that the above measures represent ad hoc accretions to the previous provisions and reflect an absence of any coherent policy on admissions or allocations which the government considers appropriate to erect. Essentially the measures are reactive rather than proactive. Those aspects of local authority practice which are considered restrictive - such as age and income barriers in Scotland - are outlawed but there is no requirement to deploy a points or group allocation system as consistently advocated. A paper presented to the Rowntree Study Group (Spicker, 1986) examined the allocation criteria used in 196 local authorities under twenty-four considerations ranging from overcrowding to lack of space. The paper illustrates the difficulty, not to mention the questionable desirability, of imposing rigid, uniform standards throughout the public sector. Despite this the criticisms levelled at some local authority practices demonstrate that the present framework of statutory control is inadequate. Allocation policies operate as a way of metering out a scarce resource amongst an over-large number of potential beneficiaries and will remain open to two major criticisms:

1. They are increasingly a welfare net designed to catch and accommodate the worst-off and most under-privileged people.
2. They take inadequate account of various forms of housing need (Hughes, 1987).

Furthermore, the processes involved in residualising council housing may encourage management practices which accentuate social division by categorising 'bad' tenants and placing them on the worst estates, and by channelling those in greatest housing need into less popular areas by dint of waiting time points gaining significant allocation preferences. While such practices are not of necessity unlawful there is an expectation of the legal system that those which are, such as discrimination on the grounds of age or marital status, will be effectively policed.

However, the courts, in both jurisdictions, have repeatedly stated their unwillingness to intervene in the exercise of a local authority discretion unless bad faith or severe unreasonableness can be proved. In addition to this general reluctance to intervene in the exercise of local discretion the dicta of Lord Porter in an English case (Shelley v LCC (1949) AC 56 at 66) were more specific:

> It is to my mind one of the important duties of management that the local body should be able to pick and choose their tenants at will.

Furthermore, in relation to the provisions requiring 'a reasonable perference' to be exercised, the categories (excepting overcrowding) lack any statutory definition and there is considerable doubt as to whether the 'duty' imposes a concomitant right in respect of those affected (Hughes, 1987).

In 1988 the Department of the Environment published a sponsored report, Queuing for Housing: A Study of Council Housing Waiting Lists (SCPR, 1988) based on a study of twenty-three English local authorities. The study confirmed the obvious - some authorities operate liberal policies and others restrictive ones, only recording on the waiting list applications which have been stringently monitored for housing need and meet local residency requirements. Most fell somewhere in between (SCPR, 1988: 3.2.1-3). One of three schemes - date order, points and group schemes - was employed by the majority in equal numbers. The majority of waiting lists showed a rapid turnover and were increasing at a rate varying from 0.5 to 26 per cent. The proportion of the applications to the local population varied from 3 to 17 per cent. Around a quarter of new lets went to elderly applicants, a quarter to single non-elderly applicants and the remainder to couples with or without children, four-fifths of applicants being either under 35 or over 55. About 40 to 46 per cent of waiting list applicants, it was estimated, were alive and still wanting council housing. The dominant reason from all types of applicant for wanting council housing was that they could not afford anything else. Although 40 per cent had thought about owner-occupation, of those who looked into it most then rejected the idea for financial reasons. While nearly half were interested in housing association tenancies, only 8 per cent had been asked about this interest by the council: 17 per cent did not know what a housing association was; 17 per cent had tried, independently, to get their names on a housing association waiting list. In comparison with the general population, the unemployment rate was higher, as was the incidence of a serious handicap or medical condition in respect of new tenancies. When waiting list applicants were compared with new tenants, the latter were found to be younger, poorer, sicker, more frequently unemployed, to have occupied more overcrowded and less 'amenable' accommodation and were more likely to have children (SCPR, 1988: 15-18). Although a question on ethnic origins was included, incomplete returns precluded analysis.

Because the survey did not analyse who got what and why and in this respect appears to confirm the view that its primary purpose was to justify the DoE's opinion that housing need was inadequately assessed by reference to council waiting lists and little else, conclusions from the survey are limited and indeed were not offered by the report itself. First, however, by comparing the waiting list profile with that of new tenancies it is clear that, whatever system is adopted individually, in general those in greatest housing need were more likely to be housed. Second, however inaccurate waiting lists might be, the fact that they were increasing, that stock was diminishing, that waiting time was increasing, and that a significant proportion of applicants had few, if any, tenure options, confirms that, irrespective of the Government's support for alternative housing, the gap between the demand for and supply of council housing is widening. Third, and no doubt partly in consequence of the above conclusions, council housing is increasingly catering for families and individuals whose particular needs are not met by residual housing stock, i.e. low-rise, one/two bedroom houses or bungalows, sheltered and community housing. Fourth, and more tentatively, by cross-reference to the 1982 PSI Survey (Brown, 1984), there remain significant differences between the profile of applicants (and new tenants) and ethnic minorities, for example, in respect of age and family size; consequently, that new build and rehabilitation which is allowed to take place in the public sector is likely to accentuate the gap between provision and ethnic minority need if it is to reflect more general needs on the waiting list. Consequently, unless local authority programmes acknowledge special needs, including those of ethnic minority applicants, it seems likely that the process of residualisation demonstrated by the SCPR report will aggravate disadvantage on racial grounds. In this light amending the allocation rules, monitoring the applications and policing the outcome are merely one side of the equation.

The courts

Because appeals to the courts are generally expensive and time-consuming, there is a strong argument for introducing a process of administrative review of housing authority decisions which may have a profound affect on tenants' lives. Such opportunity for review should not take place in a vacuum. If the Government were committed to achieving fair and rational criteria - while permitting some local discretion - it would have produced national guidelines in respect of local allocation, transfer and exchange schemes. Such guidelines, possibly published as an advisory code akin to the homeless provisions, might, in addition, create a presumption that a breach was unreasonable and subject to appeal unless the authority could demonstrate otherwise.

Ideally the government should consider the establishment of local housing tribunals, whose jurisdiction might extend to the homeless, housing benefit, repair and improvement grants, home loss payments, etc. In the absence of such a framework applicants should be afforded some opportunity for review. Indeed a back-door appeal mechanism exists with many authorities who may permit applicants to make a special case for consideration by the relevant sub-committee, but such systems are piecemeal and exclusive: they favour the well informed rather than those in greatest housing need. Accordingly, a formal administrative appeal mechanism should be a statutory and universal requirement.

Nonetheless such provision would prove of little benefit to the disappointed applicant if he were provided with merely the minimum information which the housing authority is currently statutorily obliged to supply. It would appear desirable for the tenant to require the authority to advise him of his relative position on the waiting list and the number of points which he has been allocated and which he will require, where applicable, along with the period he may be expected to wait in order to be allocated a tenancy of the housing type and in the area of his preference. Where points have been awarded or preference has been exercised on a discretionary basis, the method of calculation should also be provided. Such information - together with all personal file details whether or not provided by the applicant - would facilitate a more rational assessment by the applicant of his options and prospects. While it might in a minority of cases, lead to appeal, it might, more often, reassure the applicant that his lack of ready access to quality housing was attributable to factors of demand and supply rather than any arbitrary judgement of the merits of his application. The English case of R v Canterbury City Council ex p. Gillespie (1986) 19 HLR 7 suggests that, even within the present legal framework, the courts may intervene to quash an arbitrary decision despite its conforming with agreed policy. The facts in this case are worth brief description. Until March 1983 the applicant and her cohabitant lived as joint secure tenants in accommodation owned by Thanet District Council. They had two children. After the breakdown of the relationship the applicant left the accommodation and moved to her mother's home in the respondent authority's area. The accommodation was inadequate and she applied for housing from the defendants on 11 March 1983, a fortnight before being awarded custody of her two children. She was advised that it was not the council's policy to accept on the waiting list anyone having an interest in the title of a council dwelling house elsewhere. She then attempted to relinquish her joint tenancy but was refused because of outstanding rent arrears. On 21 May 1985 the respondent authority finally admitted the applicant to the waiting list but, through their environmental health and housing policy committee, agreed the following:

> That no applicant registered on the council's housing list be allocated accommodation while holding a joint or sole tenancy of another local authority or Housing Association property unless:
> (a) a reciprocal arrangement can be agreed with the landlord or local authority whereby the council may nominate its own tenant or applicant for rehousing or;
> (b) the case may be considered as a priority within the terms of the Housing (Homeless Persons) Act 1977 or because of violence, or a threat of violence, the applicant is unable to return to their accommodation in another area.

We have noted how in England and Wales each housing authority is obliged to maintain and publish a set of rules 'for determining priority as between applicants in the allocation of its housing' (now Section 106 of the Housing Act 1985). On application to the Queen's Bench for judicial review it was held that this policy of the council should be quashed as it constituted an inflexible rule rather than a general approach subject to exceptions which would permit each application to be individually considered: the authorities had failed to apply their minds, as they should have done, to the particular problems which the applicant asserted had prevented her from relinquishing her interest in her secured tenancy. The judgment referred with approval to the decision in Attorney General ex rel. Tilley v Wandsworth London Borough Council (1981) 1 All ER 1162, where it was held appropriate for the court to examine a resolution of a council in the exercise of its statutory discretion:

> to see whether the local authority has thereby bound itself to make future decisions in individual cases ... without taking into account some of the considerations that under the Act ought to be

taken into account. If the local authority has bound itself in that way the resolution is clearly bad.

Although the implications of this judgment are opaque and it would not be binding on Scottish courts, given the fact that there are similar obligations relating to the publication of housing rules and that the statutory provisions relating to 'reasonable preference' in the selection of tenants are almost identical, it is to be expected that courts both north and south of the border would be directed to this decision and encouraged to follow it.

Thus generally it would appear that housing rules relating to the waiting list and allocations are not to be equated with statutory rules in the sense that they must not impose a rigid requirement for observance on the local authority. It is in the nature of rules, however, that they are drawn up and applied to facilitate ready decision-making without reference, generally, to factors which they do not encompass. It would be somewhat ironic, then, for an authority to publish rules but to state that they were no more than general guidance. To state that each application will be considered on its merits to determine whether the relevant rules should be applied in a particular circumstance would go against the whole exercise - to give clear and explicit guidance to applicants regarding their expectations from the housing service. To be foolproof it would appear that the rules must either encompass every potentially relevant consideration - an unlikely and cumbersome result - or be expressed as a code of guidance imposing a collateral obligation on the housing authority to consider an applicant's case not merely in terms of the rules but also in terms of its particular merits. It is suggested that housing officers would not welcome the delegation of a discretion to disapply such rules as each applicant who felt that a particular rule was not to his or her advantage would insist on individual consideration. With the multiplicity of applicants and housing officers, inconsistency and, therefore unfairness would result.

It would appear therefore that the only satisfactory resolution of the difficulties stemming from the Gillespie decision would be for the rules themselves expressly to permit an avenue of appeal to one body - say, to the relevant housing sub-committee - on the ground, amongst others, that the rules should not be applied in the particular instance: such a mechanism would facilitate fair and even treatment of all applicants.

Given the importance of housing authority admissions and allocation procedures, systems, rules and areas of discretion, which often determine an applicant's chances of obtaining housing compatible with his or her household needs, it is important that there should be mechanisms whereby the proper performance of such areas of local authority responsibilities, particularly those which are purportedly underpinned by statutory obligation, are secured. This is mainly achieved through four sources of control: first, Central Government regulation and guidance; second, local democratic pressure - whether through councillors, pressure groups or press censure; third, the Local Government Ombudsman, and last the courts. The last mechanism - the power of the court to determine statutory appeals and to undertake judicial review of local authority decisions - may not be seen as the most effective or indeed desirable method of securing proper performance of a local authority's responsibilities. Nonetheless it is the last resort, the mere prospect of which should exert sufficient influence on local authorities to ensure responsible and sensitive management. Ultimately, the sanction of appeal to or review by the courts must be demonstrably accessible within a framework which may, for reasons of local discretion and flexibilty, remain largely administrative. The present structure, however, fails to guarantee observance of the existing statutory requirements and therefore does nothing to encourage responsible housing management. Of course there is an argument that the judiciary are - by reason of social background, training and experience - ill equipped to adjudicate housing issues and substitute their own set of social values for that of a democratically accountable local authority, and that an intrusive judicial presence in housing management will lead to defensive decision-making. To the extent that such arguments hold weight - and the absence of court decisions on issues relating to housing admissions and allocations renders any conclusion somewhat tentative - they strengthen the case for internal review of decision-making to ensure that it is fair and seen to be, thus limiting an applicant's expectations of successful appeal to the court.

From this discussion of the legislative framework applying generally to admissions and allocations in the public sector the following conclusions may be drawn:

1. Wide areas of discretion remain in the hands of the housing authority.
2. Where demand exceeds supply, such discretion is exercised to determine the match between the most suitable applicants and the best housing stock.
3. There is an absence of effective administrative and judicial review of the exercise of such discretion.
4. Applicants are not required to be fully advised of how their own needs and requirements are assessed, nor of how the housing authority exercises its discretion in practice.
5. Rules, policies, procedures and individual decisions frequently discriminate on an arbitrary basis without significant checks.
6. Those classified as 'undeserving' are likely to obtain the least desirable housing but are unlikely to be in a position to challenge the determinations made.

Despite these fundamental flaws and the history of paternalism as the predominant attitude towards applicants by local authorities, there is evidence of increasing professionalisation of housing practice and, with it, a greater willingness to come to grips with the arbitrariness and unfairness inherent in many of the systems previously adopted. Unfortunately, however, while the increased sensitivity by public authorities to the requirements of management has resulted and may in future result in significant improvements in public housing allocation, the same concern and commitment to matching housing with social need are unlikely to be at the forefront of considerations in the private sector, where commercial viability is bound to be the overriding consideration. Consequently the Pick a Landlord Scheme, which will enable tenants in the public sector to transfer to private landlords, may prove to be a particularly poor option for a number of minority groups, including blacks. Thus Haringey CRC (Inside Housing, Vol. 5, No. 2, 15 January 1988) has argued that positive steps taken by local authorities to promote equal opportunities such as ethnic monitoring, anti-racial harassment policies and other measures devoted to preventing racial discrimination would be lost under the government's new housing plans. It should be noted that the private sector has no obligation placed upon it equivalent to that placed upon the public sector by Section 71 of the 1976 Act. Moreover such processes of group and individual council house sales will promote the further residualisation of local authority housing to the extent that the increased competition for a diminishing resource will intensify housing management problems and will put pressure on local authorities to make speedy and therefore arbitrary decisions in admissions and allocations. The extent to which blacks are classified as undeserving will therefore seriously affect their chances of obtaining decent housing on an equal basis to whites reflecting similar housing need. It is with this general context in mind that the specific legislation on race and housing allocation is considered.

RACE LEGISLATION AND ALLOCATIONS

It has been noted that the 1976 Race Relations Act extended the definition of unlawful discrimination on racial grounds to include not only overt or direct discrimination on the grounds of race, colour, nationality and national or ethnic origin but also indirect discrimination: that is, the application of a condition or requirement to someone seeking a benefit which, while not apparently discrimi-natory, has the effect of discrimination because not only does the victim fail to meet the condition or requirement but also a disproportionate number of those of the same racial group fail to do so and the condition or requirement is not otherwise justified (i.e. on non-racial grounds).

Specifically Section 21(1) of the Act makes it unlawful for a person, who has the power to dispose of premises, to discriminate in the terms on which he offers the premises for disposal, in rejecting or accepting applications for the premises or in his treatment of individuals in relation to lists of people in need of such premises. 'Premises' includes land, houses, flats and business premises but not hotels (hotels are covered by the expression 'goods, facilities and services' in Section 20(2)).

Direct discrimination in allocation is likely to occur in one of eight principal ways:

1. Steering black applicants to specific areas because of staff perception that all black householders want to live in the same area or because of dispersal policies.
2. Offering poorer-quality accommodation to black appli-cants irrespective of circumstances.
3. Providing different or selective information and advice to black applicants.

4. Applying different assessment criteria to black applicants, thus affecting points allocations and/or waiting list priority.
5. In seeking to conform with management advice/policy objectives, etc., to let properties quickly, selecting offers on racial grounds on the assumption that, for example, blacks would more readily accept offers in areas of relatively high black concentration than whites.
6. In response to pressure from external groups such as the police and tenants' associations or from internal groups or individuals such as a housing lettings sub-committee, refusing to offer blacks accommodation in specific areas, or limiting or restricting such offers.
7. Through housing visitors or otherwise, applying different standards relating to housekeeping, etc., to black applicants/tenants affecting the assessment of 'good' or 'bad' applicants/tenants.
8. Operating a discretionary lettings policy which allows subjective decisions to determine or influence lettings to the disadvantage of black applicants.

It should be noted that such direct discrimination would not, in the normal course of events, either come to the attention of the applicant, nor - of necessity - be apparent to senior management.

In addition to overt discrimination in housing allocation, indirect discrimination may also take place, most frequently in one of the following ways:

1. Applying a points system which gives significant advantage to length of time on the waiting list the effect of which is to disproportionately disadvantage ethnic minority applicants.
2. Applying a points system or allocations system the effect of which is to disadvantage large households with the same effect.
3. Applying income criteria to allocations with the same effect (unlawful in Scotland).
4. Failing to advise a racial group of how to optimise choice of better housing through first or subsequent refusals. Where such failure flows from inadvertent lack of communication it is likely to constitute indirect discrimination: where, however, there is deliberation direct discrimination may be involved.

5. Allocating the least desirable housing/estates to the homeless or any other group disproportionately represented by ethnic minority applicants.
6. Applying any other requirement or condition in determining allocations the effect of which is to the relative disadvantage of ethnic minority applicants, e.g. relating to unemployment, owner-occupation, to separated families, or to 'sons and daughters' of existing residents (Smith, 1989a).
7. Operating differentiated access from different housing channels, e.g. decants, homeless, Housing Action Area or locality-based preference the effect of which is to disadvantage ethnic minority applicants.
8. Operating a referrals or nominations system in respect of housing association lets to like effect (CRE, 1989e).

While this is not an exhaustive list of potentially discriminatory criteria it demonstrates the breadth of systems used which may disadvantage ethnic minorities disproportionately, without, of necessity, there being any overt or covert intention so to discriminate. In examining the allocation system local authorities and housing associations, in particular, must first be aware of any criteria falling under the definition of a 'requirement' or 'condition' and second they must be aware of the consequences, whether intended or otherwise, of their application to the various ethnic minority groups. Hackney (CRE, 1984b) and Liverpool (CRE, 1984a, 1989e), for example, demonstrate that without monitoring local authorities are unlikely to be aware of the extent of unlawful or potentially unlawful discrimination taking place. Other studies not only confirm this pattern for local authorities with significant ethnic minority populations but also in respect of authorities with 2 per cent or less ethnic minority applicants (Thwaites, 1986). Moreover the report Queuing for Housing (SCPR, 1988) demonstrated that the majority of local authorities surveyed employed criteria, such as the requirement to be resident or employed locally, or not to be an owner-occupier, in respect of eligibility which are likely to discriminate against ethnic minorities: of course the legality of such criteria will depend on the 'justifiable' test.

If the extent of unlawful discrimination is likely to be so broad spread, how have such practices escaped the enforcement provisions of the Race Relations Act 1976? The answer to this question is simple in relation to policing but complex in legal terms.

In terms of policing, the Race Relations Act 1976 does not create criminal offences, merely civil wrongs encforce-able through two principal avenues. Any aggrieved individual may himself or herself pursue a complaint in a designated County Court in England and Wales or the Sheriff Court in Scotland. However, just as the majority of local authorities (AMA, 1985) and housing associations (NFHA, 1982, 1983), through a failure to monitor, do not know the impact of allocation criteria on ethnic minority groups, so the individual, when found with an ostensibly fair system, would find it difficult if not impossible first to suspect and second to prove unlawful discrimination. The rigorous analysis required to establish an unlawful practice may be undertaken by the CRE but it, in turn, has a limited budget, must be selective in relation to complainants whom it supports and has been advised to restrict investigations which it initiates to those of strategic importance (Home Affairs Committee, 1981a; Home Office, 1982a). Conse-quently in the absence of effective policing housing authorities have, to a large extent, not only acted with impunity but have, albeit inadvertently, benefited from a failure to monitor their own performance. Such benefits - if successful law evasion can be so described - may be at a heavy cost, in relation (1) to the disservice to the ethnic minority section of the community, (2) to the authority's own credibility and (3) to the problems and expense which eventuate from belated attempts at remedial action (CRE, 1989e).

The legal difficulties comprise three principal issues:

1. The meaning of a condition or requirement (Section 1(1)(b)).
2. The nature of proportional disadvantage (Section 1(1)(b)(i)) - the proportion of the ethnic minority group who can comply with a condition or requirement must be 'considerably smaller' than the proportion of persons not of that group, and
3. Determining in what circumstances the courts will consider indirect discrimination 'justifiable' - and thus not unlawful in relation to allocation policies (Section 1(1)(b)(ii)). These are considered in turn.

In Perera v Civil Service Commission (1983) IRLR 166 the Court of Appeal held that a requirement or condition would exist only where it amounted to a complete bar if not met. In housing terms this may be illustrated by the following example: a housing authority which had a significantly higher number of single applicants from ethnic minorities (in proportional terms) would discriminate against such applicants by imposing a condition refusing allocations to single people. Where, however, a preference - and not an absolute rule - favoured allocations to couples there would be no discrimination in terms of Section (1(1)(b) of the Act.

Because such an interpretation is arbitrary in its effect the CRE recommend that the 'requirement or condition' formula be replaced by a much wider phrase encompassing any practice, policy or situation which has a significant adverse impact on a particular racial group (see Bindman, 1985; MacEwen, 1986).

The second issue - the difficulty of ascertaining when the degree of disadvantage is 'considerable' - has been said, by one judge, to be a matter of personal opinion. This lack of clarity and resultant dependence on the whim of the judiciary is self-evidently unsatisfactory. In the United States a difference of 20 per cent in the impact of a particular situation on different racial groups is considered sufficient (Bindman, 1985). In housing terms if a requirement to have a local connection to attract additional points on the waiting list was met by 40 per cent of white applicants but only 18 per cent of black applicants, in the United States this would constitute 'a considerably smaller' proportion and would thus be discriminatory. In Britain, it appears, whether it is unlawful is dependent on what the judge had for lunch.

The third legal issue, the meaning of 'justifiable', is perhaps the most important in limiting what constitutes unlawful discrimination and allowing unsympathetic judges to nullify the purpose of the concept. Thus in <u>Ojutiku and Oburoni</u> v MSC (1982) IRLR 418 the Court of Appeal indicated that any practice which, from the employer's viewpoint, was 'reasonable' would exempt him or her from liability irrespective of how harmful the practice might be to ethnic minorities. In contrast, in the United States, the only exception to indirect discrimination being unlawful in employment cases is where the discriminatory impact is an unavoidable consequence of a practice which is essential for the business to be conducted effectively. But in this country the law does not attempt to require an objective assessment of the needs of the situation or of the impact of a discriminatory practice on those adversely affected. Mr Justice Browne-Wilkinson summed up the current position: 'The decisions of the Court of Appeal and of this Appeal Tribunal disclose a steady decline in the strictness of the requirements which an employer has to satisfy in order to show that a discriminatory condition is justified.' The willingness of the courts to interpret the concept to the benefit of a respondent together with the decision to treat justifiability as a question of fact (and thus effectively within the discretion of the courts of first instance) appears to give great latitude to local housing authorities. They may apply discriminatory practices and avoid civil action by providing some 'tolerable' reason for the practice in question.

Certainly an allocation system which was designed to match household size with house size - even though indirectly discriminating, given the uneven match of available stock and the housing needs of applicant ethnic groups - would meet the strictest criterion applied by the courts. Whether or not more value-laden criteria - for example, advantaging or disadvantaging the homeless resulting in indirect discrimination - would be considered justifiable must await either the development of relevant case law or changes in the law such as those advocated by the CRE, which suggested the substitution of 'necessary' for 'justifiable' (CRE, 1985b).

ALLOCATION POLICY:
THE LOCAL AUTHORITY EXPERIENCE

Subjective assessment

The AMA publication Housing and Race (AMA, 1985) summarises the findings of a questionnaire survey sent to all London boroughs and Metropolitan Districts after the Hackney report (CRE, 1984b). Of those responding (88 per cent), twenty authorities (33 per cent) kept ethnic records of waiting list applicants and seventeen (28 per cent) of transfer applicants but less than one-third of the authorities keeping records undertook regular monitoring of the records they kept. The monitoring which had been undertaken led to the following changes in practices and procedures:

1.   The abolition of residential/work qualifications.

2. Changes in the allocation scheme.
3. The establishment of targets in the allocation of property.
4. Increased priority for homeless people.
5. Changes in policy in respect of relatives abroad.
6. Changes in policy relating to local connections.
7. Changes in policy relating to housing association nominations.
8. Changes in practice relating to offers refused.
9. Abolition of age limits for certain estates.
10. Introduction of visits to applicants with a high number of points.

This survey emphasised that of particular significance for allocations to ethnic minority people were those factors, used by many authorities, which rely on subjective criteria (for example, social/family problems), or which could set ethnic minorities at a disadvantage (for example, length of time on the list). The report concluded that these criteria led to discrimination in allocation and considered it essential that allocation policies were closely monitored and, if they were seen to be discriminatory, the policy must be urgently reviewed. It was noted that the state of repair of a dwelling occupied does not appear to be considered a relevant factor in assessing need, in contrast with the dwelling's age.

It also emphasised that in respect of access channels by which, in particular, black groups obtain council accommodation there must be regular review to ensure that a fair proportion of the properties becoming available are offered to all ethnic groups in all localities within the local authority area. Allocation schemes should base their criteria on objective assessments, and the opportunity for subjective decisions by individual officers and officer discretion must be kept to an absolute minimum. In particular greater objectivity was required in the case of homeless applicants, where such policies as 'one offer only' might be discriminatory against people from ethnic minority groups.

Despite the existence of the AMA report an overview of policies by local authorities is problematic in that the base data produced and analysis undertaken have been for different purposes and as a result matching like with like is fraught with difficulty. While fairly constant themes and patterns may emerge it is hazardous to describe causes and prescribe remedies which are more than indicative.

Nonetheless there is sufficient congruence of opinion in studies in Lambeth (Ouseley, 1981), Hackney (CRE, 1984b), Tower Hamlets (Phillips, 1986), Liverpool (CRE, 1984a, 1989e), Nottingham (Simpson, 1981), Bristol (Davies, 1974), Bedford (Skellington, 1980), Lewisham (Ridoutt, 1980), Birmingham (Henderson and Karn, 1987) and Edinburgh (Thwaites, 1986; Taylor, 1987), for example, to suggest (1) that many, if not most, local housing authority allocation policies and practices have an adverse impact on ethnic minorities, (2) that such adverse impact is frequently unlawful and (3) that most authorities are both insufficiently aware of the impact of their policies and practices and insufficiently committed to improving their housing service to that section of the community.

The experience of many blacks, predominantly of West Indian and Asian origin, in relation to the local authority allocating system is that there is over-representation in the worst/unpopular estates, in high-rise buildings, in older properties and in those of poorer physical quality. Furthermore such disadvantage is not adequately explained by reference to the points or categories system employed, by the family size, income group or locational or house type preference stated by the applicants, nor by the availability of housing stock. Nonetheless while indirect discrimination arising from the allocation system employed may appear to be a constant factor, because of the widely differing circumstances of the housing authorities concerned, each facet of relative disadvantage is likely to have unique characteristics both in comparison with other facets and also in respect of each housing authority. Because such differences may affect recommendations to ameliorate disadvantage it is necessary to outline such characteristics in turn.

Unpopular/worst estates

Self-evidently estates may be less desirable for a number of discrete and overlapping reasons. The location, design and layout of the estate, the age, the physical condition of the property, the housing types predominating and the access to services and facilities are likely to be considered more objective considerations than a 'reputation' acquired by an estate which reflects not merely the physical characte-ristics but also the social ones: the housing of homeless families, high unemployment rates, single parents, so called 'problem' families, etc., will, no doubt, affect the perception of desirability, where prejudice and opinion are likely to be more important than fact and experience. Similarly the incidence of a relatively high proportion of black families will often evoke sufficient prejudices in the white community to stigmatise estates as 'undesirable' by a section of that community. Such observations do not deny the relevance of social factors - the incidence of crime being an obvious example - in determining the quality of life but merely illustrate problems of description. In the absence of generally accepted indices of assessment the significance of disadvantage which stems from being housed on the unpopular/worst estates should be examined at the local authority level. However where black applicants are being steered to estates with high vacancy rates, with the effect of limiting housing choice in a discriminatory manner, a housing authority will be bound to take remedial action, as such discrimination in respect of choice will be unlawful irrespective of any 'objective' indices suggesting that such estates, in physical terms, are not the worst.

The relatively high incidence of black families in such 'unpopular' estates may arise from different reasons. Over-representation of black families in the homeless persons category, in the category for rehousing from clearance or rehabilitation programmes, from within the waiting list, in single-parent households or from the unemployed may all affect access to more popular estates. Where the waiting list gives points for existing residence (unlawful in Scotland), for length of time on the waiting list or where the allocation system gives area preference to decant or transfer categories these may lead to unlawful discrimination. Self-evidently the prejudice of housing allocation staff may result in discrimination where the allocation system permits the exercise of any discretion. Moreover taking the decision from staff and placing it in the hands of a computer system may merely provide a front of objectivity. As the CRE investigation into St George's Medical School (CRE, 1988d) demonstrates, the prejudice of a computer system is no less than that of its designers or the judgments that inform the design. Even the most rigorously objective system is likely to permit some discretion and consequently unlawful discrimination is likely to occur, as in Hackney, where such staff play any role in the allocation system which is not subject to monitoring (CRE, 1984b).

Of all the factors which emerge from the various investigations into allocations to the worst estates perhaps two may be identified as the most significant. First, the existence of an equal opportunities policy and a housing race relations unit together with a real commitment to eradicating racial disadvantage is, in itself, insufficient to identify the nature of the problems to be addressed without monitoring, analysis and review. Second, the discriminatory nature of policies is site and time-specific. That is, the same policy may disadvantage the black community in one local authority but actually reverse disadvantage in another and similarly, when the characteristics of the black community and/or the housing stock available to let have changed over a period, the same policy will have a different impact on different occasions.

High-rise

As a generality high-rise allocations are less popular than low-rise flats, maisonettes, tenements or individual houses. Obviously the needs of the applicant and availability of stock will be factors constraining access but in each of the housing authorities referred to (excepting GLC Tower Hamlets, where this factor was not analysed, Phillips, 1986) black applicants were relatively disadvantaged in being allocated high-rise property within such constraints. Generally, within high-rise allocations, excepting the ground and first floor where security problems may arise, the lower the floor the more acceptable the tenancy. Because of the differing data base comparisons between local authorities are difficult: two are illustrated by way of example. The Birmingham study of Hazel Flett (1979, see also Henderson and Karn, 1987), showed that 10 per cent of black households were allocated flats from the waiting list at the eleventh storey or above in comparison with 3 per cent of white households in 1976 but such relative disadvantage was only 2 per cent in respect of 1974 allocations and was a 4 per cent advantage in 1971 allocations. Below the eleventh storey no clear pattern of advantage or disadvantage correlated to the applicants' ethnic origins emerges although, as Flett concluded, 'Almost every aspect of Birmingham's council house rationing system has, in practice, worked against black people'. Ironically, however, in terms of high-rise allocations, blacks would appear from the survey to have been better served prior to 1976, when

Birmingham was obliged to discontinue its dispersal policy following a finding of unlawful discrimination. Such a conclusion does not justify the dispersal policy, which clearly disadvantaged black applicants in many other respects. The Hackney investigation (CRE, 1984b) demonstrated that 85 per cent of black applicants on the waiting list were allocated flats between 1978 and 1979, compared with 65 per cent of white applicants: amongst the homeless category, blacks, again, were more likely to be allocated flats and more likely to be allocated flats above the fifth floor but this pattern was reversed in accommodation allocated to decant applicants although not significantly at floor level. However, within this category of decants, black applicants received a higher proportion of older and poorer-quality accommodation.

In relation to transfers the investigation showed, if anything, a marginal advantage to black families in respect of high-rise and quality although the differences were not considered to be statistically significant, and were explained by variations in the distribution of family types.

Older properties

Hackney (CRE, 1984b): 'A larger proportion of white applicants received newer properties (25% as opposed to 3%), while black applicants tended to be allocated to a particular group of inter-war properties that had been modernised but to a relatively low standard.'

Nottingham (Simpson, 1981): 'Within the allocations to houses themselves ... roughly 75% of these Asian and West Indian moves were to older, inner city properties. The majority of whites moved to "acquired" properties outside the central areas of Nottingham - properties much less likely to present substantial rehabilitation needs.'

Birmingham (Flett, 1979): 'Whilst blacks [from the waiting list] were not generally under represented in the newest stock, they were housed rather more often than whites in the oldest pre-1919 houses bought by the corporation.' In relation to allocations to clearance area residents, 55 per cent of blacks received pre-1919 housing compared with 14 per cent of whites.

The above illustrates that, in the public sector, blacks are more likely to be housed in older properties than whites. Being housed in an older property need not, of itself, be a detriment in terms of the Race Relations Act 1976: some

240

older properties may be located in popular estates, they may be structurally sound, in a state of good physical repair and they may be fitted with modern conveniences, such as central heating and kitchen appliances. Generally, however, such properties are more likely to be of poorer quality, to provide a less amenable living environment and to be more likely to require repair, rehabilitation or even clearance. At times the internal living conditions may be satisfactory but the structural defects may require attention. On occasion applicants may choose to be allocated to older properties because of location, more spacious accommodation, etc., but the studies referred to demonstrate that black families are housed in older properties more frequently than white families irrespective of preference. Similarly older housing stock may match need or be more frequently available for letting than newer stock but, again, these factors provide an insufficient explanation for the disproportionate housing of black families in such stock.

Poorer-quality stock

Given the correlation between older properties and poorer stock it is not surprising that blacks, in the majority of studies referred to, were also likely to be housed in poorer-quality accommodation. The exception, Nottingham, demonstrates the complexity of the issue. Over the period researched Nottingham's council house allocations included a number of distinct features: Asian families obtained housing on a very limited scale, mostly through the most advantageous of rehousing category - clearance. West Indian applicants were the largest group on the waiting list (42 per cent compared with 30 per cent white, 16 per cent Asian) - a low-priority gateway. Transfers represented an important avenue to better-quality housing but West Indians only received moves at just over half the rate of white households. When moved the majority of West Indians moved to post-war council estates but some 80 per cent of these were to deck-access or multi-storey complexes, which are amongst the least popular and most stigmatised parts of the council's housing stock: although a similar proportion of whites and Asians were rehoused in post-war properties, 60 per cent of such allocations were to houses. Within the allocation to houses themselves, while 22 per cent of Asians, 16 per cent of West Indians and 14 per cent of white applicants went to houses that the council had bought rather than built, 75 per cent of Asian and West Indian moves were to older, inner-city properties. The majority of whites moved to properties which were much less likely to present substantial rehabilitation needs. In Nottingham the allocations to pre-war properties (which in other cities have been the poorer ones) have coincided with allocations to more prestigious and soundly built houses or flats. It is almost a direct corollary of the differential allocations of West Indians to deck access and multi-storey properties that whites received houses at more than twice the rate of West Indian families.

Proving discrimination in allocations

It has been noted that in order to establish unlawful discrimination it is necessary to demonstrate, on the balance of probability, first that black applicants or tenants are disproportionately disadvantaged and second that such disadvantage is a consequence either of direct racial discrimination or of indirect racial discrimination which is not otherwise justifiable. The Liverpool study (CRE, 1984a: 87) illustrates both elements. The research established that within rehousing categories when such factors as family size, household type, housing need, etc., have been discounted, blacks were disproportionately represented in lower-quality accommodation and were concentrated in certain areas irrespective of choice. In the absence of any alternative explanation, direct unlawful discrimination appears as the only remaining explanation. The actual nature of such discrimination in the sense of identifying individuals who applied racial stereotyping and at what stage in the process was not established and in the absence of witnesses or admissions this is not unexpected. It should also be noted that direct discrimination may emanate from unconscious racial stereotyping of which the official(s) concerned are unaware.

The Liverpool study also concluded that the racial disadvantage was not due to any explicit racist intention in respect of housing policies but, primarily, to the unintentional racist effect of allocation procedures and practices (CRE, 1984a: 87). However, housing policies, in the absence of any racist intention, are likely to constitute indirect discrimination following from a determination to give relative priority to categories of applicants where blacks are under-represented - a factor expressly discounted by the study - while the procedures and practices are more likely to stem from racial stereotyping, i.e. direct discrimination such as that relating to perceived as opposed to actual area preference shown to have occurred in respect of allocations in Granby/Falkner (CRE, 1984a: 37).

Consequently the incidence of indirect discrimination, if occurring, was likely to arise outside the allocation systems proper, for example, in the access channel. Indeed the Liverpool study suggests this as a factor in its analysis of priority access through clearance/improvement programmes in respect of which 21 per cent of white households compared with 12 per cent of black households were rehoused between January 1977 and December 1980 (CRE, 1984a: 32). The study suggested that the council's policy concerning the designation of areas for improvement /clearance may have amounted to indirect discrimination but did not examine the issue in sufficient depth to determine whether the policy of designation was 'otherwise justifiable' by reference, for example, to the actual condition of the housing stock concerned.

The above illustrates that for both direct and indirect discrimination the identification of an unlawful discriminatory practice in allocation is dependent on a process of elimination when the investigation proceeds on the circumstantial evidence of relative disadvantage in allocations based on colour. Inevitably this is likely to prove a painstaking and lengthy process, the quality of the outcome depending on the systematic analysis of the various factors influencing the process (see Fig. 8.1). Where one factor is identified, such as access channels to better-quality housing discrimination is likely to be indirect, in which case other factors, such as the objectivity of a clearance programme, may justify the process, rendering it lawful. Where no factor is identified racial stereotyping and direct discrimination should remain as the sole explanation.

Local authority policy response

Because of the wide divergence in local authority policy and practices relating to the operation of the allocation system it will be apparent that while patterns of relative disadvantage for black tenants emerge in respect of allocation to unpopular estates and high-rise, older and physically poorer properties the reasons for such patterns are complex and not necessarily common to the local

Figure 8.1: Flow chart: race and local authority housing allocations

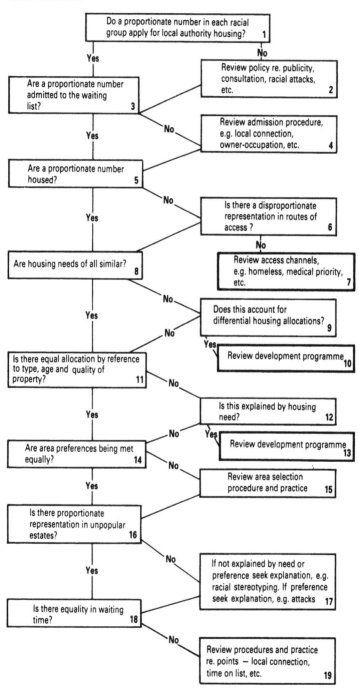

authorities concerned. Thus access to the property through the waiting list, clearance, transfers, decant and homelessness will differ, as will the points systems employed. Similarly, the area of discretion allowed to housing officers on allocation, the significance of 'good housekeeping' assessment and the nature of the housing stock itself will differ substantially from one authority to the next. Accordingly while there is a fair congruence in respect of policy responses required to ameliorate disadvantage, there is no necessary correlation between a specific detriment - such as the allocation to poorer housing - and a specific policy initiative such as staff training. However, there is substantial agreement as regards action required in respect of:

1. Ethnic monitoring of all aspects of the application and allocation process (including transfers and exchanges).
2. The need for further analysis of how the system affects black people, linked with
3. Policy review.
4. Staff training.
5. Provision of and access to information, and
6. Consultation with ethnic minority groups.

Ethnic minority staff recruitment is advocated specifically by half the reports examined. This was omitted from the Hackney report, being outwith the CRE's remit. While the Liverpool study did not specifically advise ethnic minority recruitment it did refer to staffing resources and to the need for a number of designated posts such as a Monitoring Research Officer (Race Relations), an Information/Training Officer and a Racial Harassment Officer. Furthermore the study supported the initiative already proposed by the council to rectify the overwhelming lack of black officers in the housing department (six out of 754 in 1982).

Accordingly it would be desirable for all housing departments to examine the composition of their existing workforce and to take initiatives to recruit and train - possibly in conjunction with those universities and colleges running housing courses - ethnic minority staff where they are under-represented. Such a programme will not only affect the public image of the white provider, but will undoubtedly make the whole system much more sympathetic to and knowledgeable about ethnic minority housing need. It is unfortunate that while housing authorities in England and

Wales may use Section 11 of the Local Government Act 1966 in the area of ethnic minority recruitment, attracting 75 per cent Central Government support, this avenue is not open to Scottish local authorities because of the refusal of the Scottish Office to activate the equivalent Section 11 of the Local Government (Scotland) Act 1966.

The use of Section 11 funding, as well as urban aid funding, for the purpose of employing black housing advisers is not without criticism. It has had the effect of marginalising race issues in housing - a less obvious criticism of posts which are mainstream funded. It is worth noting that the Black Caucus in the voluntary housing movement and the Federation of Black Housing Organisations (FBHO) have campaigned successfully in getting a fairer slice of the housing cake for black people (Gujral, 1986). As a result PATH National has been financed (£20,000 per annum) by the Housing Corporation to train black housing directors.

The issue of targeting was addressed directly by only one report, that relating to Lambeth, where, despite well-substantiated housing needs of the black community, many were rehoused in the worst council housing and very few in the more desirable estates in Norwood and Streatham. In February 1979 the council set a minimum target of 30 per cent new lettings to black households. Opponents, wrongly, interpreted targets as quotas (i.e. fixed numbers), as reverse discrimination and as preferential treatment despite the fact that the 30 per cent figure was below the actual percentage of blacks being rehoused by the council and did not compensate for earlier unfair treatment. By monitoring all categories of lets on a regular basis against equality targets, substantial progress was made over the three-year period reviewed. Whether or not local housing authorities choose to describe their objectives in terms of equality targets, the reality of monitoring and review must be to achieve equality in allocation which is demonstrated by percentage figures throughout the stock. In the words of a Hackney council report 'the only practical means and method of achieving the racial equality objective is to set quantitative and qualitative goals (targets) which would serve as a yardstick for evaluating the equitable nature of our allocation policy and practice'.

The Scottish Housing Authority

In advocating reforms by reference to the experience of local authorities with substantial ethnic minority populations, those with relatively small ethnic minority populations may be seduced into a state of false security. Thus Glasgow District, with an ethnic minority population of 2 per cent, has witnessed severe problems in respect of racial harassment on housing estates (Walsh, 1986), while preliminary studies in Edinburgh (Thwaites, 1986; Taylor, 1987) (1.6 per cent ethnic minority population) suggest disadvantage to ethnic minorities in respect of over-representation in high-rise dwellings (see also MacEwen, 1976). Consequently where ethnic minorities are a relatively small section of the community the nature of the policy initiative to secure equality of treatment may differ in respect of resource and staffing allocation from that of the larger authorities but the goals should not differ. Nonetheless a substantial difficulty remains in convincing the smaller, rural authorities of the necessity of examining their own practices. In 1980 a survey of the sixty local and New Town authorities in Scotland was conducted which demonstrated the following (MacEwen, 1980a):

1. Of the fifty-two responses, only Glasgow District made reference to ethnic minority needs in its Housing Plan, a local plan and an Action Area Plan and only one other authority mentioned such needs in any of these areas.
2. Only three housing authorities referred to ethnic minorities, two in relation to Vietnamese refugees, in any housing or policy document relevant to the housing service.
3. Only one authority (Dundee) had ever provided information designed for ethnic minorities: the sole example was a letter in Urdu relating to a Housing Action Area.
4. No local authority had formulated any policy in respect of ethnic record keeping.
5. No local authority, with the exception of victimisation, was aware of any difficulties experienced by ethnic minorities in respect of discrimination, disadvantage or prejudice in relation to housing.
6. Only one authority provided any staff training relating to ethnic minorities and housing.

But continuing pressure from the Community Relations Councils (both Central and Grampian Community Relations Councils have been established since the 1980 survey) and ethnic minority groups, the influence of CRE reports on housing and the more vocal advice of the Institute of Housing together with local research appear to have had some influence on practice. Glasgow District Council and Edinburgh District Council housing departments are in the process of implementing ethnic minority monitoring and both are reviewing, more generally, their admission and allocation systems. In the case of Edinburgh such review has been stimulated by a change in the political complexion of the administration.

Despite these advances a quote from one response from the 1980 survey summarised the prevalent view:

> Speaking professionally, I have yet to be convinced of the need to differentiate in this manner in housing terms, since it only seems to me to serve the argument that if we [sic] distinguish them [sic] as 'ethnic minorities' we somehow treat them differently as against other applications which of course, in our case, does not happen.

What the author failed to appreciate was that even if his assertion of like treatment was correct (which on the evidence of other authorities seems unlikely) unlawful discrimination may be and probably is taking place by applying a condition or requirement, such as points for time on the waiting list, the effect of which is to the detriment of a larger proportion of the ethnic minority population than the rest of the community without being otherwise justifiable. Such an appreciation appears to present a quantum leap in human understanding. Furthermore in a climate of diminishing housing resources, of housing officials being increasingly harassed by the more general demands for an improved housing service with much greater tenant participation, the incentives to embark on a hazardous voyage through unchartered waters are minimal. It is perhaps too easy to cast the housing official in the role of the active villain rather than the passive victim of a complex, unresponsive and cumbersome structure. Furthermore, it is apparent that ethnic minorities themselves are but one disadvantaged group in the public housing sector. A structural remedy to their disadvantage, while a necessary and achievable objective, will not ameliorate the general disadvantage suffered by other groups such as single parents, the handicapped and the homeless which in numerical terms may seem more significant.

In noting this situation, however, what is often overlooked is the fact that ethnic minorities experience double jeopardy in access to the public sector in the sense that they are as likely, if not more likely, to fall within specific categories of disadvantage in addition to experiencing discrimination on the grounds of race. As a result equality of opportunity in terms of race means no more than treatment on like terms as others even where such terms impose relative disadvantage.

It must be borne in mind, however, that in so far as race may be or is perceived as a determinant of a less favoured categorisation, such as socio-economic group, regarding admissions and allocations then the disadvantage of such groups in housing terms will result in indirect discrimination. In such instances, at least in theory, it would be for the courts to determine whether discrimination on such grounds would be justifiable.

Clearly the decisions of the courts in employment cases where justifiable equates with commercial convenience would suggest that administrative convenience would meet the judicial test now favoured. Moreover, as we have noted in Perera, membership of a social class or being employed would have to be expressed as a condition or requirement and not merely a non-exclusive preference to constitute indirect discrimination. Consequently, while untested, even where the outcome of admission/allocation policies indicated structural disadvantage to ethnic minorities on the grounds of class or other social categories the first legal hurdle - the imposition of a condition or requirement - might be difficult to prove by reference to the particular system employed. As Henderson and Karn (1987) have demonstrated such disadvantage may not be reflected in the rules but emanate from class stereotyping by officials. In such instances exceptions are likely to occur the effect of which is to remove the practice from the ambit of indirect discrimination.

CONCLUSIONS

Specific difficulties of the present law as it affects allocations have already been referred to. Thus the terms 'justifiable', 'condition and requirement' and 'significant proportion' require change to be effective in respect of indirect discrimination. Similarly Section 71 of the 1976 Act should place a less ambiguous obligation on local housing authorities to review their various functions (CRE, 1985b: 31). Although these and related proposals for change have been advocated by the CRE, the discussion of allocation policies clearly indicated that such amendments would do little to change current practice for two reasons:

1. The individual applicant would, as now, be unaware of the incidence of indirect discrimination: consequently the impetus to review practice stemming from complaints would be unaltered.
2. The housing authority's position vis à vis monitoring and review would be affected only marginally (by altering Section 71) because of the nature of policing.

Nonetheless two aspects of the CRE proposals, although ancillary rather than central to housing, might have a greater impact. First, the CRE proposed a statutory power to issue Codes of Practice, such as it currently has in respect of employment, in relation to any area of discrimination. This proposal, but in the context of its extension to housing only, has been effected by the Housing Act 1988, Section 137. The consequence of issuing such a code in relation to allocation would be to enable the courts to infer unlawful discrimination in certain circumstances where the code had been ignored. Second, there is a proposal to shift the burden of proof to the extent that where less favourable treatment is shown to have occurred the onus would be on the housing authority either to show that it was not racially discriminatory or, where it was indirectly so, that this was necessary in order to secure reasonable housing objectives (a much more onerous test than the present one).

In the present climate of opinion it is unlikely that the CRE's resources, which, following criticism, have focused on strategic investigations, would be significantly enhanced to permit effective policing by random local housing authority testing. Consequently the ability of housing authorities to evade civil action is likely to remain. Nonetheless there is a strong argument for imposing much stiffer damages - perhaps of an exemplary nature - where civil actions are successful and, more importantly, for introducing some form of class action to effect not only a remedy in respect of structural discrimination but also, because of the real financial implications of being caught, a greater willingness on the part of housing authorities to initiate monitoring and review before any threat of civil action. The experience of such class action in the United States, for example, where literally millions of dollars have had to be paid out by respondents, has proved a salutary incentive to policy review.

Recommendations for change to offset indirect discriminatory practices must be pitched at a fairly high level of generality and their effectiveness can be tested only in relation to the specific practice of a particular housing authority. A flow charge analysing allocation systems is given in Fig. 8.1. Axiomatic to this process is a thorough understanding of the impact of current local practice on the ethnic minority applicants and tenants within a local authority area. Accordingly while staff recruitment and training, the creation of a Race Relations Unit, the adoption of a comprehensive and explicit equal opportunities policy and an appraisal of the existing allocation rules together with consultation with ethnic minority groups and information dissemination may improve the chances of fairer access to a limited resource in housing, none of these initiatives will, in itself, guarantee the desired improvement without a comprehensive programme of ethnic minority monitoring in respect of each aspect of the application and allocation system employed by the individual authority. Furthermore, as the AMA and Scottish surveys illustrate, such programmes remain piece-meal despite increasing exhortations for their universal adoption: a legal requirement to monitor and review policy appears to remain the only avenue to achieve effective local analysis. Such review should be subject to CRE scrutiny, on request. As the Street Report suggested (Street et al., 1967) the relevant Minister should have power to direct changes in rules or policy on the recommendation of the CRE.

Although some twenty-five years have elapsed since the first Race Relations Act, the translation from law to practice is far from maturity: indeed local authorities are now experiencing the situation of blacks choosing less favourable local authority housing not merely as a reflection of restricted local authority access but from the real fear of white racism and harassment in particular areas (Home Affairs Committee, 1986). Often the professions have been slow to recognise changing social need (MacEwen, 1981), a factor discussed in Chapter 11, but the Institute of Housing has shown willing, producing a series of Guidance Notes on Race and Housing: the advice on allocation (Institute of Housing, 1986) is valuable and should be followed not only by the large urban authorities but also by the smaller urban and rural districts.

It has to be acknowledged nonetheless that such advice may prove especially daunting in relation to identifying racial disadvantage in housing allocation and doubly so where the black population is a relatively small section of the community. What is required, therefore, in addition to the general guidance issued, is some form of tangible and progressive checklist of action in public allocations which is user-friendly in the sense of catering for a broad spectrum of local authorities. It must be progressive in the sense of imposing a speedy check to identify major issues for analysis while simultaneously discarding peripheral issues, analysis of which may prove wasteful and inconclusive. The National Federation of Housing Associations (NFHA, 1985) has provided a checklist in respect of housing association monitoring. Based on this model, a local authority guide has also been advised (MacEwen, 1987a). A precondition of this approach, nonetheless, is a rudimentary understanding of the housing needs and experience of the local black community, who should determine the nature, focus and pace of the housing authority's approach.

Initiatives by local authority associations (such as the AMA, 1988b) and by the housing profession in producing guidance on good practice are welcomed. However, while random testing by the CRE housing section (virtually in isolation) is being actively pursued, the profession and associations must police practices regularly and impose sanctions on recalcitrant members. The evidence in council housing is incontrovertible: guidance and persuasion are a necessary but insufficient component of a strategy to eliminate unlawful racial discrimination. Compulsory monitoring should not be deferred.

CHAPTER 9

HOUSING ASSOCIATIONS

INTRODUCTION

The Housing Corporation was created by the Housing Act 1964 as a new agency for the purpose of making a contribution to housing provision which would complement that of local authorities. It was established to serve the whole of the United Kingdom although in Scotland it was responsible to the Scottish Secretary of State and had an office in Edinburgh. In the first ten years of its existence it focused upon the promotion and encouragement of cost rent housing and co-ownership schemes. From 1972 it was able to lend money to housing associations to provide housing at 'fair rents'.

A new policy was introduced by the Housing Act 1974 which was designed to channel large amounts of Central Government funds to the housing associations, which, by building new houses or improving old ones, would supplement the work of local authorities and provide a range of new housing services. In order to attract the substantial loans which, coupled with grants, would enable them to carry out these new developments, housing associations were made subject to the supervision and control of the Housing Corporation. The extensive new powers afforded to the Housing Corporation provided in the 1974 Act have now been consolidated, along with amending legislation, with effect from 1 April 1986 in the Housing Associations Act 1985. Section 75 of the Act defines the duties of the Housing Corporation as follows:

(a) To promote and assist the development of registered Housing Associations and unregistered self-built societies.
(b) To facilitate the proper performance of the functions and to publicise the aims and principles of these

Associations and societies.

(c) To maintain a register of Housing Associations and to exercise supervision and control over those registered.

(d) To act as the Secretary of State's agent with respect to applications for and payment of grants to Associations.

(e) To undertake, to such extent as necessary, the provision and management of houses, whether for letting or sale, and of hostels.

Since 1989, following the implementation of the Housing Act 1988 and Housing (Scotland) Act 1988, Housing for Wales and Scottish Homes have assumed the powers and duties of the Housing Corporation in Wales and Scotland respectively.

The primary concern of the Housing Corporation, then, is with the registration and subsequent supervision and control of housing associations, which in turn are defined by Section 1 of the Act as a society, body of trustees or company which is established for the purpose, in part at least, of 'providing, constructing, improving or managing, or facilitating or encouraging the construction or improvement of housing accommodation' and which does not trade for profit. In addition to English registered charities, the Housing Corporation may register only societies already registered under the Industrial and Provident Societies Act 1965 where they satisfy the statutory conditions laid down including first that the association does not trade for profit and second that it is established for the provision of housing to let or for occupation by members of the association only or for hostels (Section 4). Housing associations both in England and Wales and in Scotland are required to sell housing to secure but not, from 1989, assured tenants subject to certain qualifications and conditions. The Housing Corporation is enabled, also by Section 5, to set out additional criteria which must be met by housing associations in order for registration to be effected. Once registered, associations fall under the financial and general supervision of the Corporation. Associations are required to submit accounts to the Corporation; the Corporation may order an extra-ordinary audit; the Corporation's consent is required for a change in the rules of an association and for disposals of land; it may appoint a person to conduct an inquiry into the affairs of the association; it may act for the protection of an association by the removal of a committee member; the Corporation may petition for the winding up of an association; it may, subject to appeal, remove an association from the register if it appears to be no longer eligible for registration or has ceased to exist.

In addition to this supervisory role the Corporation has a responsibility for promotional activities and it has further powers to enable it to acquire land to be made available for associations and to lend money to them. The Corporation may also make grants with the consent of the Secretary of State to associations and other voluntary organisations to assist them in carrying out secondary purposes such as giving encouragement and advice to other associations eligible for registration and assisting other registered associations.

The principal advantage of registration as far as the housing association is concerned is that it is then entitled to receive Housing Association Grant (HAG), for which provision is made under Part II of the Act. Such grants are made in respect of housing associations' expenditure in connection with housing projects approved by the Secretary of State or which fall within a programme for the development of housing by housing associations. These programmes are prepared by the Housing Corporation or a local authority and approved by the Secretary of State. A project for the purpose of HAG is one for providing housing or residential accommodation, or improving or repairing such accommodation or related buildings (including for disposal). The Secretary of State, has, however, indicated that not all housing projects will be likely to be approved but only those fulfilling a priority housing need. These include projects in Housing Action Areas, projects for special categories such as elderly or disabled people, projects to meet local authority needs and shortages, and projects designed to prevent serious structural or environmental deterioration. In practice most applications for HAG are channelled to the Secretary of State by way either of the Housing Corporation, which may be lending the money to finance a project, or else the local housing authority. Section 36A (added by Section 49 of the Housing Act 1988) enables the Corporation - which includes Housing for Wales and Scottish Homes from their inception in 1989 - to issue guidance with respect to the management of housing accommodation by registered housing associations. Such guidance may relate to the demands for which provision should be made, the allocation of housing accommodation, the terms of tenancies and the principles upon which the levels of rent should be determined, to standards of maintenance and repair and consultation and communication with tenants. Of critical importance to housing associations is the amount of grant that developments will attract. For example, in Scotland it was expected that the changes proposed would effect a reduction, in most cases, from 100 per cent of approved expenditure to 85 per cent, the shortfall being made up from other revenue, most importantly that attributable to rent increases. For new tenancies housing associations will no longer charge 'fair rents' - which disregard scarcity - but market rents. Additional grants, up to 100 per cent HAG, may be payable in particular cases and reflected in lower rents, but generally the effect will be a significant increase in rent levels.

Housing associations and Housing Action Trusts (Section 72 of the Housing Act 1985, as amended by Section 70 of the Housing Act 1988) are bound to 'co-operate in rendering such assistance as is reasonable' when approached by a local housing authority for assistance in the discharge of certain statutory obligations relating to homelessness, or threatened homelessness.

Both the Housing Corporation and the local authority may act as the Secretary of State's agent for the supervision of projects and in the payment of grants. Local authorities themselves have a further interest in housing association activities; they are encouraged and expected to work closely together and in particular their tenant selection procedures should be co-ordinated with associations accepting, as a proportion of their tenants, people nominated to them by a local authority from their waiting lists. However, while such nominations will continue for pre-1988 properties, there is limited scope for nomination in respect of new schemes which are financed privately or by mixed funding.

ASSOCIATIONS AND ETHNIC MINORITIES

Types and objectives

There are approximately 2,700 housing associations registered with the Housing Corporation. Associations vary in size from the very small to those with over 10,000 dwellings; they vary in geographical scale of operation from those with a single scheme serving a village or local area to those operating nationally; they vary between associations concentrating on rehabilitation and those concentrating on new build; and they vary according to purpose - whether providing general family accommodation or catering for a variety of special needs (Niner, 1985: 33).

Although all registered housing associations are now subject to the same legislative and financial requirements, the diversity of the movement, including differences in allocation processes, can be understood only in terms of the variety of organisations which were brought together almost arbitrarily by the Housing Act 1974. Some associations originated as charitable trusts in the late nineteenth and early twentieth centuries; in the Victorian era there were limited experiments with non-profit housing of the so-called 'five per cent philanthropy' in which the return on investment was limited to the relatively modest dividend; Peabody, Guinness, Sutton and Samuel Lewis illustrate that it is the endowed charitable trusts which have survived from this era.

Second in the early 1960s the Conservative Government took a number of initiatives which resulted in the formation of cost rent and co-ownership housing societies: these bodies reflected a different philosophy of housing to let without the involvement of local authorities and public subsidies; the White Paper Housing in England and Wales (HMSO 1961) commented 'More new houses are needed to let for people who do not wish or cannot afford to buy a house though they require no help from public funds'. Fundamentally such societies and their committees created high rent schemes which developed quite distinctly from housing associations but some later developed into housing associations when they had made the necessary adjustments to their constitutions, committee personnel and working arrangements to comply with the registration requirements of the Housing Act 1974. As Niner has observed the reputation of some associations of being less concerned with meeting housing need than with selecting reliable, respectable tenants who will pay the rent and cause no problems for their neighbours may be attributable to the inherited style or philosophy of management from housing societies.

Niner identified a third group of associations sharing a very different philosophy growing after the mid-1960s as a result of the shelter campaign and the awakening realisation of the continuing problems of poverty, homelessness and poor housing conditions in British cities. Shelter's campaigning and lobbying activities contributed to the provisions of the Housing (Subsidies) Act 1967, which made a subsidy available to housing associations for the purchase and improvement of dwellings which was more generous than that available through renovation grants. In addition money raised by Shelter was used to support a few housing associations operating in inner-city areas and it was recognised that such associations had the flexibility and potential speed of response to tackle housing needs in a more sympathetic fashion than the development of local authority provision in many instances.

Self-evidently not all societies will fall into these three broad categories but such categorisation does demonstrate the diversity of philosophical approaches which underpin the housing association movement. In turn such diversity will be reflected in allocation policies and procedures in different associations. Consequently if improving the housing opportunities of ethnic minorities demands a commitment to equality of opportunity by favouring those in greatest need and those weakest in the housing market one cannot of necessity assume that all associations will share these objectives.

Despite the results of extensive research relating to race and housing in the public sector there has not been significant comparable research on housing association allocation policies (Niner 1985; Dalton and Daghlian, 1989). As a consequence little is known about the operation of association allocation processes in general, and virtually nothing about their racial or social implications. Niner has suggested that four different reasons should be considered for examining housing association allocation policies and procedures:

1. Although this sector is small in absolute terms - about half a million households in all (2 per cent of England and Wales households were in housing association tenancies by reference to the 1981 census) - it has been growing rapidly since 1974 and is still growing although at a slower rate.
2. Between 1978 and 1982 housing associations accounted for about one fifth of all public-sector construction and rehabilitation programmes and consequently this sector represents an increasing proportion of total public-sector activity; it might also be argued that, given the nature of local authority subsidy and financing,

increases in rents have moved some local authority housing out of subsidy while housing association grant and other subsidies to housing associations have remained unchanged and in relative terms therefore the increase of directly subsidised dwellings is proportionately greater for housing associations.

3. Since 1974 there have been broadly two strands in the development of housing association fair rent dwellings - new building and rehabilitation. Not infrequently it is the inner and middle rings of cities where redevelopment has not occurred, which, by definition, are ripe for rehabilitation, and it is these areas that have been the main points of settlement of ethnic minorities.

4. The attention of the housing association movement has been drawn inevitably, particularly following the disturbances in a number of cities in 1981, to the importance of its role in the housing of ethnic minorities in inner-city areas. The National Federation of Housing Associations has published two reports, the first in September 1982, Race and Housing: a Guide for Housing Associations and the second, in August 1983, Race and Housing: Still a Cause for Concern which illustrate this increasing awareness: in addition the Housing Corporation has issued circulars on race and housing, applicable in England and Wales, and the Housing Corporation in Scotland has similarly reflected the advice in a Scottish circular.

However, since Niner's report the amendments to the Housing Association Act 1985 by the Housing Act 1988 are likely to prove critical to the development of the housing association movement through the changes relating to grant, to rents and to guidance or direction, previously referred to. While public-sector rents themselves will rise, it seems likely that most housing association grants will rise more steeply. Although Housing Benefit may frequently accommodate such rises, many families are expected to be caught in the poverty trap - having sufficient income to disqualify them from full Housing Benefit but insufficient to meet the increased rents. Both part-time workers and low wage earners will be affected. As we have noted in Chapter 2, such groups are disproportionately represented by ethnic minorities.

In addition to the above reasons the fact, first, that housing associations are earmarked for taking over local authority estates and, second, that the financial arrangements may squeeze out smaller housing associations represents two issues with implications for ethnic groups.

The selection process: general

As with local authorities when housing associations are required to make decisions relating to allocation they are involved in a strict rationing process; demand for vacancies generally exceeds supply and the better the area and quality of dwelling the greater the demand. Inevitably it is the housing association which has to make the value judgement regarding suitability of applicants and in so doing when poor-quality accommodation is allocated the applicants are deemed to be only suitable for or deserving of such property. Clearly where ethnic minorities are considered, whether consciously or unconsciously, to be less deserving than other categories of applicants their eligibility for good-quality housing in the housing association sector will be affected. As we have noted in respect of local authority housing the rationing and selection processes are subject to conflicting objectives; the most significant objectives may be expressed as follows:

1. The desire to meet applicant's preferences.
2. The desire to let property as speedily as possible.
3. The desire to give greatest priority to those in greatest need.
4. The desire to build up balanced communities or alternatively to establish communities where the potentiality for conflict is marginalised.
5. The need to minimise management intervention in keeping with budget constraint.
6. The need to match available stock with household requirements (Niner, 1985).

In determining such priorities it is perhaps only the last which has universal relevance as an overriding consideration; self-evidently housing associations would not wish to house their tenants in houses the size of which would result in overcrowding and a continuing demand for transfer while, conversely, they would consider it wasteful of their resources to provide housing which was larger than that required to meet the needs of tenants. Local authority housing stock may not match housing need in respect of household size both because of historical influences on the development programme and the inherent difficulty of anticipating future population requirements in respect of housing size by reference first to demographic trends and second to social factors which have influenced patterns relating to leaving the parental home and the formation of single-parent families. In relation to housing associations the stock of individual associations and vacancies arising are frequently much more limited than those relating to local authorities. Consequently, as Niner has observed, in general terms, the available stock has a much more significant influence on the sort of household housed than the precise allocation policies or procedures operated.

Nonetheless where discretion does operate the priorities chosen differ with individual housing associations; thus it has been argued that some of the associations with links back to the 1960 housing societies may give greater priority to minimising future management problems than those founded under the Shelter initiative.

Nomination arrangements with local authorities

Although housing associations are not under any statutory obligation to enter into nomination arrangements whereby local housing authorities are able to nominate respective tenants for their schemes, in practice such nominations are often a condition of Housing Corporation finance (see DoE Circular 170/74). The actual details of how nominations are to be made and the percentage are not only left to the discretion of the two parties, i.e. the housing associations and the relevant local authority, but the decisions are largely made in the absence of guidance, whether from the DoE, Scottish Office or Housing Corporation itself. As a consequence, as has been recognised (Alder and Handy, 1987), a plethora of arrangements has resulted. Housing Corporation Circular 16/80 suggests that the generally accepted 50 per cent agreement that most associations have with local authorities should not be a hard-and-fast rule and associations should negotiate alternative suitable arrangements, particularly in the instance of special projects. Both the DoE (Circular 73/67) and the Institute of Housing (Working Together, 1983) have warned against local authorities putting extreme demands on housing associations with regard to nomination rights which, if exercised exclusively, might destroy the whole point of an independent association.

But whatever arrangements are determined between the local authority and the housing association the scope for the former discriminating in the selection of tenants will remain. The CRE formal investigation into the nomination rights of Liverpool City Council to six housing associations concluded that black nominees had consistently received poorer property than white nominees across all measures of quality. Such measures included type of housing, type of letting, type of building, age of property and whether the property had central heating and a garden (CRE, 1989e: 47). Moreover, both individually and cumulatively, housing need, area preference, household size, the year of nomination, economic circumstances, the particular housing association for which people were nominated and refusals all provided an insufficient explanation for such disparity of outcome. In addition to finding direct discrimination by the council against black applicants in respect of nominations, the CRE also found indirect discrimination in the access channels for the allocation of new build properties: blacks were under-represented in the priority categories (decant and medical cases) for which targets were set. Top priority in the council's housing programme during the period of the CRE investigation was the Urban Renewal Strategy (URS), which was expressly recognised as an important mechanism for combating racial inequality in housing as identified by the CRE in its 1984 research report (1984a). An essential recommendation of that report, the requirement to monitor, had been rejected by the council. The formal investigation concluded that black nominees to housing associations within the URS received poorer quality property and fewer new builds as a result of direct racial discrimination.

This investigation demonstrates that if there is to be equality of opportunity in local authority nominations to housing associations then closer liaison arrangements must be developed, a finding given greater importance by legislative change giving more emphasis to housing association provision (CRE, 1989e: 7).

The selection process: legal

In terms of the legal requirements, housing associations are even less fettered by legislative constraint than local authorities; there is no statutory responsibility to meet the housing needs generally in their area and no statutory duty comparable to that placed on housing authorities by the statutory provision for the homeless. We have already noted, however, that housing associations are not entirely free to choose their tenants in that they will be subject to the following:

1. First the provisions of the Race Relations Act 1976 and the Sex Discrimination Act 1975 will require allocations to be made on a non-discriminatory basis in respect of race and sex.
2. Given the general responsibilities and duties of the Housing Corporation in relation to registration and monitoring it will be necessary for housing associations to take cognisance of guidance issued from time to time by the Housing Corporation, Housing for Wales and Scottish Homes to ensure continued registration and eligibility for grants and loans.
3. In England and Wales housing associations are members of the National Federation of Housing Associations and in Scotland the vast majority are members of the Scottish Federation of Housing Associations, both of which will, from time to time, offer advice and guidance to member associations. In such instances there is no real sanction but 'peer pressure' may be persuasive in altering or modifying allocations practice and procedures.
4. As noted above, it is normal practice for housing associations where they have built with the assistance of HAG and a public-sector loan to enter into a nomination agreement with the local authority. The size of nomination quota is discretionary but 50 per cent referrals from local authorities are common and a figure higher than this is possible in respect of initial lettings. However, there is frequently a veto left in the hands of the housing association where such agreements enable a quota reference. The exercise of such a veto may, in turn, influence the nature of referrals made by local authorities.

Generally, however, associations exercise discretion in determining the channels through which they recruit tenants, whether and in what way they keep a waiting list and how they determine eligibility and priorities for their accommodation.

Housing associations frequently do not keep open lists

263

of applications and consequently may refuse an applicant in demonstrable need by reason of the fact that they do not have sufficient vacancies to be able to assist the applicant within a reasonable time. Some associations, in addition or alternatively, require reference to be made through a statutory or voluntary agency before anyone can be considered for acceptance on to the waiting list.

Although the requirements regarding applications and allocations in relation to housing associations are essentially non-statutory there are certain statutory requirements which provide a context in which these operate. In relation to local authorities we have already noted how the Housing Act 1980 and the Tenants' Rights etc. (Scotland) Act of the same year, in relation to Scotland, introduced a duty to publish allocation rules. This now applies to registered housing associations other than co-operative housing associations (Housing Act 1985, Sections 106, 114). Such rules apply to all housing accommodation and not just secure or assured tenancies. Consequently a registered housing association must publish a summary of its rules which, first, relate to the determination of priority between applicants in the allocation of its housing accommodation and, second, govern cases where tenants wish to move (whether or not by way of exchange of houses, to other dwelling houses let under tenancies by that authority or another body). In addition an association is required to maintain a full set of the rules, both substantive and procedural, to be followed in allocating its accommodation. In the Scottish context housing associations are prohibited in making allocations in the situations described in the last chapter from having regard to ownership of property (heritable and movable), residence or employment. These provisions, which originally would have applied to England and Wales by reason of Clause 27 of the Housing Bill 1979 (which would also have made criteria based on place of birth unlawful) are not so applied south of the border but no association, wherever situated, is able to let accommodation as it thinks fit without having regard to prior and published rules. Again, as we have noted in respect of local authority applications and allocations, a requirement in relation to either of these functions which restricts access by reference to place of birth might well constitute indirect discrimination in terms of the Race Relations Act 1976, Section 1. As Alder and Handy (1987) have observed, a prospective applicant for accommodation would have sufficient locus standi to seek judicial review of a decision, being within the range of beneficiaries of the statute (see IRC v National Federation of Self-employed and Small Businesses (1982) AC 617). However, it seems improbable that the statutory obligation confers a private right of action for damages upon any individual and as a consequence a remedy would appear to be restricted to a public law action by way of application for judicial review in England and Wales to the High Court and in Scotland to the Court of Session. Access to the County Court or Sheriff Court appears to be denied by reason of Cox v Thanet District Council (1983) AC 286 and by reason of Brown v Hamilton District Council 1983 SLT 397.

A prospective tenant or applicant not only has a right of access to the rules relating to the housing association; he or she also has a right of access to any information supplied by him or her to the housing association relating to the application (Section 106 (5)). Moreover so far as any information relating to the application is retained on a computerised system the Data Protection Act 1984 will require the disclosure of such information, including information which had not been supplied by the applicant/tenant, to him or her. The Access to Personal Files Act 1987 does not extend to housing associations although the Housing Act 1988 extends a duty of disclosure to Housing Action Trusts. Clearly information which is of a subjective nature and which has not been supplied by the applicant and is not on a computerised system is likely to be withheld and as a consequence such information, which may be of fundamental importance in determining the opportunity for an allocation or determining the type of allocation made, may not be open to challenge.

Constitutional requirements

Although the statutory provisions are essentially peripheral to the application and allocation process operated by the various housing associations, the constitutions of these associations are much more central to their choice of tenants. Where the constitution of an association contains a provision specifying or limiting the type of person whom it is able to assist, such as the poor, the elderly or handicapped, then, depending on how such groups are defined and depending on whether they are exclusive or otherwise, the constitution in stating such objectives may very well limit the power of the housing association to provide accommodation outside such a categorisation. If it were to do so the association would be acting ultra vires of its constitution and open to challenge. For England and Wales housing associations which are charitable and are so registered would be obliged to ensure compliance with the charitable objectives in order to continue to enjoy the benefits of charitable status.

These considerations will, of course, affect the way in which housing associations seek to attract applicants, which in turn will profoundly affect the sorts of households and needs that are in fact met. Despite this neither the legal requirements in relation to publication nor the constitutional objectives of the housing association will in themselves determine the nature of publicity and the methods whereby applications are sought.

As Niner (1985: 40) has noted, at one end of the scale there is the option of relying solely on who turns up and applies to the association: this would probably be as a result of information gained through friends and relatives or by way of proximity to existing or new developments. An example of this was Collingwood Housing Association Ltd in Greater Manchester in the late 1970s, which became the subject of a formal CRE investigation. In the CRE's view the practice of advertising by the association had been confined to notice boards which were placed on the site of its developments both during construction and after completion and by implication, given that the new developments were mostly at some distance from areas of black residence, they were unlikely to be seen by many potential black applicants. Community-based housing associations, such as many based in Glasgow, will seek to give preferential treatment to existing residents. While there may be nothing objectionable in this practice per se, its effect may be to crystallise existing patterns of segregated residential development and to reinforce racial stereotyping in relation to access to better-quality housing.

Conversely housing associations may be involved in the process of advertising in a variety of ways, for example through advertising in the local press, through providing advice and guidance for transmission by Citizens' Advice Bureaux, Housing Advice Centres, neighbourhood organisations and the rest. In addition, some housing associations have special links with a particular housing advice centre which in effect acts as their first point of contact with would-be applicants and makes initial selection decisions. The use of such advice centres and other agencies may be widespread, an example being provided by Copeck Housing Association, which between 1979 and 1981 housed applicants referred by no fewer than forty-two different bodies or individuals (Niner, 1985).

It will be seen, then, that image projection, whether by way of information, publicity or the use of referral agencies, provides an important gatekeeper prior to applications being considered.

Housing association tenants

Results of surveys of housing associations in 1985 and 1988 by the NFHA are shown in Table 9.1. These surveys dealt with new tenants only and the picture shown from the 1981 census with regard to existing tenants is somewhat different. The HMSO report Household and Family Composition which was based on the 1981 census, shows that tenants of housing associations compared to other forms of tenure fall, in socio-economic terms, between the owner-occupied sector and local authority tenants, a position confirmed by the 1985 General Household Survey (HMSO, 1987a) (see Table 9.2).

These surveys would suggest that the profile of housing association tenants is moving towards that of the local authority sector, with an expectation that new tenants are more likely to come from various special needs groups which include a disproportionate number from ethnic minorities, particularly those of West Indian and Chinese origin (Social Trends, HMSO, 1987b: Table 8.8: 141). However, the NFHA 1988 survey, although showing a marginal increase for non-white tenants, showed a fall within that for West Indians (from 7 per cent to 4 per cent) while Asians remain at 2 per cent in both 1985 and 1988 surveys. As expected the proportion of lettings to non-white households in 1988 was highest in London (31 per cent) and the West Midlands (22 per cent) but the figure of 1 per cent for the northern region remained exceptionally low.

The NFHA surveys were restricted to housing associations in England and Wales: no similar surveys identifying ethnicity have been conducted in Scotland. However, the report on a survey of four housing associations in the west of Scotland conducted by the Scottish Ethnic Minorities Research Unit (Dalton and Daghlian, 1989), by

Table 9.1: NHFA surveys of tenants, 1985 and 1988

Tenants	1985	1988
Households with a net weekly income of less than the national average	94	99
Households relying on State benefits as their main source of income	67	50
Economically active heads of households out of work	50	45
Pensioners	32	33
Single people, excluding the elderly, from an ethnic minority	21	30
Single people	48	57
Tenancies granted to ethnic minorities	11	12
Families headed by a single parent	47	48
Households qualifying for Housing Benefit	76	58

Source: NFHA, 1985 and 1988.

Table 9.2: Housing tenure by socio-economic group in England and Wales (%)

Socio-economic group	All tenures	Housing Association	Local Authority	Owner-occupied	Private renting
Professional	5.0	2.3	0.7	7.2	2.1
Managerial	15.7	8.2	4.4	21.2	11.3
Intermediate and junior	19.8	23.6	11.3	23.0	18.9
Skilled manual	30.8	27.1	36.5	29.8	29.2
Semi-skilled	15.9	19.2	25.4	11.0	18.6
Unskilled	5.1	6.6	10.9	2.6	6.3
Other	7.6	13.1	10.9	5.2	13.5

Source: General Household Survey, HMSO 1987a

cross-referencing applications and allocations, showed the following:

1. Asian households on the waiting list were more likely to have children in contrast with white households and were consequently larger.
2. Asian households were more likely to be unemployed and were heavily concentrated in unskilled manual occupations.
3. Asian households were more likely to have been lodgers, private renters or owner-occupiers, to have experienced overcrowding and lack of security of tenure.

Selecting the tenants: Provan's study of Birmingham housing associations

In general housing associations seem less waiting list orientated than local authorities (Niner, 1985). This does not mean to say that waiting lists are not generally employed but the deployment of the waiting list is more varied and dependence on it apparently more tenuous. In Provan's study of thirty general needs associations operating in Birmingham he found that two associations kept no lists whatsoever. Some new-build associations kept a list for each development and in four cases the list was closed after initial lettings were made, retaining only a few applications for re-let. Moreover the closed waiting list is not uncommon amongst housing associations, many of whom try to keep lists to a 'reasonable length' in terms of waiting time. When a list is open for a short period only a would-be applicant relying to a great extent on sound advice from personal contacts (or essentially on luck) is likely to be successful. None of these factors is likely to reflect the degree of need or the stated priorities of the association.

The other method of restricting waiting time to a reasonable period is to be highly selective in the admission of applicants to the waiting list. In this instance the criteria for eligibility and priority are likely to be of greater significance. Although the Housing Corporation in its Circular 1/75 emphasised the requirement for housing associations to prioritise satisfactory housing for those in real need this advice was not unqualified. The circular stated that it should not be concluded as a general principle that every tenant of a housing association should be someone of social, physical or financial disability: 'The object must, in part, be to arrive at balanced communities, which taken together are not going to pose problems beyond the resources of the association.' The Housing Corporation suggested the following criteria as being relevant:

1. The applicant's existing housing conditions.
2. The applicant's ability to cope with those conditions.
3. The length of time the applicant has had to put up with those conditions.
4. The other future housing prospects of the applicant. (Housing Corporation, 1975, para. 6, repeated Circular 16/80, para. 3.1).

Provan's study indicated that the following criteria were mentioned in statements of policy: existing conditions, alternative solutions, ability to cope, financial ability, building a balanced community, time in housing need, ability to compete (Provan, 1982).

Despite the existence of these criteria it seemed that the great majority of associations had no formalised system for determining either eligibility or priority. Individual cases were often examined by single officers or groups of officers (or a committee) and a decision was made whether or not to accept the application and what degree of priority to accord it. Provan found no example of a points scheme among associations operating in Birmingham and he classified all schemes as either merit schemes or merit plus date order schemes.

The description 'merit' is somewhat misleading: essentially all the description means is that an area of discretion is left to the officer or group responsible for selection and, in the absence of any formalised criteria or assessment of the relative variables involved in selection, it is unlikely that such discretion would be applied in a uniform fashion in respect of all applicants.

Housing Corporation policy on race

For England and Wales the circular on tenant selection by the Housing Corporation referred to by Provan (1/75) has been replaced (16/80) although the basic principles previously expressed are considered 'still sound'. That circular (16/80) emphasised the general need for housing associations to complement rather than duplicate local authority housing provision and to consult local authorities regarding housing need and the housing association's role when preparing housing investment programmes (HC 16/1980: 2.2). The circular advises larger housing associations to consider letting some units on a ready access basis to cater for applicants who, but for the decline in the private rented sector, would not otherwise have sought housing association assistance and might not seek accommodation through other channels. This circular also refers to the Race Relations Act 1976 (para. 3.9), refers to the CRE recommendations regarding monitoring and mentions residence qualifications and giving priority to relatives of existing tenants as potentially being 'seen as indirect discrimination'. The Housing Corporation's circular on Race and Housing (HC 22/85) quotes its board's statement on policy (1982):

> All eligible groups shall have the same opportunity. Because minority groups are over represented among those in most urgent housing need, special initiatives to assist those minority groups are likely to be necessary.

The mechanics of implementation are expressed as threefold:

1. Monitoring the activities of housing associations.
2. Promoting and developing, with the housing association movement, good practice both in housing management and in administration.
3. Legislation, policy and practice.

For detailed policy the Corporation specifically referred to the NFHA (1982, 1983) publications in this area and advised that each housing association's performance would be assessed by reference to its report in answer to questions posed in the NFHA checklist. Where responses demonstrated practice falling below the standards set the circular states that the housing corporation 'will discuss corrective action and agree remedies, to a specific timetable'. The publication Race and Housing: Ethnic Record Keeping and Monitoring' (NFHA, 1985) provides further guidance to housing associations anticipated in the relevant circular (HC 22/85). In June 1989 the Housing Corporation issued the Performance Expectations, a comprehensive guide to housing association self-monitoring. That publication together with HC 47/89

271

provides a thorough approach to race equality and equal opportunity sensitively integrated into all aspects of a housing association's functions (Housing Corporation, 1989).

In December 1986 the Housing Corporation in Scotland issued a guidance note, 'Race and Housing' (No. 22/86), which advises Scottish housing associations of the general approach expected in implementing fair housing practice and draws attention (albeit ten years after enactment) to the legal obligations of associations under the Race Relations Act 1976. The guidance states that to demonstrate compliance with the policy of the Housing Corporation, an association will need to have a system of ethnic record-keeping and to adopt a clear equal opportunity policy, both of which will be evaluated by the Corporation monitoring staff. This requirement for all housing associations in Scotland to keep ethnic records contrasts with the NFHA view, which questions the value of record-keeping where ethnic minorities comprise less than 2 per cent of the population because of the statistical significance, or lack of it, in particular situations. In both England and Wales the monitoring by the Corporation's staff is likely to be superficial because visits occur only once every two to four years and the issue of race will not be a major focus when it occurs. Moreover in Scotland it is clear that the SFHA equivocates over record-keeping and its members have not been weaned from the colour-blind approach. It is expected, however, that further guidance to Scottish housing associations may be issued following the research project into selected Glasgow housing associations (Dalton and Daghlian, 1989). More generally, the Housing Corporation in Scotland has been concerned for some time with the inadequacy of guidance provided on allocation and has embarked on a major exercise with the SFHA to provide more thorough and comprehensive guidance; a working party draft 'Tenancy Selection and Allocation' (SFHA/HC, July 1987) has been produced as a basis of consultation.

In a formal sense the guidance produced or referred to by the Housing Corporation appears to demonstrate a heightened awareness of real and potential discrimination suffered by ethnic minorities in the allocation of housing association lets. But as we have noted in the public sector the espousal of policies on equal opportunity, the advocacy of practice and the intimation of an intention to monitor are unlikely to effect change without concomitant prioritisation by both the Housing Corporation and the housing associations, efficient policing and effective sanctions. A potential weapon in this armoury of control is the power of the CRE to conduct formal investigations.

FORMAL INVESTIGATION BY THE CRE: COLLINGWOOD HOUSING ASSOCIATION

Of the thirty-seven formal investigations published by the CRE between 1977 and the end of 1986 only one related to a housing association, Collingwood Housing Association. This investigation looked into the allocation patterns of a medium-sized housing association. The report was published in November 1983, although the investigation commenced in 1978. The investigation established that the staff of the association had been influenced by considerations of race and were prepared to impose quotas for 'immigrant' tenancies in particular housing schemes. The Commission also identified that certain of the association's procedures and practices may have contributed to the low level of ethnic minority access. From 1981 new management and staff in Collingwood Housing Association cooperated with the Commission in devising new procedures and policies which were designed to ensure that local ethnic minorities got full equal opportunities to be rehoused in any of the association's housing schemes. In the CRE Annual Report for 1983 (p. 24) the Commission expressed the view that the new policies being operated by the association provided a good model for other housing associations working in areas where ethnic minorities live. The results of the investigation are summarised in Table 9.3.

The CRE concluded from the investigation that the absence of specific criteria in Collingwood's selection processes appeared to lead to a number of apparently inconsistent allocation decisions affecting some of the very few ethnic minority applicants who could be identified from the existing records. It was acknowledged, however, that, quite apart from race, apparently inconsistent decisions were likely to occur in the absence of a well structured allocation system. Despite this, concern was expressed about the small number of black tenants in the association's properties and the failure of some ethnic minority applicants to secure accommodation in view of the background evidence that considerations of race were liable to be taken into account. The housing manager had made

Table 9.3: Collingwood Housing Association: allocations

Col No.	Scheme	Ethnic minority population in area (%)	Allocations made in 1979	Ethnic minority allocations in that year[1]
1.	Trafford Ridgeway Park Estate	4.0	143	-
2.	Bury Berkshire Court Grigg Lane	2.5	41	-
3.	Bolton Glencoe Radcliffe Road	5.0	16	-
4.	Rochdale Robert Saville Court Roche Valley Way	5.0	25	-
5.	Manchester 1 Elizabeth Court Brook Road	10.0	18	- (1 Relet)
6.	Manchester 2 Eversley Court Brooklands Road	10.0	65	3 (all LA nominees)
7.	Manchester 3 Langdale Court Cheetham	10.0	38	6[2]

Notes:

[1] The number of ethnic minority allocations was gauged firstly by nationality in the application form, secondly by name and thirdly by interviewing the staff involved.

[2] Two of the six ethnic minority applicants did not take up the offer of a tenancy.

remarks indicating a predisposition to discrimination at three separate interviews and the CRE therefore concluded that if the appropriate circumstances arose, racial considerations could enter into the selection process and one or more ethnic minority applicants might fail to obtain a full and equal opportunity of rehousing. In addition, given that the housing manager had been solely responsible for the only regular training provided to the visiting and allocation staff for a number of years, it was possible that some of those staff might have been influenced by her views. The CRE, therefore, considered that the housing manager's views amounted to an arrangement whereby, in particular circumstances, some black applicants could be excluded on racial grounds from the association's housing. Although there was not evidence that any specific black person had been discriminated against in this manner such an arrangement would, in the CRE's view, be unlawful under Section 28 of the Act. However, having heard representations in the matter and having taken advice, the CRE formed the view that where the application of any prior arrangement would result in an act of direct discrimination as opposed to indirect discrimination, the practice was not covered by Section 28 of the Act. Consequently the CRE was unable to issue a Non-discrimination Notice but nonetheless Collingwood Housing Association had taken and agreed to take further steps to secure equality of opportunity in relation to applications and allocations.

Section 28(1) of the Race Relations Act 1976 provides that:

> discriminatory practice means the application of a requirement or condition which results in an act of discrimination which is unlawful by virtue of any provision of Part 2 or 3 taken with Section 1(1)(b), or which would be likely to result in such an act of discrimination if the persons to whom it is applied include persons of any particular racial group as regards which there has been no occasion for applying it.

A person contravenes this section if, and so long as, first, he applies a discriminatory practice or, second, he operates practices or other arrangements which in any circumstances would call for the application by him of a discriminatory practice. In the CRE review of the Race Relations Act

(CRE, 1985b: 25) reference was made to the Percy Ingle investigation where the Commission found itself powerless to deal with a situation where manageresses said they would not employ 'black mama types' or 'Pakistanis' where they had not issued any instructions or put pressure on anybody to that effect and the Commission was not in a position to demonstrate that unlawful discrimination had actually occurred. The Commission's attempt to formulate a 'no black' rule as an indirectly discriminatory practice under Section 28 (which need not show an actual victim) failed because indirect discrimination was held by an industrial tribunal, following the unreported case of Wong v GLC (1979) EAT 524/79, to apply only where the practice would exclude a proportion, rather than all, of the racial group in question. At the time of the Act's passage the Government view was that a practice might be so discriminatory in its effect that, paradoxically, it would not produce a victim because the practice, for example, was so well known that no one would willingly subject himself or herself to the discriminatory effect anticipated. Alternatively, an employer, for example, might, during a period of recession, give instructions to his personnel manager which enshrine a discriminatory practice which is not given effect because there has been no recruitment during the period (see H.C. Standing Committee A, 1976, Col. 419).

This investigation, therefore, illustrates the present weaknesses in the 1976 Act regarding formal investigations where direct discriminatory practices are shown to be operating but no victim can be found.

The CRE made recommendations in the following areas:

1. Policy - The association were advised to formally adopt an equal opportunity statement, to instruct all staff accordingly and to provide guidance and training to employees about the provisions of the 1976 Act and how these might affect their work.
2. Review - Collingwood were advised to review their criteria and procedures for selection of applicants for tenancies and transfers, to ensure that indirect discrimination did not take place, to reduce the scope of subjective judgment and to emphasise the use of objective criteria.
3. Monitoring - The association was advised to keep records of the ethnic origins of all applicants, local authority nominees and referrals for tenancies and

transfers in respect of the association's schemes. Reasons for decisions were to be recorded and the records were to be monitored and analysed at regular intervals to ensure that unlawful discrimination was not occurring.

4. <u>Communication with local ethnic minority communities</u> - The association was advised to ensure that the general and special needs of local ethnic minority communities were included in the consideration given and arrangements made to meet particular housing needs, to ensure that information about general rehousing opportunities was conveyed to such communities and to monitor the effect that this had on access.

5. <u>Local authorities</u> - The association was advised to ensure that each local authority which had nomination rights was made aware of all tenancies which were likely to become available.

6. <u>Responsible officer</u> - The association was to allocate to the chief executive responsibility to the management committee for ensuring compliance with the Act and specifically for the implementation of the recommendations made by the CRE (1983b).

The impact of this investigation and the CRE report on housing association practice is difficult to gauge. While none of the circulars or guidance notes issued subsequently by the Housing Corporation dealing with race and housing refers to the Collingwood investigation directly, many of the recommendations in the report relating to management, policy development and tenant selection are reflected in the advice disseminated by the Housing Corporation. The study by Niner (1985), however, demonstrates not only that the Housing Corporation advice on tenant selection (HC16/80) had not been conscientiously followed in respect of selection reflecting the needs of ethnic minority applicants but also that the Housing Corporation, at least at that point in time, had not made any noticeable impact on monitoring arrangements. Consequently while it may be unjust to conclude that the Housing Corporation had no commitment to achieving its stated objectives, there is little evidence that this area had been given sufficient priority to ensure the translation of intention into practice. As with local authority selection and allocation practice, however, it is clear that unlawful discrimination occurs and the greater the opportunity for individual discretion the greater the potential for racial disadvantage through stereotyping and direct discrimination. Moreover, the lack of individual complaints under the 1976 Act illustrates that, even in comparison with employment cases, the existing provisions of the legislation, which require neither monitoring nor the adoption of a code of practice, render appropriate counter-measures as an exhortation which is widely ignored with absolute impunity. More effective initiatives adopted by the Housing Corporation and being considered by Scottish Homes may alter this conclusion but, meantime, in many areas the promotion of black housing associations remains a tangible, if as yet limited, option for significant improvement to the opportunity for access to decent rented housing for many black families living in areas where they form a significant proportion of the total community.

BLACK HOUSING ASSOCIATIONS

Sections 25 and 26 of the Race Relations Act 1976 relate to membership of associations generally. Under Section 25 an association which has twenty-five or more members must not discriminate on racial grounds against applicants for membership or against existing members who are part of a minority racial group. The exception to the rule is provided under Section 26 and applies where the main object of an association is to confer the benefits of membership on persons of a particular racial group defined otherwise than by reference to colour. In determining whether this is the main object of an association, regard will be had to the essential character of the association and the extent to which the affairs of the association are so conducted that the persons primarily enjoying the benefits of membership are of the racial group in question. Although these two sections were clearly drawn up with associations other than housing associations in mind - for example, mutual aid societies and in particular social clubs and community associations for racial minority groups - nonetheless in the context of the housing association movement Section 26 enables an association to register rules which restrict membership to a certain racial group provided that that group is defined otherwise than by reference to colour. If the association was a co-operative (registered on rules which restrict membership to tenants or prospective tenants and which only allow tenancies to be granted to members) it

278

would appear to be lawful under the Act for the co-operative to restrict membership to a particular racial group and consequently limit the granting of tenancies to its members, all of whom would belong to that racial group. However, where an association was not a co-operative, although it could restrict membership to a particular racial group, it could not lawfully discriminate against other racial groups because, having no legal requirement to rehouse only its members, it would be bound under Section 21 not to discriminate on racial grounds in relation to the allocation or disposal of premises.

The National Federation of Housing Associations (1982: 6) then observed that in practice these two sections were unlikely to be either a help or a hindrance to housing associations requiring public funds since the Housing Corporation took the view that housing associations which receive public funds should provide housing for 'the public' and not just a particular racial group within it. It was then felt unlikely that the Housing Corporation would register any housing association or co-operative whose rules restricted membership or the granting of tenancies to any particular racial group. But even at the time of publication such opinion seemed ill-founded: Ujima Housing Association was founded in 1977 with the clear objective of providing accommodation principally for black applicants. Although its objectives, as stated, may have obscured the foci on Afro-Caribbean applicants (to such an extent that there is a current debate on the extent to which it is prepared to meet Asian housing need - see Black Housing, Vol. III, No. 2, July/August 1987) and registration was not effected until 1980 it is now clear that the Housing Corporation recognises the need to register black housing associations to meet needs which are being insufficiently met by other housing associations whether of a general or specialised nature even where they are located in areas with significant black populations. It would appear now that the difficulties experienced by black housing associations do not relate to their objectives in relation to their ability to register with the Housing Corporation but their ability to demonstrate management skills to the satisfaction of that body. By December 1987 some thirty-eight black housing associations had been registered with the Housing Corporation. Their objects included the housing of Afro-Caribbean and/or Asian tenants, some twenty exclusively so. The focus of such associations is varied, ranging from Bangladeshi widows, black single homeless and elderly Chinese, to more general categories of black families. The majority of housing associations are London-based (twenty-one) with the remainder based in the West Midlands (five), the North East (four), the North West (four), Merseyside (two), the East Midlands (one) and Bristol (one). No black housing associations are registered in Scotland. The Housing Corporation has drawn up a five-year programme for registration with a target of five black housing associations per year located in either the High Priority Areas, of which there are thirteen, or the Mid-priority Areas, of which there are eight. Such areas have been identified by reference to the proportion of black population resident and the degree of housing stress.

Despite such targeting, and the earmarking of necessary funds, it is clear that a substantial majority of black housing association tenants will continue to be housed by associations catering, predominantly, for white tenants. The Federation of Black Housing Organisations has expressed concern about the insistence, generally, by the Housing Corporation that development agreements are entered into with existing housing associations, with the obvious danger that big white brother Housing Association will be looking after number one (Black Housing, Vol. IV, No. 2, March 1988). While the FBHO recognises that in many instances such agreements are necessary, vigilance is advocated.

If anything the legal framework provided by the Housing Associations Act 1985 and the Race Relations Act 1976 might have been viewed as a restriction on the development of black housing associations. Although developmental difficulties remain, the positive progress set by the Housing Corporation demonstrates that, with changing attitudes, such restrictions are not insurmountable.

CONCLUSIONS

We have already noted that where there is statutory provision imposing obligations on housing associations and an individual prospective applicant for accommodation, for example, has been directly affected by a failure to comply with such an obligation then access to the courts by way of judicial review would result, as clearly the matter would be one of public law.

Nonetheless the substantive power to allocate tenancies does not appear to derive from statute but from the general powers of the association (Alder and Handy, 1987: 266). The case of Peabody Housing Association v Green (1978) 38 P&CR 644 is authority for the view that despite the fact that housing associations operate within the statutory context for public purposes and with public money this is insufficient to turn a landlord and tenant relationship into one of public law. Alder has argued that if the allocation of tenancies is a private law matter it is unlikely that a disappointed applicant would have a remedy in the courts, since a decision to refuse accommodation would not as such affect his legal rights. However, as Alder also noted, the case of MacInnes v Onslow-Fane (1978) 1 WLR 1520 demonstrates that the principles of natural justice and 'legitimate expectation' are not confined to public law: consequently an applicant who is placed on a waiting list may have a 'legitimate expectation' that he may be granted a fair hearing in relation to his prospects of an offer of accommodation. In England and Wales the purpose of the distinction between public and private law is confined to establishing whether challenge must be by means of the special application for judicial review procedure.

Emery and Smythe (1986) defined the term 'public authority' as including the Crown, Parliament and the superior courts (all of non-statutory origin) and the multitude of other State authorities set up by or under the authority of Parliament and endowed with executive legislative or adjudicative powers. The term 'public authority' is not, however, capable of highly precise definition. Clearly Government departments and local authorities fulfil all possible criteria: they are created by or under statute, endowed with statutory powers and funded by the public purse for the purpose of conferring what are seen to be benefits upon the community or a selection thereof. Nonetheless there are many other authorities set up by or under statute and serving important public purposes, just as housing associations which are registered with the Housing Corporation are subject to the legal constraints imposed by the Housing Associations Act 1985 and are likely to be funded substantially either by Central Government through the Housing Corporation and/or partly by local authorities themselves. Despite that, housing associations would not happily be termed 'State' or 'public' authorities because they lack one or more of the criteria mentioned. Similarly public

281

limited companies are created under statutory authority but, unlike Government authorities, exist primarily for the private benefit of their members as shareholders.

Setting aside, for the moment, the remedies available in respect of direct or indirect discrimination, there would appear to be a distinction between housing associations as non-public bodies and local authorities, as public bodies, in the potential procedures available for reviewing a decision or process where it is alleged that the rules of natural justice have not been observed: in the latter instance the only remedy would be through application to the High Court for judicial review under Order 53 or in Scotland to the Court of Session by SI 1985/500. The court may, if the application is granted, make an appropriate order (whether or not it was sought in an application itself). This includes certiorari (reduction), declaration (declarator), injunction (interdict), mandamus (implement) and, in Scotland, damages. With regard to local authorities we have noted that the English case of R v Canterbury City Council ex p. Gillespie (1986) 19 HLR 7 demonstrates that judicial review may be used to ensure that a housing authority does not apply a rule relating to allocation inflexibly to exclude appropriate consideration of a particular application on its merit. Nevertheless in relation to housing associations, given that they will not be considered a public authority and that the statutory provisions regarding the application of rules and the provision for decision-taking are sketchy, it remains to be seen whether an applicant has a legally enforceable 'legitimate expectation' in relation to the processing of his application. Whether or not in this situation housing associations are subject to the rules of natural justice in relation to applications and allocations the fact of the matter is that the procedures adopted have often been arbitrary and the criteria used for assessment frequently unknown to applicants. In this situation an individual applicant from an ethnic minority group is faced with difficulties in assessing the extent to which his application has been judged on its merits and, consequently, whether or not, in the event of his or her failure to achieve the allocation of housing which he or she might have expected, pursuing a complaint under the terms of the Race Relations Act 1976 would be justified.

In Scotland the Housing (Scotland) Act 1986 (now consolidated in the 1987 Act) may provide a new opportunity for Scottish Homes to secure not only that the new rules required under Section 8 are submitted timeously but also that, following submission of approved rules, effective guidance is given to housing associations with regard to the procedures and practices to be followed in securing not only that each application is considered on its merits but also that all applicants are properly advised of their opportunities to make representations in the process. Although Section 27 of the Tenants Rights etc. (Scotland) Act 1980 imposed upon public sector landlords a requirement to publish any rules they had governing the allocation of their houses, including rules about transfers and exchanges, this obligation was not extended to registered housing associations until the passage of the Local Government (Miscellaneous Provisions) (Scotland) Act 1981 (by the insertion of a new subsection 1(a) by Schedule 3, para. 4, in the 1980 Act). Section 8 of the Housing (Scotland) Act 1986 substitutes new sub-sections 1(a) and 1(b) to replace and strengthen the previous provisions by the introduction of a positive obligation upon housing associations to frame allocation rules. Although permissive only, these rules had in practice been drawn up by local authorities and other landlords. Some 40 per cent of housing associations had failed to make such rules (see House of Commons Standing Committee, Col. 335) and this, in the Government's view, required change. Thus the new sub-sections require those associations which have not already published their rules to do so within six months of the commencement of that section and consequently new rules should have been submitted to the Housing Corporation in Scotland by 9 July 1987. However, by January 1988 only some 25 per cent of housing associations had submitted either the new rules or the old rules to the Housing Corporation: clearly there was no perception by housing associations in Scotland that this was a matter of priority.

As Scottish Homes supersedes the role of the Housing Corporation in Scotland with the implementation of the Housing (Scotland) Act 1988 the divergence of Scottish and English practice is likely to become more significant. Policy considerations, as we have observed, may be more important in the area of housing allocation than the legal requirements. The new financial arrangements through HAG will result in rents equating market, or near market rents, as opposed to the restricted fair rent system. The financial status of the applicant will become a more critical consideration than at present, resulting in the less privileged groups, including black applicants in mainstream housing association provision, being marginalised. In Scotland, we have noted, there are no registered black housing associations. Consequently the needs of this group in Scotland are likely to become more acute. Moreover the knock-on effect of policy guidance from the south drifting north with a tartan look will be stemmed because of the severance of the interest of the Housing Corporation north of the border. It may be difficult to predict how housing associations will adapt to the new framework and how they will accommodate the interests of special needs groups but the past record of housing associations suggests that black applicants will continue to lose out and the legal safeguards against racial discrimination will continue to be peripheral to their interests in this form of tenure.

The Housing Corporation Performance Expectations (HC, 1989) formally constitute an impressive drive to ensure that housing associations, through the adoption of a 'Fair Housing Programme' using 'all available good practice guidance', through positive action and through record-keeping and monitoring, eliminate racial discrimination. Promoting equal opportunity may be characterised as a three-phase problem: recognising and defining the issues; determining appropriate solutions; and securing their implementation. The Housing Corporation in England has advanced to stage three, at least one stage further than Housing for Wales and Scottish Homes. All will be judged on how fine sentiments are translated into tangible results.

CHAPTER 10

PRIVATE HOUSING

THE PRIVATE RENTED SECTOR

Introduction

At the time of the 1919 Housing, Town Planning, etc., Act, which introduced Central Government subsidy to local authority house-building, the private rented sector was by far the largest form of tenure, with some 80 per cent of the country's housing stock being rented. The decline in the private rented sector, heralded by local authority council house provision was accelerated after World War II and by 1981 it constituted only 12 per cent of dwellings. Although there are regional variations, even in London, which has the largest proportional number of rented dwellings, this tenure had declined to 15 per cent by 1984 and this proportion will have declined still further with higher vacancy levels in private dwellings and a continuing transfer of rented stock to properties for sale. Although there has been exponential inflation in house prices in the UK, and especially in London, the system of mortgage tax relief and other benefits for owner-occupiers has encouraged a broader spectrum of society to purchase homes and this factor together with, until the 1970s, the wide availability of council housing to let at a subsidy from Central and Local Government has generally marginalised the private rented sector. As a consequence although the Rent Acts have often been cited as causing a decline in the private rented sector by reason of their effect on rent levels, rent increases and security of tenure for tenants and their successors, it is clear from research that the reason for the decline of the private rented sector is much more complex.

Moreover, in terms of condition, desirability, safety, value for money and the provision of basic amenities the private rented sector is less desirable than other forms of tenure: although the availability of improvement and repair grants has provided some incentive to the private landlord to secure the maintenance and improvement of rented property and the provision of means-tested housing benefit to the unemployed and low-paid has ensured that a proportion of the population are able to afford fair rents, the tax system generally has not favoured the private rented sector in comparison with other types of tenure.

The 1981 census confirms that households renting privately have incomes towards the lower end of the scale. The GLC's survey of privately rented housing in London showed that pensioner households were over one-third of this type of tenure, while 38 per cent of households were headed by women and nearly three-quarters of these households were either adult or pensioner women-only households (GLC, 1986b). Black and ethnic minority households form a disproportionately large and growing number of London's homeless and those in severe housing need (CRE, 1988c). Frequently there is over-representation of this group in the worst kinds of temporary accommodation, especially in bed-and-breakfast hotels and in temporary accommodation some distance from their usual places of community residence. The GLC survey (1986b) showed that some 15 per cent of households renting in inner London were from ethnic minorities and 6 per cent in respect of outer London. The 1981 census (GLC Special Tables DT1255 and DT1264) show significant variation in the number of private tenants and the proportion of ethnic minority tenants within that number over the thirty-two boroughs in inner and outer London. However, the results of the Labour Force Survey 1985 indicate that on a national level estimates of black and ethnic minority populations that are based on birthplace, as is the case in respect of the 1981 census, underestimate the ethnic minority populations by about 30 per cent, as they do not take account of black and ethnic minority peoples born in the UK.

In 1988 the Anti-racist Housing Working Group and London Against Racism in Housing, which after the abolition of the GLC in March 1986 became re-established as the London Against Racism in Housing Group (LARH), issued a report entitled Anti-racism for the Private Rented Sector. Having surveyed available data, they concluded that there was very little accurate information available about the ethnic composition of households in the private rented sector. Thus, although it is clear that there are significant regional variations as well as significant variations within metropolitan areas, as the London survey demonstrated, the dearth of more general information necessitates a reference to area-based information informing some overview of the private rented sector as it affects ethnic minorities but the danger of making generalised and inaccurate observations on the basis of such information is acknowledged (LARH, 1988).

The GLC survey showed that on average black households paid £34 a month more for their accommodation; were twice as likely to be living in housing where conditions were classified as 'very poor'; and were more likely to have renting arrangements which fell outside the protection of the Rent Acts. Some 24 per cent of black households who rented privately had one or more children, in comparison with 8 per cent for white tenants. Such local patterns or racial disadvantage are evidenced in the private sector at a national level by the 1982 PSI survey (Brown, 1984) but, as with the public sector, the extent to which such disadvantage is caused by direct and indirect discrimination cannot be assessed with any confidence. What is clear, however, is that direct and indirect discrimination in the private rented sector is extensive.

Racial discrimination in access

In the LARH report (LARH, 1988: 3.66) seven of the London boroughs surveyed had received allegations of racist behaviour by landlords and the majority of these referred to the refusal by owners or managers of 'bed and breakfast' hotels to allow black people access. In five cases the boroughs had discontinued use of the offending hotel or, as was the case with one borough, the owner of the hotel had been warned that use would be discontinued if a similar incident happened again. The other incidents of racist behaviour reported by the boroughs included the allegation that an accommodation agency had refused a viewing of a property to a black person after she had been told over the telephone that it was available.

The LARH Report (1988: 5.7-15) also provided evidence of landlords subjecting black tenants to racial abuse, harassment and exploitation through lack of knowledge and rights. The Tower Hamlets Homeless Families Campaign provided comments about three incidents of racism:

The pattern for racial harassment which takes an overtly racist form is very clearly in response to complaints about conditions and to families organising. The management of the hotels respond with physical violence. There also seems to be a growing practice of employing one or two families, often white, as cleaners and, also, to keep an eye on other families. The use of rules and regulations, often petty and unreasonable, is also part of a pattern. Some families remarked recently that 'this is not a hotel, this is a prison'.

LARH concluded that while its various surveys had not been able to provide a comprehensive picture of the race attitudes of landlords in London they did present a disturbing indicator as to the response of those with considerable power to determine whether a household was housed or not and in what conditions. The extent to which accommodation agencies and landlords adopt a positive stance towards equal opportunity in racism is evidenced by a CRE postal questionnaire, sent on behalf of LARH, to nine large institutional landlords. There were only three responses to the survey and these displayed a 'quite shocking ignorance and complacency towards racism and equal opportunities'. LARH concluded that the survey gave great cause for concern, bearing in mind that the three replies received were from so-called respectable landlords, about the attitudes and behaviour of thousands of private landlords who do not have the public prominence of the respondents.

Between 1984 and 1987 the CRE in its annual reports referred to five cases of racial discrimination in access to private rented property where the courts found in favour of the complainants: the locations were in Bradford, Nottingham, Halifax, Blackburn and London. In a further three cases in Nottingham and London a settlement had been achieved without a court hearing. Settlements, whether by agreement or by court order, ranged form £50 to £1,000 excluding costs. Although these cases and settlements reported by the CRE are not comprehensive and there are no statistics available on individual complaints pursued in the County Courts or Sheriff Courts, given the extent of discrimination evidenced by a number of surveys conducted in the private rented sector there is little doubt that a very substantial amount of discrimination in this sector remains either undiscovered or not pursued. One factor inhibiting the pursuit of a complaint of unlawful discrimination in the private rented sector in respect of refusal to offer a tenancy is the length of time taken to determine the complaint and the negligible prospect not only of a complainant securing the accommodation which was sought in the first instance but even of securing satisfactory alternative accommodation from the individual landlord or accommodation agency concerned. Moreover the settlements achieved have tended towards the lower end of the range indicated above and, consequently, the prospective financial gains from the pursuit of a successful court action in this area do not provide any incentive on their own.

Racial discrimination in conditions

In the private rented sector unlawful discrimination may occur not only in respect of a refusal to allocate a tenancy but also in respect of the terms of the tenancy, whether relating to rent or the conditions of occupancy, the quality of property offered or the relative security of the tenancy concerned. However, just as there are incomplete and unsatisfactory records in respect of discrimination by the gatekeepers to private tenancies, so in respect of the other areas where the potentiality for discrimination arises relevant data is either piecemeal or non-existent. The Francis Committee (1971) Report of the Committee on the Rent Acts provided some general information on private tenancies, particularly in London, whether in the furnished or unfurnished sector, but this information is now out of date. In comparison with owner-occupation or the public sector the proportion of rooms or furnished accommodation in properties was small: 50 per cent of furnished cases occupied one or two rooms (66 per cent in stress areas). For the unfurnished sector 11 per cent (16 per cent in stress areas) similarly occupied one to two rooms. For both the furnished and the unfurnished sector twice as many tenants in the stress areas were born abroad than in other areas. The report included a study of rent tribunal cases which showed that of the 100 examined, thirty-two concerned West Indian applicants, fourteen African, eight Irish, one Pakistani/Indian, thirty-six of 'British' origin and twenty-three other. The various tribunal chairmen approached in connection with this study indicated that nearly half of all applicants in London were black and this percentage increased to 85 per cent in east London, 70 per cent in Lambeth and 90 per cent in Brent and Harrow. The committee observed, 'these figures undoubtedly reflect the difficulties facing immigrants, especially coloured immigrants, in finding alternative accommodation'. It should also be noted that a high proportion of landlords in the furnished sector were from ethnic minorities: 36 per cent were West Indian, 9 per cent African, 5 per cent Pakistani/Indian, 15 per cent Cypriot/Greek and the remainder from non-ethnic minority backgrounds. The studies conducted for the committee also showed that 65 per cent of landlords sharing with tenants were born in India, Pakistan, West Indies or Africa and they observed, 'many landlords let to tenants of a different nationality but rarely of a different colour' (p. 468). In brief the Francis Report demonstrated racial disadvantage in respect of overcrowding and the standard of property let and high levels of dissatisfaction regarding the conditions of let as evidenced by recourse to the Rent Tribunals. The report did not attempt to correlate racial disadvantage, however, with racial discrimination.

Amongst the West Indian population the movement to council housing and housing association tenancies and amongst the Asian population a proportional increase in council housing tenancies have been instrumental in the marked shift from private-sector tenancies amongst ethnic minorities between 1971 and 1982. The GLC Housing Research and Policy Report No. 5 (GLC, 1986b) which focused on private tenancies in London found that 64 per cent of tenancies surveyed were protected but only 21 per cent of new lettings were clearly protected within the terms of the Rent Acts. Amongst all private tenancies in London occupants were three times as likely to lack basic amenities as in other forms of tenure while 13 per cent of such tenancies had resident landlords. Ethnic minorities were twice as likely to suffer from harassment from their landlord (11 per cent in the previous year) and were less likely to be protected by the Rent Acts (42 per cent) than whites (66 per cent). As Ramsaran has observed (1988: 12), against this background black people are often to be found living in the inner city, in the oldest, most crowded accommodation, which, more often than not, is lacking in basic amenities and in need of repair. And amongst black people, dependent on this form of tenure, there are groups who are more vulnerable than others: the single person with no other option but to rent because of lack of priority on council housing lists and exclusion from the over-subscribed lists of black housing associations and co-ops; the large family on a low income facing a long wait for one of the scarce larger council units who cannot afford a mortgage; and those adversely affected because of their immigrant status. The latter include migrants, refugees and since the Immigration Act 1988 and the new immigration rules the settled immigrant with a newly arrived family. Neil Stuart (1988: 8) by reference to the new council house waiting list rules introduced by Portsmouth City Council has shown the willingness of some local authorities to respond readily to the implications of the Immigration Act 1988 by disqualifying all waiting list applications where either the applicant or any member of the household has been 'domiciled outside the UK within twelve months immediately prior to the date of application': clearly such rules force low-income applicants into the private sector. It is against this background, then, of economic and social vulnerability that the Race Relations Act 1976 has to be gauged in its attempt to combat racial discrimination in the private rented sector.

Discrimination by accommodation agencies

The PEP research conducted in 1973 clearly indicated that a number of accommodation agencies did discriminate (Smith, 1977). Tests carried out in London and Birmingham indicated that 27 per cent of Asian and West Indians were treated less favourably when making telephone applications in response to advertisements offering rented accommodation. BBC testing for the programme 'Black and White' (1988) indicated a 30 per cent prospect of less favourable treatment in the private rented sector in Bristol and complaints received by the Race Relations Board between 1969 and 1977 suggested that discrimination was widespread. The two investigations carried out by the CRE in 1978/79 and 1980 in respect of Midda and DS Services and Allen's Accommodation Bureau respectively led the CRE to conclude that many accommodation agencies still blatantly discriminated against black and other ethnic minority applicants. In March 1988 the Tenancy Relations Services of Hammersmith and Fulham LBC submitted a progress report to committee which included a brief report of a survey undertaken on their behalf to test the extent to which racial discrimination occurred at the enquiry stage at some of the accommodation agencies serving the borough. Discrimination was deemed to have occurred where there was any discernible difference in either the information given and/or treatment received by the two matched applicants. Differences were indicated by whether any offer of accommodation was made; the number of addresses offered; whether any invitation to register with the agency was made, and the attitude of the agency's employee towards each applicant. The number of indices used (four) and the agencies approached (ten) allowed a total of forty possible instances of discrimination to be measured and in nine of these it was found that discrimination had occurred. The discrimination experienced included failure to offer accommodation (two), offering fewer addresses to the black applicant (two), not inviting the black applicant to register (one) and dealing with the black applicant in a perfunctory or indifferent manner in comparison with a cordial and professional attitude towards the white applicant (four).

A common experience amongst applicants was to be offered a 'company let', in other words the offer of a tenancy to be let to a company which consequently would be unprotected in terms of the Rent Acts, which do not extend to companies as opposed to individuals. As a result these would not be subject to the fair rent provisions nor to secured protection against eviction at the end of the let. This resulted in additional disadvantage to unemployed people, a category where blacks are disproportionately represented, who were effectively debarred from obtaining accommodation through these agencies by virtue of not having an employer to enter into the company let agreement on their behalf. Consequently, although one should not place too great reliance on these figures, they do tend to confirm, first that there remains a high incidence of direct discrimination in accommodation agencies and, second, that the institutional practices adopted have the effect of discriminating indirectly against black applicants.

The Housing Advice Switchboard Report on London's accommodation agencies (HAS, 1983) found that many agencies were discriminating by refusing to act for black and other ethnic minority tenants, were channelling applicants to particular types of accommodation and were stereotyping black and other ethnic minority tenants as 'holiday makers'. But clearly the picture remains fragmentary.

The CRE investigation into Allen's Accommodation Bureau in Paddington (CRE, 1980) may serve to illustrate the nature of unlawful discrimination occurring. Mr Ajao, a

Scot of Nigerian ethnic origin, called the agency in 1979 in connection with the availability of a single bed-sit and was advised that the accommodation was no longer available. After his colleagues had been advised by the agency that it was available and Mr Ajao persisted with his enquiries he was advised by the agency that there was no point in sending him along to view, as the landlord would refuse him because of his colour. In July 1980 the CRE served a Non-discrimination Notice under Section 58 of the Race Relations Act on Allen's Accommodation Bureau, requiring it to comply with the Act and to issue a policy statement to all staff and landlords, new and existing.

Just as the range of accommodation available in the private rented sector is diverse by reference to quality, condition, security, price and arrangements for letting, so in the kinds of organisations which let housing in the private sector there is a vast range from the owner-occupier with a single lodger to large multiple groups of companies with thousands of properties in their ownership. While there is clear evidence of unscrupulous landlords of individual property a growing area of concern is the financial speculator intent on buying up tenanted property, removing the tenants and selling for substantial profits at a time when housing scarcity results in soaring house values. Alternatively, private landlords may turn to squalid 'bed and breakfast' accommodation and other multi-occupied property let on terms falling outside the Rent Acts and exploiting the needs of many disadvantaged people of whom blacks and ethnic minorities are a disproportionate section. Despite the Accommodation Agencies Act of 1953 there is no formal requirement for accommodation agencies to operate through any formal system of registration or qualification and consequently the principal service which they extend, introducing tenants to landlords and vice versa, lacks any effective method of control and monitoring. Currently it would seem that neither the Race Relations Act 1976 nor the Accommodation Agencies Act 1953 constitutes an effective implement to combat unscrupulous accommodation agencies and, in England Wales, the Rent Act 1977 and in Scotland the Rent (Scotland) Act 1984 are readily circumvented in order to avoid rent control and security of tenure. Although the Housing Act 1988 has provided for a 'statutory' code of practice in housing, and a draft prepared by the CRE extends to the private rented sector, the response of accommodation agencies and private landlords to their current statutory obligations is such that there can be little confidence in a code of practice having any significant effect, particularly in areas of housing stress where a disproportionate number of black residents are obliged to seek private rented accommodation.

Bed and breakfast

An increasing number of housing authorities now spend vast proportions of their housing budget in the temporary, and often long-term, housing of homeless families in bed-and-breakfast accommodation and hotels operated in the private sector. The 1985 GLC survey of such temporary accommodation (GLC, 1986a) found that 70 per cent of London boroughs (twenty-four of thirty-three) were using such provision in December 1984 and 78 per cent of households so resident were placed by six boroughs - Camden, Brent, Tower Hamlets, Hackney, Hammersmith and Fulham and Ealing - west and north-west London being most acutely affected. This constituted a rise of more than 300 per cent from 1981. The statutory Code of Guidance relating to the Homeless Persons legislation states that such accommodation should be used for as short a period as possible but the GLC survey showed an average length of stay in respect of the six boroughs with the highest usage varying from six to thirteen months (GLC, 1986a: Table 5). Net revenue expenditure had risen in Greater London from £4.3 million in 1981/82 to £12.5 million in 1984/85 (Table 9). Including hostel accommodation, expenditure for 1984/85 was £16 million - an increase of 40 per cent over the previous year.

The changes in Central Government arrangements for board and lodging payments and hostel funding, while not finalised by the date they were due to come into effect, i.e. 1 April 1989, are expected to hit hard at homeless people under 25, especially those under 18 (Inside Housing, 1989, Vol. 6, No. 7: 3). In respect of hostels, which generated half their funding from charges and the remainder from Home Office and other support, the loss of Home Office funding for the 'core' element threatens their continuing financial viability. Some younger residents would have their income cut or withdrawn altogether through being forced to change from board and lodging payments to Income Support.

Although the impact of such reductions is difficult to assess even during 1989 when transitional arrangements protected some hostel provision, significant reductions in

Government support seen inevitable. The linkage between homelessness and the lack of supply of houses, confirmed by a DoE report (Niner, 1989), suggests that alternatives for the worst-off are not available.

The statistics on homelessness do not include figures on the applicants' ethnic origin and only twelve London boroughs, at the time of the GLC survey, kept such records: in Lambeth 40-80 per cent of homeless households in temporary accommodation are black while the figure has been estimated as 90 per cent for Tower Hamlets (GLC, 1986a: 44). Very few provide cooking and laundry facilities or twenty-four hour access to rooms: few boroughs effect regular inspections or enforce a set of standards. Use of bed-and-breakfast accommodation outside the responsible borough is widespread although most hostels and short-life property were situated within the borough. For example between May and August 1985 Tower Hamlets placed nineteen Asian families in hotels in Southend (GLC, 1986a: 44). Such accommodation provides no protection in respect of security of tenure or rent, being outside the provisions of the Rent Acts. Disproportionate allocation to permanent accommodation from the waiting list and the restriction of the number of offers by a number of housing authorities result in lengthier periods in temporary accommodation and eventually poorer permanent accommodation.

Robson (1986) provides a useful general guide to the law and practice, Bed and Breakfast in London, which, in addition to identifying relevant law, also provides advice on remedial action. However, most occupants are likely to be contractual licensees, where board is neither optional nor insubstantial, and do not even have the limited protection from eviction provided by an obligation on the landlord to obtain a court order for repossession. Similarly, in respect of rent, because the board and lodging regulations set limits on what the DHSS will pay, the actual rent will not necessarily be met by benefit payments. While the housing authority may meet the difference in respect of those whom it is obliged to house as homeless, it may seek a 'reasonable amount' from the household.

The 1983/84 GLC survey Private Tenants in London (GLC, 1986b) confirmed the over-representation of ethnic minorities in the 'unprotected' as opposed to the 'protected' private rented sector. The mean average rent at £162.75 was considerably greater than that paid by white households, £128.89, and while ethnic minorities had better access to amenities, largely explained by the length of tenure of whites, a third shared amenities, in comparison with 15 per cent of white households (GLC, 1986b: 21). Ethnic minorities, as previously noted with reference to the LARH report, are particularly vulnerable to harassment from landlords when accommodated to bed-and-breakfast accommodation as well as to refusals of accommodation itself.

Local authorities and the private rented sector

Local authorities through their housing advice services, whether focusing on the Housing (Homeless Persons) provisions in the relevant legislation or otherwise, in their tenancy relations services, environmental health services and legal services may mitigate the worst practices which come to light through tenant applications. But while they may cumulatively play an important role in the enforcement of the rights of privately renting tenants and in promoting anti-racism policies in the private sector, in a climate of stringency currently applied to local authority expenditure the opportunity for developing new initiatives in these areas is bound to be limited.

Housing Advice Centres (HACs), except where required in the provision of advice and assistance in terms of the Housing (Homeless Persons) provisions, have no statutory basis, although many local authorities have provided housing advice services for a considerable time prior to the first enactment of the Housing (Homeless Persons) legislation in 1977. As LARH (1988: 24) has concluded from its survey in London, the services provided vary greatly from one borough to the next: some provide a comprehensive service covering all types of tenure and are staffed and resourced to advise or assist on the full range of housing problems from access to council housing and financial problems of mortgage default to complex areas of housing law in the private and public sectors. Other HACs provide a service exclusively for tenants in the private rented sector and do not have a direct responsibility to advise those in other types of tenure. Unfortunately there is no common practice either in the local authority sector or elsewhere in relation to the adoption of equal opportunity policies in respect of the recruitment and training of staff or in respect of the promotion of anti-discrimination policies. Few have so far adopted record-keeping and monitoring of the ethnic origin of those using the service, which would provide an important input in policy review. Although some advice centres have used Section 5(2)(d) of the Race Relations Act 1976, which enable an employer to recruit people of the same racial group where they are providing a personal service to members of their own racial group, it would appear that these provisions are under-used. Useful guidance on improving the service provided by HACs to black and ethnic minority tenants and applicants is provided by LARH (1988: 25 et seq.).

Those local authorities which do employ tenancy relations officers (TROs) assign them a specific responsibility to deal with the landlord/tenant disputes in the private rented sector, including cases of racial harassment and illegal eviction. The deployment of TROs is perhaps best developed in London, where all boroughs provide a tenancy relations service of some kind, although a number still rely on one officer in the legal department with tenancy relations as part of his or her responsibilities: few London boroughs employ more than one or two officers to provide the tenancy relations service and there is significant regional variation in the deployment of such officials outside London. The work of the TRO is shaped by the Rent Acts. In recruitment, although most local authorities ask for some knowledge of housing law as it applies to the landlord/tenant relationship, few TROs have formal training in housing and the majority are expected to learn 'on the job'. As yet there are no regular external training courses available although the Association of Tenancy Relations Officers held its first course in April 1986 and it is understood that these are being developed on a regular basis (LARH, 1988: 28).

Environmental health officers (EHOs) have a broad statutory remit which includes the enforcement of standards of food hygiene, home and occupational safety, the prevention of infectious diseases and the restriction of noise, air and water pollution. Thus their responsibilities for ensuring the maintenance of statutory standards of public health and housing are merely one of their functions. The vast majority of EHOs are now recruited and employed directly by local authorities and it is estimated that there are some 6,000 EHOs in Great Britain with 950 employed in London. A survey by the Institution of Environmental Health Officers (IEHO) in 1986 found that fewer than 1 per cent of EHOs were from the black or other ethnic minority populations. Despite the fact that the work of EHOs includes close contact with the public and a right on demand to enter private houses or business premises to carry out their statutory duties the course content for training EHOs concentrates on the technical and legislative aspects of their work and there is a significant lack of attention to inter-personal skills and to the racial dimension of their work. Partly in recognition of these difficulties, in October 1984 the Institute and the Commission for Racial Equality published a joint report, Race and Environmental Health (CRE, 1984c), containing a number of progressive and practical recommendations for the service but there has been little progress within the majority of local authorities in the implementation of its recommendations. The IEHO has set up a working party on equal opportunities but a number of the recommendations remain unimplemented.

Although legal action arising from tenant/landlord disputes may be initiated by a local authority, its housing advice services, its tenancy relations services or its environmental health services, it is most frequently the local authority's legal officers who have to decide, on legal grounds, whether a particular case should be pursued. Consequently the attitude, training and professional skills of local authority solicitors will be an important factor in the local authority response to difficulties in the private rented sector. However, solicitors in local authorities are no different with regard to their professional background and training from solicitors in the private sector, where the institutional responses of the legal profession to racism and race relations legislation, addressed in Chapter Eleven, will be a relevant consideration. The fact that there is significant under-representation of blacks and ethnic minorities in the legal profession and that professional training both in England and Wales and in Scotland emphasises the more lucrative (or 'commercially viable') aspects of the profession is at a cost. What in financial terms are more marginal interests such as welfare law, housing law and the law relating to equal opportunities are frequently neglected. This results in local authority solicitors being ill-equipped from the aspect of training, if not from the aspect of personal professional experience, to pursue complaints relating to discrimination in the private rented sector. In Scotland the decision as to whether or not to take criminal legal proceedings will not depend on the local authority solicitor but on the local procurator fiscal, but that body of solicitors will be no better equipped in housing and anti-discrimination law than the local authority solicitor.

The rent officer

The rent officer service was established by legislation under the Rent Acts in the 1960s with the function of setting fair rents in respect of private regulated tenancies, whether furnished or unfurnished, and of accommodation provided by housing associations, trusts or co-operatives. Lodgers and licensees were excluded from their terms of reference by reason of such lettings being unprotected by the Rent Acts. Although, unlike the environmental health officer, the rent officer has no statutory power to ensure that improvements are effected to any property, he may request the landlord to carry out improvements and/or repairs as a prerequisite of setting or changing the registered rent.

In England and Wales the rent officer's services are geographically defined by borough boundaries, and staff are recruited and employed by individual local authorities. They are officers of the Crown and the costs incurred by each local authority in running the service are reimbursed by the Department of the Environment. In Scotland, rent officers are employed directly by the Crown and are located in one of the four Scottish Offices (Dundee, Aberdeen, Glasgow and Edinburgh). No formal qualifications are required for appointment and once appointed the officer's main source of training is work experience. The Institute of Rent Officers runs an annual one-week course for both new and established rent officers and its education trust publishes a manual of legislation which is annotated and cross-referenced to assist the rent officers. In Scotland there is no provision for anti-racism training for rent officers. In neither jurisdiction is there a comprehensive equal opportunity policy in respect of the service provided.

Pressure groups in the private rented sector

Although there has been an extension and growth of community action with the development of many private tenants' associations, umbrella organisations and pressure groups, frequently these bodies have been exclusively white. LARH classifies such groups as primary, the individual tenants or residents' associations as secondary, these being local umbrella groups such as the Federation of Private Tenants and Residents' Associations in Camden, and as tertiary, being the strategic umbrella and pressure groups such as the Organisation for Private Tenants, the Houses in Multiple Occupation Group and the Federation of Private Residents' Associations. Following a survey of secondary and tertiary groups, the London against Racism in Housing report (LARH, 1988: 56) found that few groups had equal opportunity policies or codes of practice for employment. The evidence showed that the majority of groups (five out of eight) did not have a specific equal opportunities policy for their organisation and none had a specific code of practice: this was particularly disturbing in view of the fact that all but one of the groups was in receipt of local authority grant aid and the absence of adopted policies and a code of practice may have meant that they failed to meet the conditions attached to the grants, calling into question the monitoring procedures adopted by the local authorities concerned. None of the groups conducted any systematic ethnic monitoring or kept records of their membership and the generally disappointing picture regarding most groups' membership from black and other ethnic minorities was reinforced by the lack of positive action on recruitment. Black members of one group said that racism gave landlords extra power over tenants and that landlords had racially abused tenants. LARH concluded that there was a clear and urgent need for pressure groups to develop specific anti-racist policies and practices and to ensure their promotion and implementation. Although three of the eight groups covered by this survey claimed to have anti-racist practices only one had a black private tenants worker: the majority had done very little and in some cases almost nothing to challenge racism in the private rented sector and a fundamental change in attitudes was required by both the groups and their funding authorities.

Conclusions

Despite the lack of systematic evidence relating to the experience of black people of racial discrimination in the private rented sector, that evidence which does exist, whether in relation to direct or indirect discrimination, demonstrates that the Race Relations Act 1976 has proved ineffective in preventing unlawful discrimination and in securing effective enforcement where such discrimination has occurred. Moreover, those institutions which are critically involved in the process and may institute measures to counteract discrimination whether by promotion of anti-discrimination training or policies or the monitoring of enforcement provisions, are ill-equipped to combat racial disadvantage in the private rented sector. In 1987 Government published its proposals for changes in the private rented sector (Cmnd 214) in order to reverse the decline of the past sixty to seventy years. The Housing Act 1988 and the Housing (Scotland) Act 1988 will seek to do this by reducing what the Government believes are unnecessary controls upon private landlords, discouraging them, in its opinion, from letting their property. In addition to private landlords being enabled to take over council properties and tenants, Government aims to transform the private rented sector by the creation of two new forms of tenure - assured tenancies and assured short-hold tenancies in England and Wales and the equivalent in Scotland with new definitions of security of tenure, grounds for possession and methods of setting rent levels. There is some doubt as to whether landlords who operate licence agreements and company lets will continue to use this form of tenure or whether they will allow their lets to become assured or assured short-hold tenancies, although the system of providing meals or attendance will not prevent the creation of any assured tenancy. There is an improvement in respect of the provisions relating to harassment by landlords but the lack of security of tenure over any lengthy period and the creation of market rents in respect of the assured tenancies, of both kinds, together with the simplified grounds for repossession are likely to increase the gentrification of many parts of the inner cities and the black and ethnic minority tenants are bound to be disproportionately adversely affected by such provisions.

Although either party to the tenancy agreement may apply to the Rent Assessment Committee to set a market rent which will be the maximum chargeable, there is every prospect that if a tenant does apply for a market rent to be set the landlord will simply refuse to grant an extension of the tenancy, which, in respect of a short-hold tenancy, may be as little as six months. It will no longer be illegal to charge a premium on entry and such premiums are bound to affect the unemployed or those on low wages. Previously assured tenancies in England and Wales were restricted to registered landlords and some form of control was at least possible: this will no longer be the case. Although tenants who are currently protected by the Rent Acts will retain most of their protection so long as they remain in their present accommodation, even moving to another room in the same building may result in the tenant losing his existing protection and becoming instead an assured tenant. Given that it is those on low incomes who cannot afford mortgages who will be forced to remain in the private sector, the availability of housing benefit will become critical in relation to the new private-sector tenures but Government has made it clear that housing benefit will be restricted to reasonable rents from April 1988, from which time a ceiling on rents will be set. Local authorities are no longer to have the power to refer high rents to the rent officer or Rent Assessment Committee in order to have the rent fixed and with the introduction of strict rules on eviction for rent arrears more and more people are likely to be made homeless. Ramsaran has concluded (1988: 10) that the new legislative provisions will not improve conditions for the vast majority of tenants dependent on the private rented sector but will simply compound the problems already facing them and increase the vulnerability of those most in need (see also Mullins, 1989).

THE OWNER-OCCUPIED SECTOR

Introduction

The 1982 PSI Study (Brown, 1984: Tables 31 et seq.) confirmed the importance of the owner-occupied sector in housing, indicating that 78 per cent of Asian known new households were in the owner-occupied sector. The percentage for the West Indian population of known new households, at 28, was significantly less than the known new white households, at 58 per cent. The Government's commitment to increasing the owner-occupied sector by means of restricting local authority new build, facilitating local authority council house sales and the differential financial contribution to owner-occupiers through mortgage tax relief will increase the emphasis on this form of tenure. Accordingly the impact of the Race Relations Act 1976 on inhibiting unlawful discrimination in the owner-occupied sector is of great importance to an increasing number of ethnic minority households. As noted in Chapter Two, the only significant exception applying to the owner-occupied sector from the provisions of the 1976 Act relate to individual private sales which are not advertised where the sale is negotiated by the individual owner-occupier with the prospective purchasers. As the number of sales falling within this category is negligible the exceptions from the Race Relations Act in respect of owner-occupied tenure are not thought to be significant. The Race Relations Board and the Commission for Racial Equality have received complaints relating to direct discrimination by individuals whether as owners or as neighbours applying pressure to discriminate but the major gatekeepers of this form of tenure are likely to be the estate agents. In Scotland solicitors, who are entitled to describe themselves as estate agents, have traditionally sold property as well as dealing with the legal side of house transfers.

Estate agents

According to a 1987 Consumers' Association survey (Which?, 1987, April: 172) nine out of ten sales in England and Wales are effected through estate agents of some description. As a consequence estate agents represent critical gatekeepers in the process of access to housing for the whole population, let alone ethnic minorities.

Although the Estate Agents Act 1979 provides by Section 22 a minimum educational standard and training for estate agents this provision has remained dormant because of the Government's view that its application would be anti-competitive. Currently there is nothing to prevent anyone being a double-glazing salesman one day and an estate agent the next. Recently a local trading standards department blacklisted an estate agent when he was found guilty of credit card fraud. It was discovered at his trial that the man had a lengthy criminal record involving many aspects of fraud, theft and shoplifting. The trading standards officer involved observed:

> it just shows that no checks exist about who sets up an estate agency business. It is a sort of negative vetting - we blacklist after something goes wrong. This man had been operating for some time as an Estate Agent handling a lot of people's money, yet no-one was aware of his past (MacDonald, 1988).

There is widespread dissatisfaction with the lack of control over the training and vetting of estate agents and also widespread dissatisfaction with the service provided to the general public. With the passage of the Building Societies Act 1986 and the Financial Services Act 1986, financial institutions have bought hundreds of estate agencies in the past few years as 'front offices' for insurance and mortgage business and this fact has contributed to, if not actually caused, a decline in professional standards. It is estimated that the financial giants own or control a third of estate agent businesses in England and Wales. In July 1988 the Department of Trade and Industry held the first of two meetings with the profession's leading representatives to try to establish a new code of conduct for estate agencies. However, until significant changes are effected the present system of control and supervision does nothing to discourage shady practices or encourage professionalism. However, Section 180 of the Local Government and Housing Act 1989 enables the CRE to produce a statutory code of practice affecting all estate agents and institutions that provide housing mortgages. The CRE sees the period of consultation and this new code as an important educative process (Inside Housing, Vol. 6, No. 6, 1989: 5) building on its recently revised guides (CRE, 1989c, d).

Currently Section 3 of the Estate Agents Act provides that the Director of Fair Trading may make an order prohibiting any person from engaging in estage agency work if he considers that person to be unfit to practise on one or more of the grounds specified in that section. These grounds are widely drafted and include a finding of discrimination committed in the course of estate agency work. The first Schedule to the Act defines discrimination as including discrimination on the grounds of gender or race. A person shall be deemed to have committed discrimination for the purposes of the Act where, inter alia, there has been a finding of discrimination in proceedings under Section 57 of the 1976 Act (the enforcement provisions relating to the provision of goods, facilities or services), where a Non-discrimination Notice has been served on the person concerned and it has become final or where he has been subject to the restraints of an injunction or order granted against him. It should be noted that while the Director of Fair Trading also has power to issue Warning Orders in respect of contraventions of Section 3 this flexibility does not extend to a finding of discrimination, where a

Prohibition Order alone may be made. As at August 1988 the Office of Fair Trading has advised that no orders have been made in respect of unlawful discrimination although such an order was under consideration in respect of the Richard Barclay formal investigation completed by the CRE in 1988 (CRE, 1988b). Accordingly, where there has been a finding of unlawful discrimination against estate agents by the courts or the issuance of an unchallenged Non-discrimination Notice by the CRE, the former being evidenced by Akbar v Morris Homes, 1985 (CRE, 1986a: 49), the Director of Fair Trading has exercised a discretion not to use his powers under Section 3. In this respect the failure to provide the Director with a power of issuing a Warning Order under Section 4 of the Act in respect of a finding of discrimination under Section 3(1)(b) is a restriction of flexibility and prevents 'a shot across the bows or a warning signal' as envisaged at the time this Private Member's Bill was introduced (Standing Committee C, 26 April 1978, Cols. 59-60). It should also be noted that both Prohibition and Warning Orders are recorded on a register open to inspection by members of the public. The offender, therefore, has no time limit within which to remedy conduct specified in the Warning Order prior to its registration. The 'warning signal' is effectively therefore a sanction which might, in some instances, be seen as heavy-handed. Consequently, it can be seen that the Estate Agents Act 1979, Section 3 and 4, as currently drafted and deployed by the Director of Fair Trading is either under-used or inappropriate and ineffective.

Although there is clear evidence of discrimination in house sales by individual sellers both directly and by inducement and of pressure to discriminate put on estate agents, as well as evidence of estate agents themselves adopting discriminatory practices to exclude house sales to ethnic minority applicants in specific areas, in Britain there is no systematic discrimination by estate agents as a written or unwritten professional code of practice, as witnessed in the United States up to the 1960s. This may, in part, be explained by the more recent settlement patterns of ethnic minorities in this country and, despite continuing degrees of segregation in residence here, the absence of any clear-cut financial incentive either to maintain emerging patterns of residential segregation or to be involved in 'block busting' - the practice of creating panic selling of houses in areas for fear of rapid change in property values by an influx of black purchasers. Gentrification of areas, and in this process the movement of whites back into the inner cities, thereby excluding blacks by reason of economic segregation, has already taken place and is not inhibited by planning legislation, whose principal concern has been regeneration of the physical rather than social fabric of such areas, nor the policies of the Conservative Government of the 1980s, whose barely qualified faith in private capital and the supposed filtering effect of improved housing provision has encouraged the process. The role of estate agents in this process of gentrification is largely peripheral and direct discrimination by the selling or purchasing agent would not result in economic gain. Accordingly the experience of the US in this respect is unlikely to be replicated in the UK. Consequently the potential for controlling estate agents is greater despite the current evidence of lack of professionalism and the unscrupulous practices associated with false property descriptions and gazumping being commonplace in England and Wales. It would, therefore, be possible for a legislative attack on such practices to include the provision of a more convincing code of practice against unlawful discrimination, with appropriate sanctions being a required rather than a discretionary response.

Mortgages

The 1982 PSI Survey (Brown, 1984: Table 49) indicates that both West Indian owner-occupiers (78 per cent) and Asian owner-occupiers (73 per cent) rely more heavily on a current mortgage than the white population (54 per cent). We have already noted that the high level of Asian owner-occupation was in part a response to the limited opportunity to find rented accommodation in the public or private sector and was often characterised by the outright purchase of relatively cheap poor-quality housing. Over the years the nature of owner-occupation among Asians has become more like that within the rest of the population and the proportion of loans and mortgages being paid off has been balanced by a decline in the proportion of those poor-quality outright owned properties. While the proportion of outright owners among Asians has remained static, amongst West Indians it has grown as loans and mortgages have been paid off. The majority of whites, Asians and West Indians are buying with money borrowed from a building society, although other sources of finance are used more commonly by Asians and

West Indians than by whites (Brown, 1984: 79). Although a higher proportion of both West Indians (28 per cent) and Asians (17 per cent) rely on local authority loans than whites (12 per cent) and the Asian purchaser is more likely to obtain a loan from his bank (12 per cent) than the white population (4 per cent) the significant dependence on building society loans in respect of both West Indian (65 per cent) and Asian (68 per cent) communities renders such societies important facilitators in the housing process. A comparison between the PEP survey in 1974 and the PSI survey in 1982 demonstrates that building societies have become increasingly important for both West Indian and Asian purchasers while all other forms of loans for house purchase have diminished in both communities over this period (Brown, 1984: Table 62).

The nature of tenure change in ethnic minority communities cannot be viewed in isolation: the transition from private rental to owner-occupation and to public housing represents, along with suburbanisation, one of the most important transformations of the housing market and residential space in Britain since World War II (Hamnett and Randolph, 1986: 122). The private rented sector declined from 7.1 million dwellings in 1914 to 2.9 million units in 1975, a loss of 4.2 million dwellings, and no fewer than 3.7 million or 88 per cent of these were sold for owner-occupation, which accounted for 41 per cent of the total growth of owner-occupation in that period (DoE, 1977: Vol. 1, Table 1.24). Critical in this process has been the role of differential patterns of investment and disinvestment and the part of institutional lending policies within this process (Harvey, 1978; Boddy, 1981; Williams, 1976). But while Central Government policy and that of the major financial institutions will be key factors in the formulation and transformation of the housing market it must also be acknowledged that large commercial landlords, while operating within this structure, will adjust their behaviour to sub-market characteristics. This behaviour, in turn, structures outcomes with respect to the renter, the maintenance of the housing sector, reinvestment and disinvestment, neighbourhood decay and the like (Harvey and Chatterjee, 1974: 32). Hamnett and Randolph in their study of flat break-up in the market in central London have demonstrated that the characteristics of particular landlords in specific sub-markets such as central London may result in substantial regional variation in tenure transition within the broader market conditions. While in general the privately rented sector is 'a predominantly small scale tenure run by late middle-aged or elderly individuals who own one or, at most a handful of tenancies' (Kemeny, 1981), Hamnett and Randolph (1986) demonstrated that two-thirds of the 500 landlords identified in the study area in central London (who owned an average of seventy flats apiece) were property companies who owned 77 per cent of the total number of flats. They calculated a 30 per cent vacancy rate in 1981 in comparison with an 8.8 per cent rate in 1971. They concluded that the vacancies arose from a combination of two forces: first, the capital gains to be derived from sales for owner-occupation and, second, the unwillingness of landlords to re-let upon vacancy, given both the financial rewards from sales and the existence of security-of-tenure legislation. Given the existence of assured tenancies and short-hold tenancies from 1980 and the wide deployment of company lets, board and lodging provision and licences, there must be some doubt as to how important this last factor has been. Whether or not the 1988 housing legislation with the revamping of assured and the introduction of short-hold assured tenancies will result in a decrease in such vacancies is uncertain but the change in succession rights to protected tenancies along with the increased capital gains to be achieved from sale must surely more than offset any 'unlocking' of vacant flats for renting purposes and result in a net gain for owner-occupation tenure. Consequently, even in central London, where the private rental sector has been most buoyant, a further diminution in this form of tenure may be expected and demands on the financial lending institutions and, in particular, building societies and banks, will continue.

In their examination of gentrification in the United States, Legates and Hartman (1986) concluded that gentrification had already touched significant numbers of minorities. It was occurring for two reasons. First, in some cities in which gentrification had already progressed through the most desirable white neighbourhoods, it was reaching into more deteriorated, primarily minority neighbourhoods. Second, minorities constitute a significant sub-population in many predominantly white gentrifying neighbourhoods, and, in a number of cities, minorities have been disproportionately displaced from racially mixed neighbourhoods. Although the nature of polarisation of segregation in English cities is distinct from that of cities in the States, the process of gentrification will have similar structural effects. Working-class residents of decayed inner-city neighbourhoods are faced with a Catch 22 situation (Schaffer and Smith, 1984). These residents are ghettoised in areas of economic deprivation and social malaise as well as physical decay, and the influx of capital and social resources is the first prerequisite for improving the quality of life. Without an influx of capital the decay will continue, yet with a large scale investment of capital in these neighbourhoods, and the fashioning of attractive communities, present residents are pushed out to housing that is not appreciably better than that which they have left. Either way they lose; if the city manages to solve some of its housing problems, it is at the expense of surrounding municipalities. The poor remain poor wherever they are moved (Williams and Smith, 1986: 222).

The bullish nature of house price rises in London and the South East prior to 1989 and its disproportionally adverse impact on low-income and black groups are not totally unrecognised by Government. Mr William Waldegrave, the then Minister responsible for housing, said,

> what we must do is escape from thinking that there is no need to take social engineering decisions. I think we have to take a deep breath and say that Government has to get involved. We are committed - within limits, it cannot be done everywhere - to try to maintain mixed communities in London and that is going to mean low-cost housing (Roof, 1988: 20).

Means testing on discretionary improvement grants, currently proposed, is projected as more sensible and sensitive targeting to those in need. However, there is little current evidence that the building societies share the view that in the longer term prices will cease to rise and there is no shortage of institutional funds either for inner-city regeneration or in respect of new build in expanding suburbs.

Building societies: the Rochdale investigation

The report of the CRE's formal investigation, Race and Mortgage Lending in Rochdale (CRE, 1985c) illustrated the importance of the operation of the mortgage market to the Asian community. Although the legal framework within which lending institutions operate has been affected by the Building Societies Act 1986 and the Financial Services Act

1986, and the general housing market will have been affected by the passage of recent housing and rent legislation, the CRE's examination of loan practice in Rochdale between 1977 and 1981 provided an indepth study of local practice which will have continuing relevance. The investigation, being prompted by concern about lending for house purchase in inner-city areas where the majority of Britain's ethnic minority communities live, as exemplified by studies by Karn (1976), Lambert (1976), Weir (1976), Boddy (1976), Duncan (1977), and Karn (1978), examined the lending policies and practices of the local authority and the building societies with branch offices in Rochdale. Research published in the 1970s had shown that Asians and West Indians were not obtaining building society mortgages to the same extent as whites. Ethnic minorities were instead using more expensive sources to buy their homes. The properties on which ethnic minorities were more likely to apply for mortgages were situated in inner-city multi-racial areas, being older, cheaper, terraced properties without front gardens and often in a poor state of repair. The studies indicated that the building societies were reluctant to lend on such properties and in some cases whole areas had been 'red lined' as unsuitable for lending. Studies in Birmingham (Lambert, 1976), Huddersfield (Duncan, 1977) and Leeds (NHPRA, 1976) suggested that, in addition to property characteristics, building societies were reluctant to lend because of the presence of a West Indian or Asian population in the area. Evidence from the Leeds study showed that the applicants for mortgages were refused on the grounds that 'immigrants' lived in the area (CRE, 1985c: 1). A more comprehensive study of building society lending in Leeds was completed for the Leeds Community Relations Council in 1981 (Stevens et al., 1981). This concluded, inter alia, that, if anything, the ethnic origin of the applicant had at least as great an impact on lending decisions as the ethnic characteristics of the area in which they were buying.

The CRE investigation was initiated in April 1979 under Section 49(3) of the Race Relations Act 1976 and, as a general investigation, its aim was to assess the extent of the disadvantage, if any, in obtaining mortgages experienced by the Asian community and, if there was any, to determine whether it was the result of direct or indirect racially discriminatory policies and practices operated by lending agencies and to make appropriate recommendations. In 1981 the population of the Metropolitan Borough of Rochdale was just over 200,000, of whom 10,500 (5.2 per cent) were living in households headed by a person born in the New Commonwealth or Pakistan. Eighty-six per cent of Asians, the Pakistanis being the largest group, lived in Rochdale town, which contained approximately half the total population of the borough: 58 per cent of the Asian community lived in the five wards comprising the town centre. Sixty per cent of the Asians fell in socio-economic group D/E (manual unskilled), in comparison with 19 per cent of the white population, with evident implications for the type of properties the Asian community could afford to purchase. In 1978, 84 per cent of the Asians in Rochdale were owner-occupiers, with only 5.6 per cent in council accommodation in comparison with 53 per cent and 39 per cent, respectively, of the white population. Over the period examined 28 per cent of Asians, in comparison with 10 per cent of whites attempted to obtain a mortgage.

With reference to property type the Asian community in Rochdale lived predominantly in old terraced properties which lacked front gardens, the CRE providing the following breakdown (1985c: 4):

1. Eighty-six per cent of Asians lived in terraced properties, compared with 45 per cent of whites. Conversely only 11 per cent of Asians, compared with 42 per cent of whites, lived in detached or semi-detached properties.
2. Forty-six per cent of Asians lived in pre-1919 built property, compared with 30 per cent of whites.
3. Fifty-nine per cent of Asians lived in properties which lacked front gardens, compared with 25 per cent of whites.

In 1976 Rochdale Local Authority embarked upon an extensive area improvement programme in areas of poor housing in the borough. A series of community-based Action Areas had been deployed, utilising Housing Action Areas, General Improvement Areas and Environmental Improvements. Forty-five per cent of Asian applicants for mortgages through the Support Lending Scheme wanted to purchase properties in Improvement Areas, compared with 4 per cent of white applicants. Between 1977 and 1981 building societies provided 72 per cent of loans in Rochdale, the local authority 14 per cent and bank lending 7 per cent: other lenders including insurance companies, solicitors, relatives, etc., provided the remaining 7 per cent.

At the lower end of the property market and of particular significance to purchasers was the operation of the local authority home loan scheme but by 1981, owing to the cutback in public expenditure this source of finance had been restricted by the local authority to loans made to applicants threatened with homelessness. Largely in substitution for the home loan scheme and as a response to the limits placed by Central Government upon the mortgage lending of local authorities in the mid-1970s, the Support Lending Scheme was introduced. The scheme was introduced nationally in 1975 by agreement between the Government, the building societies and the local authorities, with the purpose of helping into home ownership 'people of modest incomes and limited capital who were often seeking to buy older, cheaper houses especially in the inner urban areas' (DoE, 1978). A survey conducted by the Commission during its investigation (PAS, 1981) demonstrated that the greater the proportion of Asians in specific polling districts the less the proportion of loans provided by the building societies and the greater the proportion provided by banks and by the local authority (CRE, 1985c: Tables 5-7). Although the investigation identified two instances where direct discrimination was thought to have occurred the principal reason for the investigation was to uncover practices and policies the effect of which was indirectly to affect Asian applicants adversely and the Commission concluded that the following three fell into this category:

1. The policy of not lending on properties without a front garden.
2. The policy not to lend on properties below a specified purchase price, and
3. Policies not to lend on properties in particular areas of Rochdale.

In respect of the first the survey showed that 58 per cent of Asian mortgage applicants in Rochdale applied for loans on properties without front gardens, compared with 20 per cent of white applicants. A similar analysis of support scheme applications for the period 1978-79 showed that 84 per cent of Asian mortgage applicants applied for loans on properties without front gardens, compared with 22 per cent of white applicants. Since the principal building societies operated practices of not lending on such properties Asian applicants would be considerably more likely to be refused a mortgage than white applicants. The Commission also examined whether or not this practice was 'otherwise justifiable' and would not therefore be unlawful in terms of the Act. It had been argued by one building society (the Provincial Building Society) that properties without gardens were likely to lose their value in comparison with those with gardens. Having examined the experience of other lending agencies and having sought the opinion of professionals such as valuers, estate agents and building society managers, the Commission concluded that properties without front gardens would not incur greater risk and this was confirmed by the experience of other building societies which had afforded loans on such properties, where no greater incidence of repossession or financial loss occurred during the period of study. Accordingly those building societies which used such rules of thumb and did not ensure that the circumstances of each application were assessed were likely to be in contravention of the Race Relations Act. Moreover, if a local authority were to accept the requirement of a building society participating in the Support Lending Scheme for properties with front gardens then the council would be knowingly aiding an unlawful practice and would be contravening the Race Relations Act itself.

Similarly the Commission demonstrated that the practice of refusing to lend on properties below a certain value had a disproportionately adverse affect on the chances of Asian applicants obtaining mortgages. On the question of justifiability, the Commission had examined the experience of the Halifax Building Society in lending extensively on properties priced below £5,000 during the period 1977-81, which showed that the society did not incur any greater repossession problems and no actual losses during the period. In like vein the local authority lent substantially on the cheapest properties and did not incur losses, despite the fact that the council was providing a social service rather than a commercial one. The CRE also noted that all but one of the individual applicants who had been refused by the societies on the grounds that the property price was too low later obtained mortgages from the local authority or from the Halifax on the same property. In the light of such findings and taking on board the commercial, prudential and statutory considerations applying to building societies the Commission found that this practice was not justifiable in the context of lending in Rochdale and that such

313

discriminatory practices were therefore unlawful where practised by a specific building society and, moreover, a local authority's acceptance of such instructions from a building society would constitute knowingly aiding an unlawful discriminatory practice.

In respect of lending in particular areas of Rochdale there were economic, religious and cultural constraints upon the Asian community's choice of district in which to live, resulting in their disproportionate representation in the inner areas: where building societies had policies or practices of not lending in improvement areas or in multi-racial areas, then the Asian applicants would be disproportionately affected. The Commission examined the Halifax pattern of lending over the period 1977-81 within the inner multi-racial areas of Rochdale and concluded, with the exception of one area (Sparth Bottoms), which lacked sufficient data, that in view of the fact that there were no repossessions or losses the practice of applying a requirement or condition that properties in the inner areas or in improvement areas should not be considered for a mortgage constituted indirect discrimination which could not be justified in the context of Rochdale. It was noted that a decision to refuse loans on properties located in multi-racial areas on the grounds of the racial composition of the population would constitute direct discrimination where the question of justifiability would not arise. The Commission also noted that any policy not to extend loans to joint family applicants, while not significant in Rochdale, would be so in other areas of the country such as Bradford, Birmingham or London. Such a policy would be indirectly discriminatory and on the evidence available would not be justifiable on economic, commercial or grounds other than race.

The Commission made a series of recommendations to lending agencies (CRE, 1985c: Chapter 5) which were intended to be a comprehensive set of guidelines for good practice in the field of race and the provision of finance for house purchase. The agencies directly involved responded positively to the recommendations and consultations and discussion followed between the CRE, the Building Societies Association and the Local Authority Associations, amongst others, to ensure that the lessons learned from this investigation were applied more generally.

The only legislative proposal stemming from the investigation was that the Chief Registrar should be given authority to take action where a building society had been proved to have committed racial discrimination, similar to the authority provided to the Director General of Fair Trading under the Estate Agency Act 1979. Given that building societies have not only refused outsiders access to their data but have also failed to carry out or publish their own analysis of the impact of lending decisions on racial minorities (Henderson and Karn, 1987: XXI), some supervisory mechanism is clearly desirable but in the light of the failure of the Estate Agency Act to have any impact on the practices of estate agencies in respect of equal opportunity policies and the promotion of good professional practice, such legislative change could not have been expected to produce significant dividends. Recent legislation heralds a change in direction for many building societies although they may well continue to see themselves as financiers and facilitators rather than as housing managers (Cole and Wheeler, 1987: 67). However, in the debate leading up to legislative change the focus was on a narrow 'financial and commercial framework' and issues of race, gender, discrimination, homelessness and debt were not addressed (Hawes, 1987). Specifically provisions enabling Prohibition Orders to be taken out against discriminating building societies were overlooked in the Building Societies Act 1986, the CRE failing to make representations in the matter. However inadequately local authorities may have addressed issues of racial discrimination in housing, both legal and political accountability allows some prospect of improvement: in contrast the building societies have few legal responsibilities reflecting the needs of their clients, particularly in respect of low-income groups, of whom ethnic minorities form a disproportionate number. The bottom line of financial gain is an inauspicious base on which to build social responsibility and accountability.

The Building Societies Association

The Building Societies Association (BSA), as its title suggests, is a voluntary association of member building societies. As such it operates as an advisory umbrella body with limited sanctions against members. The guidance which it does offer, by way of circulars, marketing briefs and memoranda, tends to be couched in advisory rather than directory tones and the extent to which the advice is followed by individual societies is often a matter of local discretion. Following the CRE report on Rochdale the BSA issued a circular (No. 3425 of 27 November 1986) for the purpose of reminding societies 'of the law relating to race relations, to inform them of the CRE's findings and recommendations in the report and to give guidance to those societies who may wish to adopt them'. This circular reiterated the CRE recommendations relating to the formal adoption of an equal opportunities policy and appointment of a senior officer, the monitoring of the ethnic origin of all mortgage applicants, the review, by building societies, of lending policy and criteria for assessing mortgage applications to ensure that discrimination was not occurring and the provision of guidance and training to staff concerned with the allocation of mortgage finance on the implications of the Race Relations Act 1976 and the equal opportunities policy. The circular also made reference to the desirability of providing information about lending criteria in different languages, the BSA referring to its own pamphlet entitled 'A Guide to Savings and House Purchase', which it had issued in various languages. Further advice was offered in relation to the viability of special initiatives with local authorities and the Local Authority Loan Guarantee Scheme (Section 442 of the Housing Act 1985), where the CRE had suggested that the society should use the scheme in situations 'where the security is in the valuer's opinion marginal and where, without the safety net of the guarantee, the society would refuse to lend'. The circular also referred to regular consultation between the local authority and the building societies and other lending agencies in Improvement Areas, focusing on the progress of improvement strategies. In this regard the CRE had welcomed agency services which were comprehensive schemes to help people, especially the elderly, to improve their homes, including employing the builder and monitoring his work and arranging necessary loan and grant finance.

The BSA has not carried out any follow-up to this circular and is unaware of the extent to which member societies have implemented the various recommendations. Consequently building societies are largely left to their own devices and the BSA cannot be seen as an effective vehicle for implementing policy change to counteract disadvantage demonstrated by the CRE in its Rochdale report or by the patterns of disadvantage in respect of mortgage lending demonstrated by the PSI survey (Brown, 1984).

Conclusions

The Race Relations Act 1976, by dint of Sections 1, 20 and 21 in particular, provides a fairly comprehensive network of statutory controls to outlaw direct and indirect discrimination in the owner-occupied sector. This clearly covers the vast majority of services provided by estate agents, solicitors, building societies and other institutions concerned with land transactions and lending for acquisition. Indirect discrimination, as the courts have acknowledged, is, however, notoriously difficult to establish and, where relevant information is retained by these institutions, the individual purchaser is unlikely to be aware of the nature of disadvantage to which he may be subject, and even less likely to be able to establish a prima facie case in respect of indirect discrimination when access to information is denied. Although CRE investigations and research reports may lead to improved practices the ability of the CRE, as the principal enforcement agency, to police unlawful discrimination in the owner-occupied sector is clearly deficient. Such deficiency, which may no doubt be attenuated by improved practices relating to strategic investigations, is on such a scale in respect of resourcing as to negate the potential threat of being 'found out' and if found out of being subject to substantial financial penalties. Observance of the law, therefore, depends on the goodwill of the individuals and institutions concerned and that of the professional or commercial associations to which they may belong. While goodwill may have a commercial value, its social value goes largely unrecognised in respect of racial discrimination: few participants pay more than lip service to a concern for equality of opportunity.

The extent of unlawful discrimination in the owner-occupier sector is not known and not readily estimated. There is evidence of direct discrimination, and the 1982 PSI survey (Brown, 1984: Table 37) demonstrates a continuing belief by Asians, West Indians and whites that ethnic minorities get worse treatment than whites from building societies, banks and estate agents; this belief is more widely held, paradoxically, by whites in respect of estate agents than by Asians and West Indians and contrasts with the white perception that West Indians and Asians are marginally better treated than whites by council housing departments, an area marking the sharpest division of opinion between black and white. However, the studies on building societies in particular have demonstrated the existence of large-scale indirect discrimination which has, largely, gone unchallenged. While some lending institutions have examined their practices in the light of the Rochdale inquiry the regulatory institutions and professional associations have exercised insufficient control over their charges or members to secure change or to police bad practice. Unless there is a legal requirement to monitor and allow public access to the results, it would appear that the law will remain solely of declaratory effect. While there is little evidence that commercial investment in discrimination, as evidenced in the United Sates in respect of realtors, has affected patterns of racial segregation, the extent of racial disadvantage in housing, when considered by social class, economic group and even area of residence, suggests that ethnic preference provides an inadequate explanation of disadvantage in this sector. Some studies (e.g. Davies, 1985) have questioned not only the statistical evidence provided by Rex and Moore (1967) in respect of their seminal study of Sparkbrook but also the theories of housing, class and racial disadvantage to which it, in part, gave rise, and there can be little doubt that Asians and other ethnic groups have on occasion and in specific locations been able to benefit from the owner-occupied housing market. Such studies, however, do not outweigh the substantial evidence of racial discrimination which remains unchallenged by the legal system.

If the law is to be effective in this area, it is the enforcement provisions which must be addressed: the proposed statutory code of practice will effect improvements but the evidence of the employment code indicates that it will be insufficient, in itself, to ensure significant change. To secure equality of opportunity in the private as in the public sector, ethnic record-keeping and monitoring must become a statutory requirement, particularly in respect of the major building societies and estate agents. Some form of accountancy to which the CRE and the public has access is necessary on the part of the governing institutions. Penalties must be salutary and not, as now, frequently derisory. Minimum statutory awards are required as the courts cannot be relied upon to adjudge the social injustice at which their leniency has connived. Without a solid anchor of legal sanctions, the CRE's promotional activities are cast adrift and float aimlessly in the swirling political currents.

Part III

EVALUATION AND RECONSTRUCTION

CHAPTER ELEVEN

INSTITUTIONAL RESPONSES

INTRODUCTION

The purpose of this chapter is to provide a rough sketch of the institutional responses to racial disadvantage in housing, with a focus on Central Government and related agencies and professions which play a critical role in the housing process. Two factors, however, must qualify the utility of this approach. First, as the first chapter has attempted to emphasise, ideology, attitudes and the general climate of opinion will be significant factors not only in moulding formal responses to race issues but also in shaping the manner and extent of their effect. Accordingly a description of what has taken place may prove a useful indication of future action, but without an explanation of the reason for past responses, whether explicit or implicit, anticipating developments may prove hazardous. Second, because it is clear that racial disadvantage in housing is correlated both directly and indirectly with other areas of disadvantage such as education and employment, any analysis of institutional responses which is confined to housing will be partial. Moreover in both Central and Local Government it would be foolhardy to suggest that the political decisions of the Cabinet and the administrators in power are given effect by a neutral, value-free, Civil Service and Local Government officer cadre: their self-interest will frequently lie in neutralising innovation so far as it is seen as disruptive of the status quo without set-off benefits. To an extent, however, this chapter has already been written. The description of black experience regarding public and private housing, homelessness and access to housing finance portrays an accumulation of institutional responses but here the purpose is to seek a vertical explanation of decision-making to be read in tandem with the horizontal analysis provided in the previous chapters.

GOVERNMENT

While it may be argued that the battle against institutional racism in housing has been ineffectively fought by both Labour and Conservative Governments, in terms of political ideology, at least, there is much clearer alignment of the former than the latter in the formal promotion of equality. Moreover, there is some evidence of this on the statute books. The Race Relations Acts of 1965, 1968 and 1976 were enacted by Labour Governments, as were the Equal Pay Act 1970 and the Sex Discrimination Act 1975: the last, admittedly, was an improvement of a Conservative Government Bill published in 1973. While both political parties have pandered to the immigration lobby - the Labour Party schizophrenia if not xenophobia enshrined in the Commonwealth Immigration Act 1968 - the Conservative Government, as evidenced by the Commonwealth Immigration Act 1962, the Immigration Act 1971, the Immigration Act 1988 and, by association, the British Nationality Act 1981, has more consistently adhered to a view that the stricter the controls on black immigration the better the prospects for improved race relations. Mrs Thatcher has stated that the neglect of the immigration issue was driving some people to support the National Front whom she wished to attract to the Conservative Party (Layton-Henry and Rich, 1986: 75). She claimed that people were rather afraid that this country and the British character might be swamped by people with a different culture, that people had a right to be assured about numbers and that there should be a prospect of an end to immigration, except in compassionate cases. The juxtaposition of black immigration control and good community relations was implanted in the 1979 Conservative manifesto (Conservative Central Office, 14 February 1978) and there it continues to reside, albeit such hypocrisy has not gone unnoticed. As Bernard Levin observed (The Times, 14 February 1978) 'If you talk and behave as though black men are some kind of virus that must be kept out of the body politic then it is shabbiest hypocrisy to preach racial harmony at the same time.'

The 1979 Conservative manifesto comprised an eight-point programme to reduce immigration from the New Commonwealth but neither this manifesto nor subsequent ones suggested in what way, if any, community race relations would be improved. Accordingly while there was no intimation that the Race Relations Act 1976, somewhat reluctantly acceded to by a decision of the Conservative Opposition to abstain at the time of its passage through Parliament, would be repealed there was no commitment of substance to achieving racial equality. As a result the Conservative Government of the 1980s has been in a defensive, reactive position, handing over the initiative for analysis and policy development to others such as the House of Commons Home Affairs Sub-committee on Race Relations, the Commission for Racial Equality or ad hoc bodies such as Swann (1985) and Scarman (1981) reporting on Education for All and The Brixton Disorders respectively. Such a policy vacuum enabled the Conservative Government to placate the wets, as the liberal wing became known, and the right simultaneously: the former by periodic pronouncements from the front benches that the Government believes in equal opportunity and is opposed to unlawful racial discrimination as defined by the Race Relations Act 1976, and the latter by its unwillingness to initiate policies, of its own, which give practical evidence of such commitment. As a result, so far as the Conservative Government did have a policy on race relations, it was not to be found in any comprehensive statement but in piecemeal responses to reports and recommendations initiated by others. This is not to suggest that the Government was implacably opposed to ameliorating the circumstances of black British but it is striking that it neither fully implemented the recommendations of the various reports which it had to address nor suggested or implemented alternative strategies. A major exception was a confidential report, 'It Took a Riot', prepared by Michael Heseltine, then Secretary of State for the Environment, for the Prime Minister after the Brixton riots of 1981 had transformed inner-city policy from a minor item on the Cabinet agenda to one of urgent importance (Layton-Henry and Rich, 1986: 89). This report, reinforced by Scarman (1981), led to the appointment of a special Merseyside Task Force under Heseltine and an increase in the Urban Programme allocation from £202 million in 1981/82 to £338 million in 1984/85. Voluntary sector schemes to combat racial disadvantage were doubled to £15 million and the number of ethnic minority projects under both the traditional urban programme and partnership schemes expanded. In truth, however, it was the fear of further civil unrest rather than any commitment to alleviating racial disadvantage which was the impetus for the injection of funds and focusing on ethnic minority provision. This aspect had been a concern of the Urban Programme, introduced under Harold Wilson in 1968, but it had lost this hard edge, so far as it was evident, through the 1970s, suffering substantial general criticism as an area-based approach (Batley and Edwards, 1978) and specific criticism vis-à-vis the opportunities afforded to ethnic minority interests (Stewart and Whiting, 1983).

THE HOUSE OF COMMONS SELECT COMMITTEE ON RACE RELATIONS AND IMMIGRATION

Although reports of Select Committees are not usually debated in Parliament, they are frequently substantial and authoritive documents which require Central Government departments to issue responses to various recommendations. The House of Commons Select Committee on Race Relations and Immigration was established in the 1968/69 session of Parliament (November 1968) and in the period of ten years prior to it being superseded by the Home Affairs Sub-committee it issued eight substantial reports, the second of which was on housing (House of Commons, 1971) and the antepenultimate of which was on the organisation of race relations administration (House of Commons, 1978). The evidence adduced from the Home Office in respect of the latter sheds some light on how that department then considered its function: 'The role of the Home Office is to take the lead within Government in promoting and developing concerted policies - based, as far as possible, on common principles. These principles are at present under review.' When asked what these principles were the official giving evidence suggested that they (the Home Office) were perhaps a little sanguine in implying that there were any common principles, but that recognition that racial problems were rooted in the wider problems of urban deprivation might be one of them. As Banton (1985: 82) has observed, when the Home Office (1975) produced its White Paper, Racial Discrimination, it had advanced no further in the search for such principles. The official accepted that the Home Office had very little control and not much knowledge of what was going on and did not have the capacity to do more although it had some 300 reports and recommendations touching on education, housing, social services and employment which it was co-ordinating in an attempt to evolve common policies for Government departments.

324

The 1971 report on housing by the Select Committee (House of Commons, 1971), which contained some forty-six recommendations had to wait until September 1975 for a formal response from the Secretary of State for the Environment (HMSO, 1975a). The response focused on three issues: the intention to improve the housing circumstances of ethnic minorities through general policies which were not made specific, record-keeping by local authorities and dispersal. All other recommendations were dealt with in an appendix.

The general approach to improving housing which would, directly or indirectly, improve the housing situation of ethnic minorities included policy on area rehabilitation, on public house-building and management, on housing association initiatives, on a revised subsidy system, on inner-city initiatives and on the greater security of tenure afforded by the Rent Act 1974. Government believed (para. 12) that the effect of its housing measures and of those of Local Government, whilst designed to benefit the community as a whole, would give increasing priority to the areas of greatest housing need and would therefore accommodate the needs and problems of ethnic minorities 'which they share with others'. Specific action against discrimination - and it shared the view of the Select Committee that there was no evidence of this in respect of public-sector allocation - would be made possible by a strengthening of the legislation on race relations.

On record-keeping the Government rejected the adoption of a national system which 'could mean that statistics collected at considerable cost in staff time, in a sensitive area of social policy, would prove of little practical use'. The onus of record-keeping was firmly placed on local authorities - as far as they deemed it appropriate and useful - as part of the authorities' wider arrangements for understanding and dealing with the housing and social problems of the area, and for housing management purposes. While Government suggested that when information was collected it should be capable of indicating the proportion of the population in various forms of tenure and in various qualities and types of estate as well as being capable of being analysed 'to establish better understanding of its [the proportional distribution of ethnic minorities] background causes' (para. 19), there was no indication that the Department of the Environment would monitor local authority record-keeping or its analysis.

On dispersal of ethnic minorities as a matter of policy, which the report had advocated, the Government gave qualified agreement as to the benefits - the opening up of economic, social (including housing) and educational opportunities - which might, potentially, improve race relations, but apart from removing the obstacles to dispersal it did 'not at this stage consider it appropriate to go beyond this statement of policy, or to issue further guidance to local authorities on this issue' (para. 28). To the Government the most important question was not that of concentration versus dispersal, but of

> improving the housing, environment, educational and employment opportunities of the inner urban areas where large numbers of coloured families will continued to live for some time to come, whatever the rate of movement out; and [of] providing the opportunities so that coloured families, along with others, will in practice be able to move out from the inner cities.

Although the general tenor of the response to the Select Committee report on housing was positive and a number of specific initiatives already taken by the Department of the Environment were identified two cardinal points of Government policy on race and housing were implicit rather than explicit. First, while ad hoc initiatives by Central and Local Government might prove desirable, the thrust of measures to counteract racial disadvantage was area-based and not specifically targeted to benefit ethnic minorities directly: general improvement programmes directed at areas of social stress would benefit all residents and it was only where such residents were disproportionately drawn from ethnic minorities that such groups would be a focus of concern. Second, while local authorities should assess the general needs of their communities in a systematic fashion and should include ethnic minorities in such assessment and resultant programmes, Central Government would not be responsible for providing an explicit framework for such activities, nor would it be concerned, directly, with monitoring the extent to which local housing authorities were successful or otherwise in identifying and meeting ethnic minority housing and related needs. Despite such avoidance tactics there is little doubt that the Select Committee was a major vehicle for the making of Government policy, which, without it, might have appeared even less convincing. Moreover it forced Government to focus on issues it might otherwise have preferred to ignore and, through a process of accumulation of evidence, created a much improved information base on which Government could ground its future policy review. It must also be recognised that just as Government is capable of putting a positive gloss on responses to which it is not unequivocally committed so, conversely, it may formally reject proposals yet, virtually simultaneously, take measures to secure their adoption. Accordingly, the formal responses to the Select Committee reports will provide a partial view not only of what Government is doing but also of how policy change should be anticipated.

THE HOME AFFAIRS COMMITTEE

This committee (HAC) was one of fourteen all-party committees established by Parliament in 1979, replacing the previous Select Committees which had been established in an ad hoc fashion mainly during the 1960s and 1970s and which were predominantly departmental or subject based (Nixon, 1986). The purpose of the reformed committees was broad: 'to examine the expenditure, administration and policy of the principal Government Departments, and associated bodies ...' (para. 1 of Standing Order 86A, 25 May 1979). These committees were appointed for the full term of a Parliament and they were given substantially increased powers regarding access, information, attendance of witnesses and the production of records and papers. The HAC was one of only three committees given power to appoint a sub-committee and, as expected, it appointed a sub-committee on Race Relations and Immigration at its first meeting in December 1979. This provided a continuum with the previous committee solely concerned with this area from its inception in 1968. The HAC has not issued a report which deals specifically with housing but the 5th Report (session 1980-81) on racial disadvantage contained a series of recommendations relating to Central and Local Government responsibilities which were of direct or indirect concern to housing provision (HAC, 1981b). None, however, with the exception of the recommendations relating to

Section 11 of the Local Government (Social Needs) Act 1966, advocated legislative change. The report criticised Department of the Environment project control of housing schemes under the old cost limit system (revised from 1 April 1981) and the Department of the Environment, while accepting that value for money might discourage the creation of larger housing units by conversion, pointed to the possibility of higher eligible expense limits for such conversions in Housing Action Areas and the greater flexibility permitted by the new regime. The report advised that housing authorities should examine their allocation and transfer criteria and procedures and should ensure that ethnic minorities were aware of available housing facilities (paras. 87-8). The response by Government was predictable: 'it is for local authorities to determine their council housing allocation policies in the light of the needs of their area and the resources they have available to meet them'. It acknowledged that 'such policies should be consistent with the general duty placed on local authorities by Section 71 of the Race Relations Act 1976, and also Section 21 of the Act' (Home Office, 1982b: 24). Hardly a clarion call to rigorous action.

In respect of the internal organisation of the Department of the Environment the HAC recommended that the department should create a specialist unit concerned exclusively with the racial disadvantage aspects of its responsibilities (para. 43): evidently this would improve the information, the focus and guidance proffered on housing issues. The Government responded somewhat blandly and with disingenuous honesty that it did not see the need to establish a separate unit within the Department of the Environment 'since this would be divorced from the work of particular policy divisions, with whom the prime responsibility must lie' (Home Office, 1982b: 17). Responsibility 'for taking a general view ... as it affects the Department's whole area of interest' was allocated to the Inner Cities Directorate. These illustrations of Government's response together with a refusal to establish a body with general oversight of research into race relations and disadvantage and a refusal at that time to introduce comprehensive monitoring of the Civil Service workforce provide an indication of profound lack of political will and commitment to substantial measures to combat racial disadvantage. The previously negative and defensive stance which, it is argued, characterised Government's response to this report might well be challenged on the grounds that Government's response to the HAC report acknowledged a commitment 'to a multi-racial society in which there is full equality of opportunity irrespective of colour, race or religion' and that ethnic minority communities 'who were here or settled here are part of this country and part of its future'. Moreover there was a recognition that special measures, on occasion, were required particularly by Central and Local Government to overcome racial disadvantage. But in stressing that 'an overriding objective ... is the creation of a stronger and more prosperous economy' as the most potent means of controlling racial disadvantage, Government clearly signalled that the creation of equal access to housing or any other aspect of public and private provision was not a priority and this conclusion was borne out by the consistent trimming of the various recommendations made by HAC in its report, and the absence of alternative or complementary strategies to achieve the stated aims of equality of opportunity.

The HAC report on <u>Racial Attacks and Harassment</u> (HAC, 1986) and the Government response (Home Office, 1986) must qualify any conclusion that the influence of HAC is consistently marginalised. Clearly there is evidence here, as in responses to other reports emanating from HAC, that the Government is capable of responding speedily and sympathetically to specific issues. This fact, however, does not detract from a view that the impetus to such positive responses, as indicated in respect of the previous Select Committee reports, is derived from considerations other than the achievement of a racially just society, such as the containment of urban crime. Inevitably there is a sense of frustration when members of the committee feel that its reports have not been properly treated by Government: one observed, 'sometimes you feel you do a lot of work and prepare the report, and to all intents and purposes, it tends to be forgotten' (Nixon, 1986: 152).

The purpose of the Select Committees is 'to probe, to expose, to enquire, to challenge', in the words of John Wheeler, then chairman of HAC's sub-committee, and in this HAC has been generally successful by identifying weaknesses in departmental policies and exposing Government inaction on a number of important issues. Moreover in respect of Government policy towards the repeal of the 'sus.' laws, towards the need for an ethnic question in future censuses and towards modifying urban aid, health and social services provision and Section 11 funding HAC's sub-committee has proved of value. These factors, together with the assemblage and exposure of a wealth of evidence relating to racial disadvantage has, in Nixon's view, enabled the work of HAC to prevent race relations from being of only marginal interest to either Parliament or Government and has helped to sustain that interest as a continuing item on the political and administrative agenda. That it has not had a greater effect on the formulations of proactive Government policy may, consequently, be attributable to the inherent weaknesses of such committees and to Governmental and administrative resistance, rather than the reports that HAC has issued. From 1988 HAC has not appointed a Sub-committee on Race Relations and Immigration, apparently owing to the weight of its other responsibilities: as a result there is little prospect of Government being advised by this means of the substantially adverse impact of the current programme of legislation on housing on ethnic minorities.

Whether or not there is truth in the assertion that Government consciously uses Royal Commissions, Select Committees and the like as a ploy to defer, evade or avoid issues and decisions it would prefer not to address, there can be little doubt that their deployment deflects the immediate responsibility of the departments and Ministers concerned for devising comprehensive strategies in the area under consideration. Without a developed system of accountability and monitoring Government may rationalise both association with and dissociation from the recommendations it receives and control the extent to which it is subject to further scrutiny. In the area of race and housing, Government has never established firm objectives, has never established a programme to secure that legislation or policy aims are periodically reviewed and has never established criteria with which the Home Office, the DoE, the Welsh Office or the Scottish Office are to gauge the success or otherwise of their own functions and responsibilities. Perhaps one of the major failings of HAC has been its neglect of procedures and processes adopted by Government which enable substantive issues to be by-passed.

GOVERNMENT DEPARTMENTS

The Home Office

Nixon (1986) has suggested that the Home Office retains responsibility for race relations by default. 'It is an area in "policy space" which is occupied by many bureaucracies and where none seeks sovereignty.' In a formal sense, however, there is no ambiguity concerning the Secretary of State for the Home Department having overarching responsibility for race relations both in terms of his responsibility for the Commission for Racial Equality under the 1976 Act (albeit the Act conforms with the conventional fiction that there is only one Secretary of State who is not named) and in terms of his assumed responsibility 'to form an overall view of the race relations situation and to ensure that the Departments with specific responsibilites are developing their own policies on a co-ordinated basis' (HMSO, 1980). In written evidence to the HAC sub-committee (June 1980) the Home Office described Division II as 'broadly concerned with the co-ordination and assessment of the Government's policies for combatting racial disadvantage', including 'inter-departmental liaison on the implications of Government policies generally on race relations' (HAC, 1981b: EV: 165, para. 27). However, the HAC sub-committee found it 'disturbing' that the Home Office did not see itself as a co-ordinating department and that it rejected the notion that there should be anybody in Government 'in charge of monitoring the race relations performance of all the different Departments'. Moreover HAC found in its report on Racial Disadvantage (HAC, 1981b) that there was no effective machinery at ministerial level for co-ordinating Government policies impinging on racial disadvantage. On the basis of constitutional convention the Minister then responsible for race relations - Timothy Raison - refused to advise whether a Cabinet committee existed which included this function in its merit. HAC concluded that if there was a Cabinet committee or its equivilant it was ineffective; if there was not, one should be established (HAC, 1981b: para. 37).

The Home Office Advisory Council on Race Relations, which was established in 1975, meets about three times a year and includes representatives of the CRE, the CBI, the TUC, the four Local Authority Associations operating in England and Wales, the National Association of Community Relations Councils (NACRC) and twelve members appointed by the Secretary of State from ethnic minority communities. COSLA (the Convention of Scottish Local Authorities) and SCRE (the Scottish Council for Racial Equality) are not represented, although Ministers and/or officials from Government departments (including the Scottish Office) will attend from time to time - depending on the subject on the agenda - in addition to the Home Secretary or Minister of State, Home Office, with responsibility for community relations, who chairs the meeting. The division responsible for race relations has twenty-four staff (at all levels, 1988) and two Community Relations Consultants who, inter alia, have a special responsibility for training. Divisional responsibilities include responsibility for the race relations legislation, relations with the CRE, Section 11 grants, the Home Office ethnic minority business initiative and 'other general advisory duties' (Harnett, 1988). However, the Advisory Council has no staff of its own and has no parallel committee of officials. Despite its then impending revamp, HAC remained 'sceptical of the value of a high-powered talking shop deprived of effective power'. A matching vacuum was found at official level and HAC recommended the establishment of an inter-departmental committee chaired and vigorously led by the Home Office - which should break with inhibitions about interference with other departments. HAC insisted, 'some provision must be made at official level for inter-departmental co-ordination of Government policies for combating racial disadvantage.

The Home Office, however, rejected the recommendation that it should adopt a more rigorous co-ordinating role and generally remained obdurate on the essentials (Nixon, 1986). No standing inter-departmental committee of officials has been established although, following the HAC report on Racial Attacks and Harassment (HAC, 1986) an ad hoc committee was established with a specific remit in this area. Its report was published in 1989 and it thereafter disbanded (Home Office, 1989).

Department of the Environment

Although the Department of the Environment (DoE) does not have overall responsibility for the implementation of the Race Relations Act 1976 and the promotion of equality of opportunity, given its size (5,500 staff in post at April 1987), its budget (£6.6 billion of £13.5 billion total gross expenditure, 1986/87 out-turn) and its broad responsibilities, its influence on the formation of Government policy and practice on race issues is of critical importance. Moreover through its responsibilities for Local Government, housing, planning, and the inner cities, the DoE determines the framework of local authority, private housing, and housing association operations.

However, in common with other Government departments, the DoE has no strategic plan or overarching policy by which it, or others, may gauge the extent to which it meets the requirements of the legislation or to which others, such as local authorities, New Towns, and the Housing Corporation, for whom it is responsible, meet such requirements. While the absence of policy planning, monitoring of performance and review of objectives is not unique to the field of race relations - a similar vacuum is to be found in respect of sex discrimination, equal pay, and provision for the disabled - it must surely reflect, implicitly if not explicitly, either the lack of priority and political and administrative commitment to achieving equal opportunity or a blind indifference to its own performance.

In respect of housing the DoE (Sherman, 1988) summed up its policy as follows:

> The Department encourages the promotion of racial equality and prevention of racial discrimination in all fields of housing. One of the ways this is being done is through the CRE's code of practice on housing to which we are giving statutory force. Local authorities are also asked to keep ethnic records of applicants and monitor their application practice so that they can see whether they are fulfilling their statutory obligations under Section 71 of the Race Relations Act 1976. The Department also supports the development of local authority staff training and co-funds the Local Authority Race Relations Information Exchange.

> Within the Urban Programme, the Department's policy with regard to ethnic minorities is that due priority be given to projects designed to benefit disadvantaged minorities, particularly through the provision of work and training. Ethnic minority involvement in projects is established at the outset, by the requirement for applicants to say:

333

(a) If the project will be run by an ethnic minority group.
(b) What proportion of the project benefits will be going to ethnic minorities.
(c) If the project is specifically directed to ethnic minority needs.

This information enables local authorities to sum up the benefits both aimed specifically at, and which flow incidentally to, ethnic minorities ...

The production of a good practice guide (from research undertaken by Brunel University) will show local authorities the best way to deal with problems of racial harassment. Within both housing and ICD (the Inner Cities Directorate) there is a co-ordinating point on ethnic minority issues... see Home Office, 1989.

ICD have a co-ordinating role with DoE on ethnic minority issues and other policy advice to other divisions and outside bodies concerned as required. The Department liaises closely with the Home Office on ethnic minority matters and is occasionally asked by the CRE, usually through the Home Office, to offer advice and comment on a variety of race issues pertaining to the Department's work. When appropriate a DoE minister attends meetings of the Interministerial Advisory Committee on Race Relations which is convened by the Home Secretary about three times a year.

At the time of the Home Affairs Committee report on Racial Disadvantage (Home Affairs Committee, 1981b: 43) the DoE had no staff exclusively concerned with racial disadvantage: in 1976 Government told the former Select Committee, 'a specialist race relations unit would not be appropriate'. In 1981, H2 Division in the Housing Directorate carried the major responsibility for race in the DoE but the committee concluded that these arrangements were insufficient to ensure 'that the particular concerns of ethnic minorities are articulated in policy formulation and administration throughout the Department': the committee recommended the creation of a specialised racial disadvantage unit within the DoE.

Government rejected this proposal because it 'does not see the need to establish a separate unit ... since this would divorce the work from particular policy divisions, with whom the prime responsible must lie' (Home Office, 1982b: 17). However, it conceded that a division would be responsible for a 'general view of racial disadvantage as it affects the Department's whole area of interest' and allocated this to the division within the Inner Cities Directorate although no staff have race issues as their sole responsibility.

Inevitably, given the absence of set objectives within the department and of yardsticks by which performance may be measured, the work of the ICD is largely responsive. While the need to gear the Urban Programme, for which the division has a direct responsibility, more closely to perceived ethnic minority needs is reflected in the increase in the proportional number and resourcing of ethnic minority projects, even in this area there is not any Central Government assessment of need and setting of targets to ensure that they are met. Such assessment, albeit with guidance from the DoE, is done by local authorities. Consequently the criticism made by Stewart and Whiting (1983) that the progress is response-led and consequently still dependent on local initiatives, frequently divorced from comparative need either within the local authority area or nationally, remains and the structures within the DoE are not geared to redressing resultant imbalances.

The involvement of the ICD in housing issues for which it has no direct responsibility is even more tenuous. The Housing Division of the DoE is not in any sense answerable to the ICD for the implementation or monitoring of equal opportunity objectives. Consequently the 'general view of racial disadvantage' afforded by ICD appears to be confined to a consultancy role when thought appropriate by the division concerned. Even where policy guidance is issued - such as the monitoring of public housing allocation and racial harassment - the fact that the relevant housing division does not monitor the extent to which it is given effect is not within the ambit of IDC's 'general view'.

The DoE acknowledges that race relations are not given a high priority under the Conservative Government and it is apparent that the Civil Service is bound to reflect this in determining the administration of its own responsibilities: where initiatives may be viewed as operating in politically sensitive areas, self-preservation may result in inertia or

335

trimming to avoid conflict. Self-evidently research findings may provide an objectivity for action and remove initiatives from the party political arena but the DoE does not and has not carried out any research in the area of housing, race and law (Gooday, 1988).

A report on ethnic origin surveys (OMCS, 1987) of non-industrial civil servants showed that 9.4 per cent of the respondents employed in the DoE in Greater London were from ethnic minorities. This percentage compared favourably with the Home Office (9.2 per cent) but unfavourably with the DHSS (20.7 per cent) and the overall percentage (12.8 per cent) for Government employees. Those at Principal level or above (Grade 7) of ethnic minority origin were 1.8 per cent in the DoE compared with 3.6 per cent in the DHSS (the Home Office could not produce a statistically significant number at this grade) and contrasted with 7.7 per cent of white Government staff on this grade employed in Greater London. The report blandly concluded that the variations between white and ethnic minority are significant and cannot be fully explained by differences in age and length of service. What the report did not conclude, but what the returns suggest, is that there is widespread unlawful racial discrimination in the Civil Service in respect of recruitment and in respect of promotion and that the DoE and Home Office, amongst others, reflect this pattern.

On the basis of this report, which is suitably anaesthetised to avoid public criticism, the nature of Government's workforce is such that it is unlikely to be sensitive to the needs of ethnic minorities. Clearly had the DoE in its workforce mirrored the composition of the general population at all grades, it would be in a stronger position, on the basis of knowledge, awareness and sensitivity, to secure that the Race Relations Act 1976 in its impact on housing was not so readily written off by political scepticism if not aversion.

If political scepticism towards the promotion of equality of opportunity is a current fact of life, it could be argued that the absence of a coherent plan and strategy for implementation may protect ad hoc projects and proposals, such as those to be found within the Urban Programme, on the basis that ministerial intervention in minutiae is, first, less probable and, second, when it does occur, it may not adversely affect other ad hoc activity - a more likely occurrence where threads of policy are woven together. It must be acknowledged that the DoE does take valuable initiatives in the area of race and housing from time to time but whether these would have been inhibited by political decisions within a broader and more cogent framework is a matter of conjecture. What is not conjecture is that, even in the absence of a plan, there has been no attempt to analyse the cumulative effects of the department's work on ethnic minorities (whether in comparative terms they have benefited or disbenefited in the area of housing from the various initiatives over a period of time). As a result the DoE is ill-equipped to judge the effect, let alone the purpose, of what it does and what it asks others - such as local authorities - to do. The absence of any criteria of assessment of the department's various responsibilities ineluctably means that it does not know what is worth protecting and what is not other than by informed guesswork and intuition.

The Scottish Office

The Scottish Office comprises five divisions (or National Departments), the Scottish Home and Health Department (SHHD), whose responsibilities largely match those of the Home Office, the Scottish Education Department (SED), the Scottish Development Department (SDD), equating with the Department of the Environment, the Industry Department for Scotland (IDS) and the Department of Agriculture and Fisheries for Scotland (DAFS). The Scottish Office, through the five departments within it, is responsible for a substantial range of subjects, by breadth and depth, including those which have significant relevance for race relations such as Local Government, health, education, housing, economic development, inner-city policy and the urban programme, police and prisons. The Home Office, however, retains responsibility for race relations and immigration and some functions, such as social security, are administered through 'British' departments such as the Department of Social Security. It should also be noted that the separate legal system for Scotland, guaranteed by the Treaty of Union in 1707, may have implications for race relations not only in respect of the relevant legal procedure which is to be applied but also in respect of common law and statutory provision. As an example the Local Government (Social Needs) Act 1966 - which enables 75 per cent Central Government grants to be given to local authorities in

England and Wales for special staffing needs in respect of New Commonwealth immigrants - does not apply to Scotland: separate provision is made, however, by an equivalent Act (also Section 11 and also 1966) which the Scottish Office has consistently refused to operate. Further distinctive provision is exemplified in respect of the law relating to public housing allocation and racial harassment.

Given the relatively small numbers of Scottish residents of NCWP ethnic origin, the issue of race relations and racism has not been seen, by successive Governments, as having any political priority. This is evidenced by the fact that the Scottish Office does not employ one civil servant with an exclusive remit on race relations, has no post designated as that of a race relations adviser, whether shared with other responsibilities or not, and has no advisory committee or body whether at political or official level to provide advice or guidance in implementing Central Government policy.

Inevitably those aspects of the work of the Scottish Office which should have accommodated a race dimension have, in general, failed to do so. Thus the Morris Committee report on Housing and Social Work (Morris, 1975) and the guidance Assessing Housing Need (SDD, 1977b), while valuable in themselves, ignore the fact that Scotland is a multi-ethnic society in failing to mention let alone analyse the racial dimension to the subject area considered. Some changes have, however, been effected. Following representations from the Scottish Council for Racial Equality (SCRE), the Scottish Office intimated in July 1980 that Mr Russell Fairgrieve, then an Under-Secretary of State, had been assigned an umbrella responsibility for race issues in Scotland. Similar appointments have been made subsequently. The administrative support for the Junior Minister is provided by the Urban Renewal Unit, previously under the Scottish Development Department and now responsible to the Industry Department for Scotland, reflecting the responsibility of the Inner Cities Directorate in the Department of the Environment who similarly administer the Urban Aid Programme. Such political and administrative arrangements are little more than a 'post box' facility: it has been made clear at meetings between the SCRE and the Minister that they will not result in policy guidance being issued either to departments within the Scottish Office or to local authorities themselves (MacEwen, 1980b).

The only 'research' on race undertaken by the Scottish

Office, through its Central Research Unit, related to a demographic profile of ethnic minorities in Scotland. In 1988 the Scottish Office commissioned a research project comprising a survey of ethnic minorities in the four major cities, Glasgow, Edinburgh, Dundee and Aberdeen. It was a condition of this survey that it eschewed attitudinal questions - such as the experience of discrimination and racism. While modelled on the PSI survey <u>Black and White Britain</u> (Brown, 1984), and utilising questions from the National House Dwelling Surveys, neither of which - despite their misleading titles - extended to Scotland, it is clear that the Scottish Office, at least at the political level, did not want to gauge, from that survey, whether or not any findings of racial disadvantage - clearly predicted by other surveys - are related to racial discrimination. Moreover, it is also clear that, at present, the administrative structure in the Scottish Office relating to race issues is incapable of a proactive stance on race. Consequently, unless the Scottish Office looses its milk teeth on race and grows a veritable set of gnashers, there is every prospect that the survey report, due in 1990, inadequate as it will be, will be allowed to collect dust as soon as it is published. The SCRE has advocated the establishment of an advisory committee to the Secretary of State - perhaps equating with the Home Secretary's Advisory Council or, alternatively, the Scottish Office Committee on Scotland's Travelling People, established in June 1971. The CRE, establishing its presence in Scotland with an Edinburgh office for the first time in 1986, has supported this proposal and a response from the Scottish Office is currently awaited (June, 1990).

Within the Scottish Office administrative structure, or more accurately in the absence of any coherent structure for race issues, each department is responsible for issuing policy guidance on implementing legislation. One circular has been issued, following the passage of the Race Relations Act 1976, on the issue of race which merely summarises the statutory provision in relation to local authorities. But the SDD, which, until the inter-city responsibilities were transferred to the IDS, had comprehensive responsibility for issues relating to Local Government and housing, has not issued advice or guidance to local authorities in this area. It is clear that the absence of a focus on race within the Scottish Office has affected the quality and extent of input from Scotland not only in respect of projects and agencies operating in Scotland but also those operating at a national level. Thus the inquiry into racial attacks and harassment (Home Affairs Committee, 1986) received no submissions from the Scottish Office, did not ask it to give evidence, and did not mention it in its findings and recommendations. The effect, whether by design or otherwise, is to confirm the marginalisation of race relations on the political agenda in Scotland, and to justify the continuing evasion of responsibility by the Secretary of State for Scotland. On occasion, however, some aspects of Central Government policy impinging on race relations have been addressed. The Central Committee on the Curriculum, an advisory body to the SED, has considered the issue of multi-cultural education in relation to the Swann Report (1985). The SHHD, in respect of its responsibility for the police, has considered the relevance of the Scarman Report (1981), in relation to community policing, and several others, culminating in the report of an inter-departmental group in respect of racial attacks (Home Office, 1989). Such deliberations have almost inevitably, given the lack of administrative coherence within the Scottish Office, provided further illustration of the Government's overall concern for race issues being dictated by reaction to external pressure and focused on containment and social control. Within such a framework of priority, housing has rarely surfaced. Consequently issues relating to public-sector allocation, racial attacks in the public or private sector, access to mortgage finance, discrimination by estate agents, access to improvement or repair grants and the effect of Housing Action Area declarations on ethnic minorities have not attracted Scottish Office concern. As a result the impact of the Race Relations Act 1976, in respect of discrimination in housing in Scotland, operates in a political and administrative vacuum. The Urban Programme, although largely of secondary importance to housing, has no specific ethnic dimension, as guided by SDD circulars to local authorities, in contrast to those emanating from the Department of the Environment. And the Scottish Office, as already noted, refuses to implement Section 11 of the Local Government (Social Needs) (Scotland) Act 1966, which would enable supplementary staffing of local authorities where particular housing needs of ethnic minorities have been identified. Overall, then, the picture is driech - a tedious, unremitting Scottish gloom - and seldom relieved by chinks of fleeting and unsuspected light. There is no ambiguity, therefore, about the fact that if discrimination in housing is to be combated through the vehicle of the 1976 Act, the divisions of the Scottish Office will not be found near the barricades.

The Welsh Office

Although the Welsh Office enjoys a broad array of umbrella powers similar to the Scottish Office two factors qualify their comparability. First, the legal system in Wales is shared with that of England and, second, Wales has, in many respects, the status of a region. The number of Welsh Acts or Parliament are limited and frequently reflect a perception of regional need such as the Welsh Development Agency Act of 1975. Unlike Scotland, there is no separate legislation for Wales relating to Local Government, planning or housing and this legislative and administrative congruence with England results, inter alia, in the frequent promulgation of joint circulars by the DoE and Welsh Office.

While the proportion of the Welsh population of ethnic minority origin is relatively small there are significant concentrations in Cardiff, 4 per cent of whose population is of ethnic minority origin, and elsewhere.

Within the Welsh Office, Central and Management Services 2 is the division which has overall responsibility for ethnic minority matters and it will liaise with the Home Office, which, formally, retains responsibility for race relations in terms of the Race Relations Act 1976, as and when necessary, for example, on the scrutiny of Section 11 funding under the Local Government (Social Needs) Act 1966. Mr Wyn Roberts, the Minister of State for the Welsh Office, under Mr Peter Walker (the imported Secretary of State), was a member of the Advisory Council on Race Relations.

The Housing Division would liaise with the DoE on race issues, such as the issuance of joint circulars: the Welsh Office has not issued any circulars on this subject.

No official in the Welsh Office is exclusively concerned with race relations. In the Central and Management Services Division race relations form 'a minor part of the work of a Grade 7, Grade 5 and an EO' and also form 'a very minor part of the responsibilities of an EO in [the] Housing Division' (Padfield, 1988). A Principal (Grade 7) attends meetings of the Welsh Consultative Committee of the CRE but the issue of housing has not been discussed recently (November 1988).

Clearly the Welsh Office has not taken and is unlikely to take any initiative in respect of race and housing. The

Home Office and DoE make the running, or stroll, and the Welsh Office, generally, keeps in step. Given the low priority which this issue attracts, the small number and low level of staffing responsible, the lack of research activity and the willingness to perceive issues in numbers, the Welsh Office looks sideways or upwards to the 'English' subject departments or downwards to such as Cardiff City Council as those primarily responsible for dealing with race relations and the 'problems' which may occasionally surface.

THE COMMISSION FOR RACIAL EQUALITY

The statutory framework of the CRE has been described in Chapter Five and its role in providing advice and legal assistance and in conducting formal investigations has been illustrated in subsequent chapters. The purpose of this section is to describe the CRE as an organisation and to outline, from that perspective, its influence regarding one of its two other functions, its promotional activities - its work to encourage the adoption of good housing practice. Its remaining function, the responsibility to advise the Secretary of State for the Home Department on the functioning of the Race Relations Act 1976 in achieving its objectives is addressed in the penultimate chapter.

Figure 11.1 and Table 11.1 set out the organisation of the CRE and the allocation of the £10.5 million budget for the year 1986. The CRE has five divisions: Employment; Education, Housing and Services; General Services; Field Services; and Legal. Self-evidently it is the second of these which has prime responsibility for housing in respect of promotional activity in the private and public sector so far as national strategies are devised and co-ordinated. There are, however, major qualifications to this allocation of responsibility. The Field Services division, working through regional offices in Birmingham, Manchester, Leicester, Leeds and Edinburgh, is responsible for all aspects of promotional work at the local level, which will, not infrequently, draw on the specialist staff, whether in housing, employment or education, employed in London. The Legal division is responsible for legal representation and advice: this includes all complaints and investigations. Consequently, so far as these inform promotional work, this division will be involved. The General Services division conducts and co-ordinates all research which in turn will

Figure 11.1: Structure of the CRE

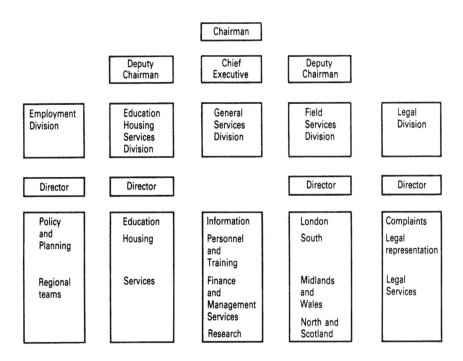

Source: CRE Annual Report 1987: 11 (CRE, 1988a)

have implications for promotional activity as well as informing the Commission, more generally, on the effectiveness of the statutory provisions. The Employment division in conjunction with Field Services will be involved in promoting good employment practices - such as those advocated in the 'statutory' code on employment - which will be relevant to local authorities and other housing agencies. One third of the CRE budget (£3.3 million) pays the staff salaries: the authorised complement is 226 (at 1986), of which 205 permanent staff were in post. The

Table 11.1: Analysis of expenditure by CRE, 1987/88

Expenditure	£	%
Staff salaries	3,456,842	32
Grants to CRCs, including discretionary grant aid	2,920,625	27
Grants for projects, self-help groups project aid and bursaries	1,927,782	18
Overheads - accommodation, travel, staff training, printing stationery equipment, etc.	1,700,879	15
Information services, publications and conferences	365,893	3
Research	109,356	1
Legal and professional costs, including external training	465,454	4
Total expenditure	10,946,831	100

Source: CRE Annual Report 1987: 10 (CRE, 1988a)

Education, Housing and Services division has thirty-two staff - marginally more than Legal Services (twenty-six) and Employment (thirty-one) but significantly less than Field Services (fifty-five) and General Services (sixty-one). Of the thirty-two staff in the Education, Housing and Services Divisions, eight are concerned solely with housing. Following administrative reorganisation consequent upon the Home Affairs Committee report on the CRE (HAC, 1981a), which recommended the allocation of specialist staff to regional offices, if necessary in place of the current fieldwork staff (para. 40), officers in the regional offices have been given functional responsibilities reflecting the specialist divisions in London. Accordingly CRE 'housing' staff will include some regional officers in addition to those in London.

In 1986 there was a change in the thrust of the Commission's work. The earlier emphasis on establishing, through investigations and individual complaints, how discrimination occurs and on proposing ways to prevent it is now accompanied by a new emphasis on outcome (CRE, 1987d). Increasingly, the CRE argues, the questions asked of any set of practices or of an institution have been less concerned with how they have come to be in the situation they are in, or what procedures have or have not been adopted in the past, and more concerned with what can be done, within what time-scale, to put things right. The strategy for formal investigations is to concentrate on areas where there is a real prospect of effecting change quickly. Moreover in housing and other aspects of service provision, where there is evidence of discrimination, the concept of targeting is applied for the purpose of devising a plan of action for which the individual, authority or company is held accountable.

Ultimately responsibility for this and like policy changes lies with the Commission rather than the staff. Members of the Commission, including the full-time chairman and two part-time deputy chairmen, are appointed by the Secretary of State for a five-year renewable term but are neither servants nor agents of the Crown (Schedule 1, para. 2, 1976 Act). In terms of Section 43 of the Act the number of commissioners is to be not less than five and not more than fifteen but this may be amended by order of the Secretary of State. The number of commissioners is (1988) now fifteen, one-third of whom are members of an ethnic minority group. Past appointments by the Secretary of State have been criticised not only because of their 'political' nature but also on the basis of racial bias - neither the CRE nor its predecessors the RRB and the CRC has had a black chairperson - and lack of knowledge and expertise, particularly in respect of non-ethnic minority appointees. Although the CRE has, not infrequently, been severely critical of Central Government, such appointments have been used to justify accusations of the CRE being an Uncle Tom which, however misplaced, undermine its credibility, especially with black and ethnic minority organisations. Thus the decision in April 1980 to replace four of the seven black commissioners and only one of the eight white commissioners created the impression of a black purge, resulting in calls for the potential black nominees not to accept places on the Commission (Layton-Henry, 1984: 142).

The Commission has power to appoint committees and currently deploys a Field Services Committee, a Complaints Committee, an Employment Committee and a Committee for Education, Housing and Services. However, while the Chairman and the Deputy Chairmen may play a key role in devising policy, the other commissioners, through the committees and otherwise, are more involved in monitoring and advising on such policy as opposed to determining its formulation. Consequently, although some commissioners may play a role in policy formulation through their regional responsibilities, as a generality policy development is more likely to be generated by the full-time staff. In this respect, the various directors and section heads will be key personnel. Such allocation of responsibilities, albeit implicitly, was endorsed by the report of the Home Affairs Committee (HAC, 1981a) which observed that 'The essentially non-executive and supervisory functions of the Commissioners should be emphasised' on appointment.

In its promotional activities in housing, the CRE will inevitably experience a 'gut response' similar to the criticism attracted by its other activities. As a quango, Government may use the CRE, as other quangos, to deflect or avoid criticism while, not infrequently, pursuing contradictory policies itself (Gregory, 1987). Those who view the CRE as an unwarranted intervention into market forces, those who see it as a cosmetic token to disguise inactivity, those who see it as a mechanism of social control and containment and those who see it as a potentially powerful vehicle for social change will all view its record from a different perspective. Inevitably such perspectives must be placed in an historical context. The CRE assumed many of the professional staff of the defunct Race Relations Board and Community Relations Commission but perpetuated this historical division by the administrative organisation, which separated law enforcement from promotional activity. The Home Affairs Committee criticised this as one aspect of the overall 'lack of coherence' which, in its view, characterised the work of the Commission. The first chairperson was David Lane, an Old Etonian and Conservative MP for Oxford who had had previous dealings with the Race Relations Board regarding immigration and nationality as a Junior Minister in the Home Office and he did little to appease a far from quiescent black lobby by reiterating links between good race relations and strict immigration control (Coote and Phillips,

1979). The CRE was accused of racism in internal staff disputes (Rasul v CRE (1978) IRLR 203). Despite the significant strengthening of the legislation in respect of indirect discrimination, the CRE had experienced an inauspicious inauguration. By 1980 it had commenced forty-five formal investigations but by 1981 had seen only ten to a conclusion. It was accused of having moved off in all directions without any clearly defined priorities or objectives, having bitten off more than it could chew, having concentrated on 'small fry', of being amateurish and having failed to provide useful ammunition which should have been the cornerstone of its promotional activities (see McCrudden, 1987: 227-66).

> Even one single completed investigation into practices of a major employer or provider of services would have had a greater effect than those small fry (Home Affairs Committee, 1981a: para. 43).

The Home Affairs Committee was at pains to stress the links between law enforcement and promotion, the latter being the 'ripple effect' of the former. In its view the Commission's promotional activities should be largely confined to the publishing, and the incorporation into codes of practice and the like, of the results of the enforcement of the 1976 Act through investigation and court cases (para. 80); much of the promotional activity should more properly spring from Government departments (para. 81); and the promotional work should be solely dictated by the need to eradicate racial discrimination (para. 14).

In its response Government (Home Office, 1982a) first accepted that the Commission should not be a shadow race relations department; second, while accepting that the CRE should be and was reviewing the relationship between its promotional and investigation work in order to align the two more closely, Government stated that to confine promotional work to the eradication of unlawful discrimination would be to define the CRE's role too narrowly; and, third, albeit by footnote, referred to the Commission's view that it did not regard its promotional programme as over-ambitious or vague. The CRE had argued that the development of programmes takes place firmly in the context of what results can reasonably be expected or aimed for and this remains a guiding factor in regular reviews of progress. The CRE argued that a considerable amount of responsive work is often demand-led, as illustrated by the requirement to respond to the Home Affairs Committee's own inquiry, the Scarman inquiry, and an increased number of requests for advice from a variety of institutions.

However, as formal investigations and research projects were completed, the relationship between enforcement work and promotional work became more apparent: thus the Hackney, Liverpool and Rochdale investigations and research provided the basis for a promotional drive into housing allocation by other housing authorities both directly and through associations of local authorities and into mortgage lending through the building societies and the Building Societies Association. Much of the promotional work of the CRE, nevertheless, has evolved round a process of establishing networks and of developing dialogue with key personnel in umbrella organisations which have the potential for influencing their members to adopt good practice with a view to changing rules, procedures and processes which have been shown to disadvantage blacks. Thus from 1980 a Standing Committee of local authority associations including the Association of Metropolitan Authorities, the Convention of Scottish Local Authorities and the Association of London Boroughs, under the Local Government Training Board, has been convened. This body has considered various reports on housing issues and will consider the Code of Practice on Housing once it is finalised. Unfortunately the pace and commitment of this approach depend, essentially, on the goodwill and receptivity not only of the various umbrella bodies addressed but more importantly on the willingness of members to examine and adopt the practices advocated. The CRE has no statutory authority to require the adoption of Codes of Practice. Until recently, the Code relating to Employment, which, in terms of the Act, has required approval by the Secretary of State, was the sole Code of Guidance having statutory recognition by reason of Section 47 of the Act. The benefit of adopting such a code, in legal terms, is that the employer may establish a legal presumption that his revised practices meet the requirements of the Act to establish a defence under Section 32(3). Consequently the adoption of the code may distance the employer from unlawful discriminatory acts of an employee, even in the scope of his employment. While this special defence is unusual in employment law in allowing the employer to escape indemnifying a claimant (CRE, 1985b:

29), it has failed to provide employers with sufficient incentive to adopt the code on a wide scale.

It is naive to suppose that a majority of organisations in the interests of equality of opportunity, the optimisation of resources and the promotion of sound professional practice will court the antagonism of their employees to innovation and change - if not to the objectives themselves - without sanction or incentive. Of course where discrimination has been exposed or where political or social pressure has been brought to bear, the code may be adopted. The vast majority of housing authorities, however, are not subject to such exposure or pressure and their priorities will lie elsewhere. In these circumstances the promotional work of the CRE, whether or not riding on the ripple of investigations and successful complaints, will, however well targeted and monitored, have little effect on housing practices in the public or private sector. This is not to deny that opportune intervention, following adverse publicity in a County Court or Sheriff Court or following elections and a change of political commitment at local level or consequent to the appointment of a key official, will not individually and cumulatively be valuable. But these opportunities depend on sensitive local knowledge and resourcing beyond those available to the CRE and its 'crime' prevention work, as with its enforcement role, is unlikely to affect the general climate of opinion. Since the vast majority of people in this country are white and are most unlikely to have experienced racial discrimination they have no personal reference point for the location of this phenomenon. An irrational but understandable reaction is to deny the significance and extent of such discrimination, particularly when, in tandem with the lack of first hand knowledge, there is a lack of awareness about the impact of stereotyping which, from cradle to grave, may well go unchallenged and be reinforced by white peers, white Government and white press. Racial discrimination is, then, characterised by the majority as individual, atypical acts motivated by explicit racial intent. It is a minority who see or sense the mundane underlying assumption regarding equality which becomes embedded in institutional practices and translated into blacks systematically achieving poorer housing, by type, by age, by location. But the church of reason is poorly attended and the preaching of the CRE is unaccompanied by any Government organ or chorus of popular tunes to convert the infidel. Moreover, the concept of participatory Local Government, of seeking a broad spectrum of public involvement in the formulatory stages of decision-making, is not widely practised and as a result the everyday experience of all tenants, for example, including blacks, is not harnessed as a management resource to inform future practice. Certainly the Bains (1972) and Paterson (1973) reports on Local Government management under the reformed systems introduced in the 1970s have facilitated a move away from the paternalistic and technocratic committee structures. Policy and Resource Committees may now set goals within which racial equality objectives are more readily housed both in terms of structures - Race Relations Committees, Race Relations Units - and of policies - on equal opportunity in employment, or ethnic monitoring of service provision. While it is not possible to assess the significance of such change or the receptivity of Local Government to the promotional work of the CRE a number of considerations indicate that it is likely to have been marginal.

First, the manpower and regional structure of the CRE impose severe constraints on its outreach activities. Scotland, as we have observed previously, was served from Manchester until 1987. The South West is served from London and Wales from Birmingham. There are only three field officers to cover activities in Greater London. Inevitably this affects promotional work whether in the public or private sector of housing which a concentration on local authority associations, the National Federation of Housing Associations, the Scottish Federation of Housing Associations, etc., merely accentuates.

Second, while the CRE recognises the restrictions imposed by its lack of resourcing and acknowledges the contribution of the local CRC as a complementary sister organisation it has failed to optimise these arrangements in its promotional activities. Following the harsh criticism of CRCs by the Home Affairs Committee (HAC, 1981a: part VI) and the frequently poor liaison with their paymasters, the CRE has introduced annual work plans and area profiles which are designed to facilitate an overview of local community work, the setting of objectives and the monitoring of results. However, the PSI report on the CRCs concluded that there was a mismatch between the planning of work programmes and the work actually done and the CRE staff were insufficiently equipped to fulfil the development and assessment role (PSI, 1988). In the opinion of the PSI the present regional structure of the CRE is ill suited to the task of enabling CRCs to be fitting partners of community groups, local agencies and the CRE itself - the regions are too few, too large and too diverse, with too few field officers over too wide an area. For greater effectiveness in the field, the CRE required radical decentralisation, devolution to manageable territories and the transformation of the field officers' role to one that is counselling and advisory. After the CRE review of the funding and role of the CRCs in 1979/80 and the abandonment of various alternative models then proposed, the CRE 'has given every sign of having reduced its expectations of CRCs' (PSI, 1988: 120). Such clearly discouraging signals from the CRE have affected the capacity of the CRCs to work as local agents for the Commission's promotional activities.

Third, the concentration of specialist staff in housing and other fields in London has undoubtedly affected the professionalism of promotional work in the various regions. While regional staff are now assigned subject specialisms, line management within the regional structure may impede the development of specialist skills necessary to ensure effective dialogue between specialists in the local authority housing department and the regional officers. Such lack of professionalism was noted by the Home Affairs Committee and while the new arrangements and future changes within the CRE following the Pliatzky review (CRE, 1986a) may result in improvements, these will take some time to be translated into more effective promotional activities.

Fourth, these structural weaknesses will not only affect inter-personal contacts at the local level but also the impact of publications on housing disseminated by the CRE. The Home Affairs Committee (1981a: para. 77) noted that several witnesses had praised the Commission's promotional work in housing and the committee was impressed with the professionalism of the literature on housing which had been produced while noting that the outcomes in terms of changed practices among local housing authorities or housing associations may have been slight. Clearly the value of such publications will be limited by the lack of a defined and co-ordinated structure of information dissemination and follow-up. Substantial advances have been made in this area but there are few signs that the CRE is making the best use of the CRCs to launch relevant housing publications. There is little doubt that the CRE did not trust the local CRC to act as a partner in such ventures, or alternatively, the regional officers were too keen to secure absolute control of related promotional campaigns.

COMMUNITY RELATIONS COUNCILS

In the immediate post-war period Churches, welfare groups and concerned individuals in many parts of Britain made spontaneous moves to found local committees to promote the welfare of settlers from the 'New Commonwealth' (Hill and Issacharoff, 1971). Such local committees became formalised and funded by reason of the Race Relations Act 1968. This Act enabled the newly established Community Relations Commission, which had assumed the responsi-bilites of the National Committee for Commonwealth Immigrants (NCCI), appointed in 1965 by the Prime Minister for the purpose of co-ordinating and developing these local liaison committees, to provide support to its local sister organisations, now called Community Relations Councils. This approach had been heralded by the 1965 White Paper (Immigration from the Commonwealth), which made the first attempt to define the work and experience of the thirty voluntary liaison committees as providing 'the necessary focal point in their areas for the co-ordination of effort and a channel for the exchange of information, ideas and experience' (Home Office, 1965). Above all they hoped to create a 'climate of mutual tolerance in which the stupidity of racial prejudice cannot survive'. The main aim of the Community Relations Commission, now subsumed within the Commission for Racial Equality, was to encourage the establishment of, and assist others to take steps to secure, harmonious community relations and to co-ordinate on a national basis the measures adopted for that purpose by others.

From their inception, Community Relations Councils (CRCs) had no statutory basis and accordingly were not directly involved in the pursuit of complaints under the Race Relations Act 1968, which remained the sole responsibility of the Race Relations Board. However, by the 1970s it was realised that CRCs were no longer the local community expressing goodwill to the new immigrants but a group of white and black people struggling to promote racial equality in the public and private sectors. As the PSI report observed, a major problem foreseen for CRCs as they struggled to move with (or in some cases resist) the tide of social change was their entrapment by the unchanging expectations of some of the more powerful interests with which they had to deal (PSI, 1988). Local authorities in particular saw real utility in supporting voluntary organisations which sought to promote harmony and diffuse conflict. The House of Commons Select Committee on Race and Immigration was advised in 1975 by the Association of County Councils that if it were not for the presence of CRCs 'the local authorities would undoubtedly have to employ additional staff to deal with community relations problems'. Since the Race Relations Act 1976, however, by which Section 71 placed a specific statutory responsibility on local authorities regarding the elimination of racial discrimination and promotion of good relations, there has been a greater willingness on the part of many local authorities particularly in large urban areas to adopt a more interventionist stand in relation to the promotion of equal opportunity by the establishment of race relations committees and race relations units and the appointment of race relations advisers. Moreover, as we have noted elsewhere, the emerging opportunities for involvement in local politics and the growing confidence of black organisations, whether in housing, employment or education, have stripped the CRC of its previous implicit role as a power broker with Local Government. In addition changing attitudes amongst the black and Asian communities, particularly the young, with regard to the ability of the local CRC to effect the attainment of equality of opportunity, along with sectarian conflict and competition not infrequently marginalised the role and status of the CRCs. These factors along with a continuing ambiguity about whether they should be in the harmony business as opposed to rights enforcement have increased uncertainty as to their future. In 1988 the Commission for Racial Equality embarked on a series of consultation exercises throughout Britain seeking advice and comment on the PSI report Community Relations Councils: Roles and Objectives. This report, in concluding that CRCs were necessary and desirable, sought a shift in attitude on the CREs part to enable the local CRCs to become more fitting partners to community groups, local agencies and the CRE itself. Accordingly it seems likely that while the structure and roles of local CRCs may undergo change, their existence is not threatened and their partnership with the CRE is likely to be strengthened (MacEwen, 1988).

353

Part III Evaluation and reconstruction

The importance of CRCs may partly reflect their number. In 1987 the CRE provided grant aid for the employment of full-time staff in eighty-four CRCs out of a total of ninety-six. Although they are not the only organisations concerned with this area of work or supported by Section 44 funding by the CRE, the fact that their funding constitutes about a third of the CRE budget demonstrates their significance in financial terms.

With reference to the legal requirements of the Race Relations Act 1976, local Community Relations Councils are likely to be influential in two ways. First, through their casework, they may refer complaints to either the CRE or other agencies providing professional or legal advice. In addition they may deal with matters informally themselves. Unless staff are legally qualified they may not, as yet, represent complainants in the County Court or Sheriff Court with regard to a housing matter. Second, through their promotional work, they may seek to secure the observance of Section 71 by housing authorities and, more generally, seek to secure the compliance of housing associations and private-sector landlords, estate agents, building societies, etc., with the requirements of the Act.

Casework

As Table 12.5 shows, in the region of a quarter of applications for assistance received by the CRE annually constitute referrals from the local Community Relations Councils. However, given that the total number of housing complaints received by the CRE per annum has varied between forty-three (1987: 3 per cent) and 112 (1979: 11 per cent), it seems likely that in any one year the CRE will receive fewer than thirty housing complaints in total from all CRCs throughout Britain. While CRCs are not the only agency making referrals to the CRE in respect of individual complaints they are the most significant. There is no published account as to why the number of housing complaints referred to the CRE by local CRCs is so low, and neither the National Association of Community Councils, which operates principally in England and Wales, nor the Scottish Council for Racial Equality, which has a similar role as a co-ordinating body in Scotland, has yet sought to explain these figures. None of these bodies, however, has contended that discrimination in housing is negligible or that it is diminishing. Clearly, then, the CRCs are not acting as an effective agency of the CRE in such matters. Not infrequently, rather than referring a complaint of unlawful discrimination to the CRE, a local CRC may choose to investigate the matter itself and may negotiate informally with, for example, the local housing authority, a solution acceptable to the complainants concerned. While this approach may be both legitimate and justifiable, the absence of national figures as to the frequency of this approach and its strategic implications has a serious impact on the ability of the CRE or, indeed, any independent agency to arrive at an objective assessment of the impact of the Race Relations Act 1976 in the area of housing.

Promotional work

Just as there are no national statistics collated in respect of CRC casework there is a matching vacuum of information at a national level with regard to the CRCs interest in promotional work on housing although the CRE is able to monitor the housing work of local CRCs through the annual approval of work programmes; such programmes are internal documents for CRE assessment and there has been no attempt to collate the information provided by these programmes nor, being the subject of negotiation between the CRE and local CRC, are they readily and generally available. As a consequence a view of the promotional work of local CRCs with local authorities, housing associations, estate agents, building societies, etc., will either constitute an extrapolation from a selection of the work of local CRCs or be confined to local case studies.

Inevitably, local case studies will give a flavour of local CRC work but will not give a reliable overview, given the wide diversity of CRCs themselves and the differing priority they are likely to give to housing. In addition it has to be recognised that much CRC work may be opportunistic in the sense of being able to respond to changes in Local Government, both at a political and at an administrative level. The work programme of Lothian Community Relations Council (LCRC) for 1988-89 illustrates the local CRCs' concern with housing. The programme anticipates continuing liaison with Edinburgh District Council regarding its implementation of a new policy in relation to racial harassment on local authority housing estates, extending embryonic links with housing associations in order to develop allocation systems which meet ethnic minority needs and seeking the implementation of a report by SEMRU

(1987) which included a series of recommendations relating to the monitoring of applications and allocations for local authority housing, including exchanges and transfers and housing associations referrals. From its inception in 1972 Lothian Community Relations Council has not been involved with work in housing with estate agents, building societies or private landlords at the structural level, although casework may have involved the CRC in dialogue with such bodies. Moreover, again since its inception, LCRC has had few if any links with East Lothian, West Lothian or Midlothian District Councils, whether in respect of their housing policy or otherwise, and it is probable that this comparative neglect of the more rural areas within the geographical responsibility of a CRC will be reflected in the area profiles submitted and approved by other Community Relations Councils.

In respect of LCRC's work with Edinburgh District Council in relation to housing it is clear that the CRC's influence on policy development has come much later than one might expect. The Home Office/Community Relations Commission report Urban Deprivation, Racial Inequality and Social Policy (Home Office, 1976) outlines some of the most pressing problems facing ethnic minorities in Britain and highlighted housing, in addition to employment, education and social services, as a particular area where local groups were keen to strike up relationships with Local Government. As Jacobs (1986: 107) has observed, a number of local authorities were thereafter willing to concentrate resources into these problem areas where, in particular, education and housing were seen as giving rise to 'special needs' in comparison with the rest of the community. The early 1980s were a period conducive to the improvement of group/ Government relations over housing policy. In September 1975 the Labour Government issued Race Relations and Housing, which, in commenting on the 1971 House of Commons Select Committee report on housing, observed that people in housing need should have and feel they have a better opportunity than they had had so far of access to decent housing, whether public or private; and of sharing in the improvement of housing conditions. 'It is the Government's policy to secure this' (DoE, 1975: para. 5). However, no advice was given to the Scottish local authorities by the Scottish Office in respect of how improvements should be effected and the dialogue which then took place in some local authority areas with both ethnic minority groups and local CRCs bore little fruit in Scotland. Indeed in 1979 Edinburgh District Council (EDC) refused to pursue training opportunities for its housing officials and it was only after a change of the political complexion of the local authority in the 1980s, which resulted in the establishment of a Race Relations Sub-committee, that an opportunity for review of the stance over housing arose. Clearly at that stage the impact of the Race Relations Act 1976, both in respect of the general requirement not to discriminate unlawfully in housing provision by reason of Sections 1 and 21, and in respect of the duty to promote good race relations by dint of Section 71, had not affected the climate of opinion at the local level and had not evinced a positive response to approaches by the local Community Relations Council for a review of housing allocation policies or others affecting the local ethnic minority community. The apparent opportunity for intervention had been the change in administration, the establishment of a Race Relations Sub-committee, the sponsoring of a three-volume <u>Ethnic Minorities Profile</u> by SEMRU, the political decision to devolve some housing responsibilities to area teams and a general review of housing allocations initiated in 1987. Cumulatively, these provided a substantial opportunity for the LCRC to re-emphasise its interest in ensuring that the housing authority adopted a policy much more sensitive to ethnic minority needs than it had in the past as it was clear that these renewed overtures would be sympathetically received and would coincide with the housing authority's own view as to how best progress was to be achieved. Although in the case of Edinburgh District Council the influence of the local CRC in effecting policy review may not have been the determinant factor, it should also be borne in mind that the CRC, by securing a virtually monopolistic representation of ethnic minority interests on the Race Relations Sub-committee, influenced the formulation and direction of EDC policy through its own internal structures. Accordingly while EDC may well have instituted a policy review in order to reflect black interests without the aid of the CRC, it is unlikely that the pace and depth of the review would have been similar. It could be concluded that LCRC met the test of 'additionality' to which the PSI study (1988) refers and this beneficial influence of CRCs on the formulation and reformulation of Local Government housing policies has been reflected in other studies such as those on Wolverhampton Council for Community Relations (Jacobs,

Part III Evaluation and reconstruction

1986: 107 et seq.), on Nottingham CRC (Simpson, 1981) and Lambeth (Ousley, 1981).

Strategies for change

In 1981 the PSI conducted a study into local authority provision for ethnic minorities which highlighted the difference between authorities but also the similarity between local situations and the general willingness of black groups whether through the CRE or other organisations to enter into closer liaison with Local Government. This study argued for the Department of the Environment to play a leading role in developing strategies which would produce greater co-ordination of central and local efforts to implement policies designed to combat racial disadvantage more effectively which would lead to a lessening of the discrepancies between authorities and better co-ordination of race-related policies (Young and Connolly, 1981). The status and influence of the CRC are likely to be as much a reflection of local political attitudes as of the calibre of their personnel and the political commitment of their Executive Committees. Given the areas of local discretion and the fact, as we have already noted, that there is a total absence of Central Government guidance on the legal and policy implications of the statutory provision relating to equality of opportunity in housing, a piecemeal approach is almost inevitable. Certainly the CRE and the local authority associations have the opportunity of promoting standardised practice but they lack both the status and the authority to secure their general acceptance. Despite this, however, there is a strong argument to be made for the CRE in particular to review its current policy to ensure a much better co-ordinated approach to implementing and monitoring its guidance on housing matters. An obvious solution to the central/local dilemma in race relations work is to impose a more rigid structure on the relationship between the CRE and the CRCs: various models of reform were put forward by the CRE for discussion in 1979, including central employment of local CROs, but these were generally rejected because of the loss of local autonomy and the recognition that such loss would be to the detriment of the input of voluntary members and more importantly local ethnic minority groups. The current PSI proposals to harness this somewhat unwieldy horse suggest a 'framework of operations' within which a locally negotiated package

358

between the funding bodies, i.e. the CRE and local authorities and CRCs, should take place on an annual basis. While this and complementary proposals to harmonise the interests of the CRE and CRCs may result in greater cohesion what remains an unanswered question is whether the revised framework will ensure effective implementation and policing of the provisions of the 1976 Race Relations Act at the local level where the CRE and CRCs have a tenuous, largely unco-ordinated and haphazardly monitored impact. After thirteen years of the working of the 1976 Act it has become abundantly clear that Central Government will not issue effective guidelines to secure compliance with the spirit of the legislation in relation to housing in the public sector. Although the CRE response to the PSI report on the CRCs proclaims 'A New Partnership for Racial Equality' (CRE, 1989a) and spells out the mechanics of how this is to be achieved with sense and sensitivity, especially in respect of communications, joint planning and linked promotional activity, it is less forthright on law enforcement. Given the statutory role of the CRE in this area, such caution may have been inevitable but the litmus test of the new partnership will remain its ability to secure broader and deeper compliance with the 1976 Act: here judgment must be deferred.

THE PROFESSIONS

While a political commitment at Central and Local Government level may be a necessary condition of a comprehensive approach to meeting the objectives of the Race Relations Act 1976 in respect of equal access to housing, particularly in the public sector, the attitudes of the professions and their influence on their members may either frustrate the political will if it is present or, whether it is present or not, provide an ambience to promote such objectives. Consequently, any commentary on the Race Relations Act 1976 must acknowledge the potential influence of the key professional association in housing, namely the Institute of Housing, and the four legal institutions - the Law Societies of England and Wales and of Scotland, the Bar Association and the Faculty of Advocates. Other professionals including social workers and environmental health officers will often play an important part in the housing process but their impact on law and its implementation in respect of race and housing are likely to be less direct than that of housing officials and lawyers.

The Institute of Housing

The Institute of Housing is an embryonic professional body for people working in housing. It was formed in 1965 when two separate housing organsations merged, the Society of Housing Managers (founded 1916) and the Institute of Housing Managers (founded 1931). The Institute was incorporated by royal charter in 1984, and while this may be symbolic of its growing influence, of greater significance is the fact that a professional qualification in housing, with the expansion of university and college postgraduate courses in the past five years, has only recently become widely recognised as a desirable if not necessary attribute of a housing official. The Institute has 8,500 members (IoH, 1987b: 7) drawn mainly from local authorities and housing associations. Some members work for building societies, property companies, developers, Government departments, private landlords, universities, polytechnics or various housing groups but it is the first category - local authorities and housing associations - which dominate both the membership and the direction of institutional concerns. As the main qualifying body for housing management in the public sector it is charged, in terms of its royal charter, to 'promote the science and art of housing, its standards and ideals and the training and education of those engaged in the profession of housing practice'. The education and training programme is run by a professional administrative staff in London and Edinburgh.

Even within the local authority sphere membership is varied, as is exemplified by the membership within English district councils, which progresses from the nominal (Hereford, Leominster, Mole Valley, Teesdale, Tynedale, Thanet, West Somerset - all one member) through the significant (Nottingham forty-two, Portsmouth twenty-nine, Middlesbrough thirty-one, Leicester thirty-three, Hull thirty-three, Ipswich thirty-four, Bristol thirty-four) to the impressive, principally the metropolitan districts, where Leeds, Liverpool, Manchester and Birmingham all have over seventy members, and the London boroughs. Although the Institute is currently taking positive steps to recruit black members, including those of Asian origin, currently there is clearly a significant under-representation of Asians, amongst others, in its membership. Thus in London authorities, where membership totals approximately 1,000, less than 3 per cent are of Asian origin and the vast majority of that 3 per cent constitute student members. Generally student membership represents over 60 per cent of the total membership and this state of affairs reflects the need for access to professional training and education not only in England, as the figures above demonstrate, but also in Wales and Scotland, where a similar lack of qualified staff can be found in a significant number of local housing authorities.

Thus even if it were suggested that a professional housing qualification (MIH) embodied a coherent set of values concerning housing policy and practice the wide-spread lack of such qualifications throughout the various ranks of housing officers in England, Wales and Scotland would modify the influence of such values. Nonetheless, the role of the Institute of Housing in providing its own examinations and in recognising those offered by others, together with its provision and encouragement of post-qualifying education and training and the promulgation of professional codes of conduct is likely to be of increasing importance in developing a more unified approach to housing issues, including that of race. This does not imply that the diversity of housing tenure and conditions and the similar diversity of Local Government approaches has been or will be harnessed within a straitjacket of professional practice, but rather that an increasing commonality of training and shared values and experience will influence the shape of change and will carry with it the potential for promoting or undermining equality of opportunity in housing.

The influence of the Institute of Housing in such processes stems then from three principal sources, professional education, post-qualifying training and guidance on good professional practice. In turn these sources are likely to reflect the dominant values of the Institute's officials and membership.

In respect of professional education more than 2,000 individuals sit the Institute of Housing's own professional examinations each year and annually more than 300 complete the qualification and become eligible for membership, while there are also in the region of 200 graduates each year from the sixteen recognised courses (as at (1987) which offer both full and part-time qualifications approved by the Institute. Such graduates, from approved courses, become eligible for membership on completion of the practical experience requirement so far as it is not otherwise met on the approved courses. Within both the Institute's own course structure and those approved course structures are elements on race and housing although there is no requirement for this subject area to comprise a specific course or to comprise an approved syllabus. As a result the extent to which equality of opportunity in housing is covered is limited within the Institute's own course structure and will vary from one course to another so far as these courses are provided by other institutions: there is therefore no guarantee that any student has studied, for instance, the policy implications regarding racial attacks on housing estates or the effect of indirect or institutional racism on housing allocations and transfers. The Institute has regularly organised courses on various aspects of race and housing and, indeed, has made a significant effort to promote such courses from 1987, and these courses have frequently involved the CRE. But attendance will frequently reflect the interests of members (and their employers) rather than, of necessity, the discrete needs of the profession. Nonetheless, in 1987 the Institute announced a programme of equal opportunities dealing with such fundamental issues as selection of staff, staff development, ethnic monitoring of the organisation's procedures and racial harassment. Clearly then the Institute is conscious of its responsibilities in this area and while one might criticise a programme for being belated and insufficiently comprehensive it does incorporate positive initiatives.

While the Institute has encouraged PATH local and national schemes, the objective of which is to bolster the number of ethnic minority housing professionals, the continuing dearth of ethnic minority students in housing, generally, suggests the need for further positive action programmes, including career guidance work at schools and special scholarship provision in mainstream university and college courses to redress this imbalance, including incentives in Scotland, Wales and the North East of England.

In 1985 the Institute established a working party on race and housing which produced drafts on: (1) monitoring, (2) consultation, (3) allocation and (4) recruitment. These drafts then became Professional Practice Notes 2, 3, 6 and 7 respectively: a further draft Practice Note on racial harassment was prepared but the working party was disbanded on the establishment of the Local Authorities

Housing and Racial Equality Working Party, whose report on racial harassment was issued in 1987.

The Law Society of England and Wales

The Law Society is the governing body of solicitors, of which there are nearly 53,000 holding practising certificates, the majority being in private practice, but a growing number estimated at over 5,000 being employed in Central and Local Government and in commerce and industry. In July 1988 the Law Society published a survey of its membership: less than 2 per cent were members of ethnic minority origin (618), 440 being Asian, 86 Chinese, 70 Afro-Caribbean and 22 African: 22 per cent of ethnic minority solicitors are sole practitioners in comparison with 9 per cent of the whole profession (Independent, 28 July 1988).

It is led by a president elected annually and a council of seventy solicitors drawn from England and Wales: the Society is responsible for the education and training of solicitors and sets the standards of professional conduct to which all solicitors must adhere. In addition to representing the long-term interests of solicitors through dealings with Government, the general public and the media it provides the profession with a range of services such as further education, specific advice and information, publications, recruitment, etc., and funds a number of special interest groups concentrating on particular sections of the profession such as trainee solicitors, Local Government solicitors, women solicitors and solicitors in commerce and industry.

The Law Society has set up a Race Relations Committee and a Housing Working Party on the Civil Justice Review which reported in 1988. The Race Relations Committee has the function of monitoring the race developments affecting the profession and ensuring equal opportunities for all solicitors. Specifically its remit is:

1. To examine all areas of potential racial discrimination within the profession and to make recommendations on its avoidance and in particular in relation to:
   (a) Difficulties in obtaining articles.
   (b) Difficulties in obtaining posts as assistant solicitors.
   (c) Problems of obtaining partnerships.
   (d) Involvement of solicitors from ethnic minority groups in local law societies.

(e) The instruction and briefing of counsel from ethnic minority groups.

2. To consider complaints of discrimination, whether direct or indirect.

3. To examine and make recommendations concerning monitoring in the profession, to obtain statistics on ethnic minorities, and to consider the results of such monitoring.

4. To encourage the profession to be aware of the problems of racial discrimination and to encourage other committees of the Council and of the Law Society to consider such problems within the context of their own business.

5. To liaise with the Race Relations Committee of the Bar, and

6. Through the Professional Public Relations Committee of the Council to make such reports and recommendations to the Council as the Race Relations Committee may consider necessary or expedient.

The Race Relations Committee published a report (July 1988) which contained statistics as to the ethnicity of solicitors and made recommendations for amending the Race Relations Act 1976 in regard to the narrow exclusive relationship between solicitors and barristers. It has advocated that a code of practice be introduced pertaining to racial activities and that the Council of the Law Society make racial discrimination a disciplinary offence referable to the Solicitors Complaints Bureau. In May 1989 it made representations to the Lord Chancellor. These stressed that the effect of building societies doing conveyancing was likely to lead to a reduction in the number of small legal firms where black solicitors tend to work (Runnymede Trust, 1989b).

The need for a Race Relations Committee with broad terms of reference was emphasised by Goulbourne (1985). Entry to the profession is dependent upon examination and articles:

The evidence to date indicates clearly that the ethnic minority candidate who is unable to obtain articles through chance contacts or in areas of ethnic minority concentration and is therefore thrown on the market is unlikely to obtain articles

until he/she has passed the Law Society's finals.
(1985: 7)

The Housing Working Party on the Civil Justice Review was established in response to the consultative paper on housing in January 1987 with a view to preparing responses to the consultative documents on housing issued by the Lord Chancellor's Department as part of the civil justice review. Apart from the ongoing remit of the Race Relations Committee the Law Society has endorsed the appointment of an ethnic minorities officer whose main task will be to try and recruit more ethnic minorities into the profession.

The Law Society of Scotland

The Law Society of Scotland as at 31 October 1987 had 7,797 enrolled solicitors of whom 6,500 held practising certificates. The Scottish Law Society has similar objects to those of the Law Society, described above. Its council has appointed some thirty-nine committees, none of which deals exclusively with race relations or ethnic minorities. The Society has a staff of sixty-five, none of whom is black. The only report produced by the Law Society of Scotland on the subject of race relations was that of the Law Reform Committee in response to the CRE report Review of the Race Relations Act 1976: Proposals for Change: that report (LSS memorandum, 8 November 1986) was prepared by a sub-committee lacking ethnic minority membership.

The Society does have a code of professional practice but racial discrimination is not mentioned (LSS, 1989). The Society has issued no guidelines on race, has no policy in respect of recruitment of ethnic minorities, does not monitor entry to the profession and has never run any training course (PQLE) on race issues. Following the publication of an article on this topic in the Law Society Journal (MacEwen, 1981) and the responses to a questionnaire circulated to all members with that issue, it supported the establishment of the 'Equal Opportunity Policy and Law Group' by a contribution to an inaugural conference but there is no other evidence of any promotional work whatsoever. The Law Society was notable for its absence at a law seminar held in Edinburgh by the CRE in January 1989 for the purpose of promoting new initiatives regarding delivery and equal opportunity in the profession.

Part III Evaluation and reconstruction

The Bar

The Royal Commission on Legal Services reported in 1979 and found that 'the evidence shows clearly that barristers from ethnic minorities are less successful than others in finding seats in Chambers...' The report observed:

> we were made aware of the fact that there are a number of sets of Chambers whose members are drawn exclusively from ethnic minorities. The Commission was particularly concerned that, with the increase in the number of qualified solicitors from ethnic minorities, if the present pattern of events were repeated the results would be that firms of solicitors composed exclusively of members of ethnic minorities would set up in practices in areas with a substantial minority population ... there would be a clear division on racial lines in the practice of the law and, to some observers, in the administration of justice itself. (Benson, 1979)

The report recognised that a failure to remove even the appearance of discrimination from the legal profession reduced the confidence of every sector of the public in the fair administration of justice and placed the responsibility for counteracting the then state of affairs squarely on the governing bodies of both branches of the profession in recommending the establishment of Standing Committees of both the Senate of the four Inns of Court and the Law Society. Goulbourne has observed that the Commission failed to be specific in identifying the barriers facing black solicitors and black barristers to entry and progress in the profession and, second, it failed to place responsibility on individuals, firms and chambers. Following the recommendations of the Commission, and in response to pressure from black groups and barristers, the Senate established a working party which became the Race Relations Committee and this reported in August 1984. Although it had received no formal complaints of racial discrimination it concluded on the basis of evidence before it that a black barrister faced a particular difficulty in respect of the unwillingness in chambers to treat him as a serious contender for a vacancy. The committee was concerned about the de facto segregation of chambers, finding that out of 210 black barristers 164 were tenants in fourteen sets of chambers which had five or more black tenants. Only thirty-four sets of chambers (excluding the fourteen which had more than five black tenants) had a black tenant. In its view the remedy for this state of affairs lay in the hands of established white chambers. Various recommendations of this committee were adopted by the Bar at the AGM in July 1984, which included: the monitoring of the award of scholarships and financial assistance by the Inns of Court and by sets of chambers; the continuing dialogue with the Law Society, the Barristers Clerks Association and prosecuting authorities about means of eliminating direct and indirect discrimination against black barristers; the submission by all chambers of an annual report stating the numbers of white and black barristers obtaining privileges and tenancies; the implementation of proposals submitted from the CRE on 21 June 1984 and the formulation by the Bar Council of an amendment to the Code of Conduct providing that it would constitute professional misconduct for a barrister to cause or permit racial discrimination.

The Lord Chancellor in 1988 announced the appointment of two black barristers as QCs. The current total of black QCs is now two of 641 (Daily Telegraph, 1 April 1988).

Goulbourne concluded in her discussion paper on minority entry to the legal procession (Goulbourne, 1985) that the various initiatives then taken by the professional bodies marked an important step towards attacking racial discrimination in the profession but that very little would be achieved unless there was an individual commitment on the part of all practitioners to the elimination of racial discrimination. The general attitude was one of disbelief when the issue of racial discrimination in the legal profession was raised and this lack of awareness may well be due to the perception of ethnic minorities as immigrants. When pressured into thinking about the issue of racial discrimination lawyers immediately became defensive, arguing that with the competitiveness of the profession and the pressure of work they were not in a position to change the system.

In Mr Justice Steyn's review of the work of the Bar's Race Relations Committee (RRC) for the period October 1987 to December 1988 reference was made to an independent professional survey commissioned by the Bar Council, the report of which was produced in 1989. He noted that the Bar is exempt from the provisions of the Race

Part III Evaluation and reconstruction

Relations Act 1976 - in so far as barristers do not provide a service to the public or a section thereof in terms of Section 20 - and that, similarly, there is no legal remedy when a solicitor discriminates against a barrister but emphasised, pending necessary amendments to the 1976 Act, that the fact that the detailed Guidelines on discrimination (now in the Fourth Edition of the Bar's Code of Conduct) were 'non-mandatory ought not to mislead anyone' (Steyn, 1988: 10): failure to observe the code 'will clearly be cogent evidence that the Bar has behaved in an unacceptable fashion'. With reference to the RRC's investigation to determine whether black barristers obtained a fair share of prosecution briefs from the Crown Prosecution Service, he advised that the RRC regard the circumstantial evidence as overwhelming in respect of racial disadvantage: the RRC could not be satisfied with the undoubted goodwill, integrity and fairness of the Director of Public Prosecutions; it would be looking for corrective action.

Such forthright assessment is surely encouraging and suggests that the Bar will respond positively to the report of the professional survey and seek to secure necessary changes in the profession.

Advocates in Scotland

The Royal Commission on Legal Services in Scotland reported in 1980 but, although one of its members subsequently became a member of the Commission for Racial Equality, its report (Cmnd 7846) failed to make one mention of ethnic minorities in relation to either the Law Society or the Faculty of Advocates. The Faculty of Advocates is part of the College of Justice in Scotland, which exists primarily to provide a body of specialists in written and oral advocacy, its members having an exclusive right of audience before the Supreme Courts of Scotland. The Faculty is administered by the Dean, Vice-Dean, Treasurer and Clerk and Keeper of the Library, all of whom are elected annually from amongst practising advocates. There are presently in the region of 200 practising advocates, none of whom is black. It has no race relations sub-committee or working party, it has no equal opportunity policy and had, until 1989, never given any consideration to its exclusively white nature.

Judges

A study by Michael King and Colin May on the recruitment of black magistrates led to the disclosure of very fundamental problems in the system of recruitment of magistrates generally (King and May, 1985). This concern was reiterated by Dr Wilfred Wood in an address at the NACRO annual general meeting in 1986 (The Times, 7 November 1986). In referring to a further study, Black People and the Criminal Justice System (NACRO, 1986), Dr Wood pointed to the fact that black defendants were less likely to be granted bail, were ten times more likely to be stopped and searched and were more likely than whites to be prosecuted rather than merely cautioned. He said, 'there are many people who feel that the law does nothing to protect their rights to decent housing and fair opportunities for suitable employment'. In his opinion many seemed to have nothing to gain from the order of society and nothing to lose by flouting it. Black peoples' general lack of faith in the justice system was illustrated by comments made by Judge Abdella (Guardian, 21 August 1984), who observed that it was becoming increasingly difficult to enlist the assistance of the black public in order to form routine identification parades. Some action has now been taken in respect of the white image of the Bench. Thus in 1986 4.57 per cent of the 1,379 appointments made to the office of Justice of the Peace were black (Daily Telegraph, 25 March 1988), representing a marginally greater proportion than the 3.92 per cent of blacks within the age range 35 to 54 from which appointments were made. Nonetheless, by 1 January 1988 only 455 of the 23,730 magistrates were from ethnic minorities, representing 1.9 per cent in contrast to the 4.3 per cent of the general population who were black and the 14 per cent of the prison population who fall in this category.

A NACRO briefing paper published on 13 June 1988 (New Law Journal, 1988: 415) referred to a study of juvenile sentencing in Hackney, north London, between 1984 and 1986 which found that young black people were given custodial sentences nearly twice as often as young whites. A 1985 survey of 117 youth custody trainees by the South East London Probation Service found that Afro-Caribbeans had fewer previous convictions and were less likely to have previously been the subject of probation or supervision orders than whites. Yet 29 per cent of the trainees were

Afro-Caribbean as against 4.7 per cent in the local population. In 1986, West Midlands Probation Service found that in a sample of 222 cases, 25 per cent of the black offenders got custodial sentences compared with 21 per cent of the whites; 30 per cent of blacks convicted of burglary received custodial sentences compared with 25 per cent of white burglars. Probation officers' recommendations for non-custodial treatment were followed in 58 per cent of white offenders' cases but in only 43 per cent of black offenders' cases. A study of juvenile sentencing in north London in 1985 and 1986 found that black Afro-Caribbean youths given prison sentences had fewer previous convictions than their white counterparts but received longer sentences. Cumulatively these figures do not show that black people are more prone to crime than white people. Taking into account the age distribution of the black population, there being more young black people, and young people generally commit more crime, and also the disproportionate representation of black people in areas associated with crime, the crime figures themselves do not suggest a greater propensity by black people to offend than by white people. What they do suggest is more rigourous policing of black youth, less leniency regarding bail, less reliance on the probation service regarding sentencing and a greater inclination by the Bench to impose custodial sentences (but see Walker et al., 1989). Accordingly blacks experience direct and deep-rooted institutional racism in the legal process and this experience is bound to affect their judgment as to the objectivity not only of the police but also of lawyers and the courts. Since the same personnel within the professions are involved in both criminal and civil jurisdiction any argument for distinguishing the approach by the profession in one as opposed to the other will be seen as facile and unconvincing.

Membership of the 'professional' Bench is drawn almost exclusively from barristers and advocates. Inevitably, therefore, the extent to which racial discrimination is tolerated within these branches of the profession will be reflected in both the membership and attitudes of the Bench. In 1988 there were ten Law Lords, twenty-seven Lords of Appeal, twelve judges in the Chancery Division, fifty-three on the Queen's Bench and seventeen on the Family Division: not one of these judges, nor of those in the Court of Session or shrieval bench in Scotland, is black. The Bench, then, has been a white male coterie and so it will remain until the feeder professions themselves effect change and race becomes a legitimate reference in judicial appointments as it has been in relation to lay magistrates.

Moreover while we have noted that both the Law Society and the Bar Council have taken recent action to diminish the extent of institutional disadvantage and discrimination in the profession, such steps have, generally, been a response to external criticism either in official reports such as the Benson report in 1979 or from pressure groups, including the Association of Black Lawyers. The Legal Action Group (Guardian, 7 January 1985) accused the Law Society of hypocrisy over black lawyers regarding its unpublished report in 1985 which blandly claimed there were no problems in respect of black solicitors and concluded that there was no need for ethnic monitoring nor of special places for black solicitors on its council or committees. This conclusion flew in the face of the report of the Senate to the Inns of Court and the Bar in 1984 that many predominantly black chambers complained of few briefs from large firms: the response, therefore, of the professions has been seen to be largely a reactionary one, limiting damage and responding to some of the worst criticisms by others rather than being creative and innovatory by way of initiating positive action from an analysis of its own weaknesses. Obviously it may be argued that whether or not corrective action is motivated by self-appraisal or is a response to external criticism is immaterial; what is material is the extent and nature of the action taken. But the fact that they have been reactive indicates a lack of awareness, an insensitivity and a lack of conviction about the corrective action on the part of the professions which, in turn, may dictate a cosmetic response to the perception of a cosmetic problem. Imagery becomes a substitute for reality.

CONCLUSIONS

The most striking aspect of Central Government's institutional 'response' to race issues relating to housing is its continuing ad hockery whether in respect of policy formulation or in respect of the process of implementation. Whether the refusal of the Home Office to be actively involved in developing 'Government-wide' policies on race is a genuine reflection of the historical independence of other

Government departments, or is merely a device to off-load responsibility, or is a conscious or unconscious attempt to marginalise the impact of issues relating to race relations is, perhaps, less important than the consequences - the virtual absence of any coherent strategy to secure the effective implementation of the Race Relations Act 1976.

That Government has recognised the need to supplement the legislation by a 'more comprehensive strategy' and that an effective strategy to deal with problems of 'deprivation and disadvantage' is evidenced in the White Paper on Racial Discrimination (Home Office, 1975: 6). But, in the context of civil rights retrenchment under the Reagan administration, it has been observed that anti-race discrimination law is not, despite its continuing presence on the statute book, a permanent statement of political commitment (Crenshaw, 1988). While the absence of a comprehensive strategy statement may imply that there is no strategy in fact, this is clearly not a necessary conclusion. Conversely critics have argued (Layton-Henry and Rich, 1986; Solomos and Jenkins, 1987) that in a period of 'recessionary politics' the ethos of individualism and the rigorous control of public expenditure together with the political manipulation of the 'immigration question' and the primacy of social control constitute an unstated strategy of minimalist support for anti-discrimination measures.

Of course, it is arguable that, within the framework of a statement of Central Government policy objectives on equal opportunities, there is advantage in 'devolving' responsibility to the various Central Government departments, in the instance of housing the DoE, the Scottish Office and the Welsh Office, who are best placed to secure that such policies are implemented, periodically reviewed and, when found wanting, revised. But these departments operate in a policy vacuum. Within the DoE it is clear that the Inner Cities Directorate focuses its responsibility for ethnic minorities on the Urban Programme and provides little useful guidance or support to other divisions. The structures in the Welsh Office and Scottish Office are even less convincing: in the latter case the IDS, charged with a nominal responsibility for race issues, has no locus in housing. Whether or not the establishment of a Scottish Assembly (or a Welsh Assembly) would, by instilling a greater sense of accountability, stimulate some sensitivity and awareness to policy, process and practice is open to question but it is difficult to escape from the conclusion of hopeless ineptitude. It is argued, repeatedly, that, as far as housing is concerned, Central Government has withdrawn from day-to-day supervision and control: within the public expenditure limits and fiscal controls on capital and revenue, income and expenditure, local housing authorities are provided with broad areas of discretion to develop those policies they consider best in the light of prevailing local circumstances. But there continues to be a plethora of Central Government advice and guidance to local authorities on planning, on pollution control, on building standards, on joint venture schemes and on mainstream housing provision itself: why is equal access to housing unworthy of Government's attention? Again, had there been convincing evidence that local authorities had addressed this issue effectively, a lack of central guidance would be understandable: we have seen that such evidence is exceptional.

In respect of the 'race relations industry', the CRE and CRCs, the former has yet to demonstrate an effective and sustained linkage between formal investigations and promotional work. Although internal reorganisation, 'regionalisation' of specialist advice, including housing, and a review of priorities and practices may prove beneficial, there appears to be considerable scope for further co-ordination of promotional work. In particular the partnership between the CRE and the CRCs is frequently tenuous and fails to optimise opportunities for local initiatives. The new partnership arrangement, the new name of CRCs (Racial Equality Councils) and their new constitutions offer a sharper focus on promotional work if not on enforcement. Clearly, however, the stature of the CRE and CRCs affects the impact of advice given to local housing authorities: their characterisation as vested-interest lobby groups is indirectly supported by their lack of presence in Central Government policy guidance. While not wishing to suggest that such groups are not important in themselves, this characterisation, in ignoring the statutory responsibilities of the CRE and its Government agency status, may appear to justify a local authority's decision to ignore good practice guides which are proffered.

Given the CRE's annual budget of £11 million per annum, it is apparent that without Central Government support and Local Government co-operation its impact and that of the CRCs on whom it is financially dependent, will be limited in respect of housing in the public sector and negligible in respect of the private sector. In this situation, therefore, the initiatives of the professional cadre of lawyers and housing officials assume greater substance. Certainly the housing profession has addressed racial discrimination belatedly, frequently hesitantly and often with the appearance of reactive containment. But, despite such shortcomings, in comparison with the legal profession it is light years ahead: however, the extent of its membership and the control of its members through dint of their predominantly employee status may limit the benefits of the guidance issued and in preparation. Consequently the ethos and attitudes which pervade the employing agency may clash with and countervail against the emerging sensitivity of the housing professionals.

The legal profession is something else. Current initiatives by the Law Society and the General Council of the Bar reflect, at least, a growing awareness that opportunities for ethnic minorities within the profession have been restricted. External pressure, including the Benson report, has resulted in new initiatives for England and Wales but these are limited and ignore the potential contribution of the profession, not merely in respect of membership, to the effective implementation of the Race Relations Act 1976. Whether the wide gulf between black and white lawyers regarding racial problems, acknowledged by the Race Relations Committee of the Law Society in its 1989 report, would be materially improved by the adoption of its recommended Code of Practice on Racial Discrimination is debatable. But the mere recognition of issues and recommendations for change represent a welcome advance (Runnymede Trust, 1989b), confirmed by the committee's representations to the Lord Chancellor in May 1989 to the effect that the proposed reform of the legal profession would have an adverse impact on black solicitors and clients.

In truth, however, both Law Societies are predominantly trade unions, protecting the working conditions of their members, and wearing the 'public interest' cloak only on the few occasions when the climate of public opinion demands protection from the chilling winds of reality. Their contribution to welfare rights including public-sector housing is far from impressive despite the emergence of Neighbourhood Law Centres of which there are now some fifty-six in England and Wales. In Scotland only one such centre exists, in Castlemilk, and the Law Society of

Scotland has, until recently, been resolutely opposed to the establishment of other centres and the extension of the areas of law in which the one centre may operate.

The Law Centre Federation has (September 1988) applied to the CRE for funding for lawyers to focus on anti-discrimination work. They were to be based in London (two), Sheffield and Glasgow but differences regarding case referrals, management and finance have yet to be resolved. Generally, however, the extent of anti-discrimination work undertaken by the network of law centres throughout England and Wales is not monitored. Evidently many centres undertake significant casework on nationality and immigration as well as racial attacks and harassment. The Law Societies and, for England and Wales, the Lord Chancellor's Department which funds six such centres, have no policy in respect of this area of work. Consequently strategies for its development reflect a piecemeal approach led by local initiatives. Unlike national organisations, such as the CPAG, law centres generally do not deploy test cases where alternative strategies including political pressure may have equal if not better chances of successful resolution of a complaint without the financial and strategic risk of failure. It may be unfair to expect the legal profession to play a dominant role in developing strategic approaches to such issues but it remains open to criticism for its failure to secure, through professional training, entry opportunities and mid-career development, a professional ethos which encourages the sympathetic involvement of lawyers in such areas.

Inevitably judges reflect the attitudes and perceptions of the cadre of barristers, advocates and, occasionally, solicitors from which they are drawn. While rogue antipathy from the Bench to race relations legislation may not be subject to stringent controls, currently the extent of such antipathy together with the ignorance and insensitivity displayed in judicial pronouncements indicates that ethno-centricity if not racism is institutionalised. Such a proposition is clearly true in criminal cases, where analysis of sentencing demonstrates discrimination against black offenders. In civil cases, inevitably, the evidence is much more tenuous. Cases on homelessness, referred to in Chapter Seven may be said to be merely illustrative: it has to be admitted that extracts from judgments may be quoted to support a counter-proposition. But in the absence of detailed analysis it would seem more likely first that judges reflect the ethnic insularity of the community as a whole and the legal profession in particular: second that the 'criminal record' cannot, in logic, be severed entirely from civil matters so far as the ethos of the Bench is concerned: and third that the limited 'training' which may be provided is insufficient to counterbalance these factors.

The situation in Scotland also illustrates the geographical compartmentalism of racism. Clearly the Bench and the Faculty of Advocates in doing nothing to offset racial exclusivity see themselves as blameless victims of institutional apartheid while the Law Society of Scotland, so far as its collective conscience recognises racial discrimination, limits the relevance of training, access and ethics to areas of substantial black settlement.

The various proposals for radical change in the legal profession in England and Wales announced by the Lord Chancellor in January 1989 while emphasising consumer choice and the need to break down protective practices fail to identify the advantages which would accrue to special interest groups, including ethnic minorities. There is a danger that such proposals, and those for Scotland, will attract the most vociferous opposition not from such concerns, but from the professional vested interests which are threatened. The rejection of proposals for a Housing Court in favour of a simplified 'Housing Action' and the proposal to off-load areas of legal aid to advice agencies (including aspects of social welfare and housing) makes predictions about the structure, let alone the practice, of future legal services in respect of race and housing difficult. What is less uncertain, however, if the past record of institutional responses as outlined in this chapter is viewed as a social barometer, is that racial discrimination in housing will remain incidental to the processes of change: any gains will be fortuitous and, without a sea change in institutional awareness, will be unsustainable.

CHAPTER TWELVE

ANTI-DISCRIMINATION LEGISLATION: TOWARDS AN EVALUATION

INTRODUCTION

Although there has been no systematic analysis of the impact of the Race Relations Act 1976, the evidence of discrimination and racial disadvantage, previously alluded to, in both the public and the private field, together with evidence of continuing residential segregation, suggests not only that discrimination is pervasive and continuing but that, with the exception of unlawful advertisements, which were effectively policed with the introduction of the 1968 Act, the scale has not diminished. However, before attempting to draw even tentative conclusions as to why this should be the case it is necessary to disaggregate the major functions of the Commission for Racial Equality enabled by the 1976 Act, namely the investigation of individual complaints and the deployment of formal investigations, the Commission's promotional work having been considered in Chapter Eleven. Thereafter the Act itself is examined with regard to real or perceived weaknesses both in substance and in process.

INDIVIDUAL COMPLAINTS

The CRE policy on providing assistance has been essentially pragmatic rather than strategic. If the referral indicates a good prima facie case, then, subject to existing commitments, it will be supported: until recently there has been no requirement to restrict support under Section 66 to those referrals which have a particular strategic value. However, because the resources of the CRE have been halved in real terms in the past ten years there is increasing pressure to restrict support. Consequently if the number of complaints pursued is to be maintained, let alone increased, it will be necessary for a larger proportion to be assisted through CRCs, advice agencies or solicitors. Given the lack of legal aid available for representation before tribunals - a requirement for all employment cases - and the current proposals to tender out particular areas of law now qualifying for legal advice and assistance, effectively removing solicitors from the first port of call in such areas, there will be increasing pressure on CRCs and advice centres to fill this gap. Currently Neighbourhood Law Centres and CABs have little experience in this area of law, do not monitor ethnic origin and are not involved in strategic planning with the CRE in respect of complaints relating to racial discrimination. Because such changes in the support offered to complainants by the CRE are imminent, it is of critical importance to digest the experience of the CRE at this juncture. It is to be regretted, however, that the information available is frequently inadequate and particularly so in respect of housing.

Between 1978 and 1988 (see Tables 12.1 and 12.2) the CRE received an average of 1,063 complaints from complainants seeking assistance which the CRE is enabled to give under Section 66 of the 1976 Act. The lowest number of complaints was received in 1980 (779) and the highest number in 1988 (1,440). The CRE Complaints Committee dealt with an average of 934 complaints per annum, varying from 672 (in 1980) to 1,403 in 1988. On average 16 per cent of applications for assistance were refused, while initial assistance only was provided in 49 per cent of applications: 27 per cent of applications were given substantial assistance either through CRE representation or through legal representation provided through the CRE: withdrawals run at about 9 per cent. With reference to Table 12.2 only 7 per cent of applications, on average, relate to housing, the majority, 63 per cent, related to employment matters. Although Tables 12.1 and 12.2 show some variation in respect of complaints received between 1978 and 1988 in general there is a remarkable consistency both regarding the overall numbers and in respect of the percentage relating to employment and non-employment matters. With reference to housing complaints variation is greater, the highest number received being in 1979 (112) and the least in 1987 (forty-three). But the figures for 1983 (seventy-two) 1985 (eighty-five) and 1988 (seventy-eight) suggest that there is no particular pattern in respect of housing complaints. According to the 1981 census approximately 43 per cent of

Table 12.1: Applications made to the CRE for assistance under Section 66 of the 1976 Act in respect of complaints relating to unlawful discrimination, 1978–88

		A	B	C	D	E	F	G
	Year	Complaints received	Dealt with by committee	Number not assisted	Assistance only	CRE rep.	Legal rep.	Withdrawn
1	1978	1033	952	99	317	371	137	28
2	1979	986	868	52	381	255	111	69
3	1980	779	672	84	321	185	29	53
4	1981	864	682	110	214	189	118	51
5	1982	956	771	148	323	98	114	88
6	1983	994	1,022	90	630	55	126	121
7	1984	1,202	1,189	324	507	46	185	127
8	1985	1,150	1,112	219	600	90	88	115
9	1986	1,016	1,032	169	566	116	78	103
10	1987	1,271	1,275	191	700	170	97	117
11	1988	1,440	1,403	261	800	151+	103+	88
12	Totals	11,691	10,978	1,747 (16%)*	5,359 (49%)	1,726 (16%)	1,186 (11%)	960 (9%)

+ The total of 254 for columns E and F combined is quoted in the 1988 Annual Report: those figures (151 and 103) are an estimated disaggregation.

* Per cent of total complaints dealt with by committee, rounded to whole numbers.

Source: CRE Annual Reports 1978–88

the New Commonwealth population live in London and the proportion of complaints at 46 per cent over the period 1978 to 1988 (Table 12.2, last column) approximates to what might be expected. For the same period 25 per cent of complaints originated through the Community Relations Councils and although it is not possible to state with any confidence how many of such complaints would have been made direct to the CRE in the absence of a local Community Relations Council it is apparent that they play a significant role in referrals: it is not known, however, how many individual complaints are pursued with the direct assistance of the local CRC without reference to the CRE.

The results of individual complaint cases fully assisted under Section 66 of the Act and reported to the Complaints Committee are summarised in Table 12.3. Although the lapse of time between the decision to provide representation and the reporting of determinations by the court to the committee may result in a different total for each of these two categories in any given year, it would be expected that these totals would even out over time, taking into account cases withdrawn in the interim. However, the totals for cases reported consistently exceed cases represented over the five-year period analysed (Table 12.3). This may be explained only partially by the number of withdrawals in respect of complaints only receiving initial assistance from the CRE. The totals given in Table 12.3 (for employment and non-employment cases) exceed those represented consistently over the four-year period analysed (Table 12.1, E and F, 7 to 10). But in any event the published figures in the CRE's annual reports do not readily lend themselves to analysis. This is borne out by Table 12.4, which provides a breakdown of housing complaints received in the same five-year period 1984 to 1988 but does not include success/failure rates, as these figures are not available. However, the figures for success or failure in Table 12.3 show a significant and increasing success rate in respect of both employment and non-employment cases where settlement on terms is included with the success rate following on hearing. In isolation the hearing success rate, ranging from 17 per cent to 27 per cent in employment and 21 per cent to 29 per cent in non-employment cases, it may be argued, is low but there is evidence of an increasing proportion of successes following on a hearing. A CRE survey (CRE, 1986b: 18) showed that applicants with the benefits of representation in race cases before industrial tribunals were successful in

Table 12.2: Type and source of Section 66 complaints 1978-88

	A	B	C	D	E	F	G
	Year	Complaints received	Employment	Non-employment	Housing	All from London/South	All from CRCs
1	1978	1033	699	334	55	415	255
2	1979	986	619	367	112	402	275
3	1980	779	458	321	79	364	207
4	1981	864	547	317	67	410	213
5	1982	956	595	361	+	458	242
6	1983	994	567	427	72	464	*
7	1984	1,202	765	410	67	591	351
8	1985	1,150	734	402	85	563	232
9	1986	1,016	619	380	59	493	272
10	1987	1,271	827	428	43	572	349
11	1988	1,440	982	449	78	679	299
12	Totals	11,691	7,412	4,196	788	5,411	2,961
13	% of total	100	63	36	7	46	25

+ (+71 average)   * (+266 average)

Note: The difference between the column B and the sum of columns C and D consists of 'out of scope'.

Source: CRE Annual Reports 1978–1988

Table 12.3: Results of individual complaints cases fully assisted (Section 66) as reported to the Complaints Committee during the year 1984–88

| | Employment | | | | | Non-employment | | | | |
	A 1984	B 1985	C 1986	D 1987	E 1988	F 1984	G 1985	H 1986	I 1987	J 1988
1 Settled on terms	38	46	33	59	67	7	12	12	12	7
2 Successful after hearing	18	38	16	39	38	5	15	9	7	3
3 Dismissed after hearing	52	54	29	48	52	12	11	9	5	5
4 Total	108	138	78	146	157	24	38	30	24	15
5 Settlement/hearings success rate (%)	52	61	63	67	67	50	71	70	79	67
6 Hearings only success rate (%)	17	28	21	27	24	21	39	30	29	20

Source: CRE Annual Reports 1984–88

Table 12.4: Housing complaints received by the CRE 1984-88 (%)

	Year	A Total	B London	C South	D Birmingham	E Leicester	F Manchester	G Leeds
1	1984	67	41(61)	12(18)	3(4)	4(6)	6(9)	1(1)
2	1985	85	40(47)	15(18)	4(5)	13(15)	8(9)	5(6)
3	1986	59	24(41)	9(15)	3(5)	4(7)	7(12)	12(20)
4	1987	43	22(51)	7(16)	6(14)	3(7)	2(5)	3(7)
5	1988	78	26(33)	28(36)	6(8)	5(6)	6(8)	7(9)
6	Total	332	153(46)	71(21)	22(7)	29(9)	29(9)	28(8)

Source: CRE Annual Reports 1984-88

over 25 per cent of cases whereas those without represen-tation were successful in only 5 per cent of cases. Of all race cases heard before the industrial tribunal during the survey period the Commission provided representation in only 29 per cent. Survey data on race cases going before designated County Courts or Sheriff Courts, either in respect of non-employment matters generally or in respect of housing cases only, is not available and comparisons, therefore, between the relative success rates of CRE-assisted complainants and others cannot be made.

The number of housing complaints received by the CRE, averaging 7 per cent over the eleven-year period from the inception of the 1976 Act, is surprisingly small in both absolute and comparative terms. Recent testing of discrimination in the housing market was conducted for the BBC in its series of programes entitled 'Black and White' (14 April 1988 et seq.) which, in comparing the treatment received by a black applicant and a white applicant, demonstrated that, in the Bristol area, less favourable treatment might be expected in 30 per cent of applications for bed and breakfast and 18 per cent of applications for accommodation on the basis of colour, purely on initial inquiries. Accordingly, and in keeping with the previous PSI surveys, there is a continuing high expectation of discrimination in housing and, given that housing along with employment and education is one of the most important areas of an individual's living experience, it is apparent that the low figures referred to are not indicative of an absence of discrimination but are much more likely to reflect either a reluctance to complain or, in the absence of testing, the absence of relevant knowledge by a potential complainer which would justify the pursuit of a complaint.

There is great variation in the number of complaints originating through CRE offices: in Birmingham the variation has been between 4.5 per cent and 14 per cent; in Leicester between 6 per cent and 15 per cent; in Manchester between 4.5 per cent and 12 per cent and in Leeds between 1.5 per cent and 20 per cent. Such regional variation is not confined to the housing sphere. Thus, in employment in the year 1987, of the fifty individual applications for assistance alleging discrimination by trade unions, thirty-nine (78 per cent) originated in Birmingham, compared with one (2 per cent) originating in London.

With reference to complaints referred to the CRE by local CRCs the variation is between 20 and 29 per cent.

Table 12.5: Individual complaints referred by CRCs 1984-88 (%)

Year	A Total	B London	C South	D Birmingham	E Leicester	F Manchester	G Leeds
1 1984	351(29)	13(9)	51(14.5)	43(12)	81(23)	87(24)	76(22)
2 1985	232(20)	10(4)	30(13)	40(17)	53(23)	59(25)	40(17)
3 1986	272(27)	13(5)	55(20)	30(11)	49(18)	60(22)	65(24)
4 1987	349(27)	5(1.5)	71(20)	121(35)	35(10)	66(19)	51(14.5)
5 1988	299(21)	16(5.3)	60(20)	52(17)	70(23)	54(18)	47(16)
6 Total	1,503(25)	57(3.8)	267(17.8)	286(19)	288(19)	326(22)	279(19)

Note: Column A: percentage of total complaints received by CRE during period. Columns B to G: percentage of total CRC – referred complaints during the period.

Source: CRE Annual Reports 1984-88

However, amongst the regions it is noteworthy that reference from London CRCs is consistently low, varying between 1.5 per cent in respect of 1987 and 9 per cent in 1984 as a proportion of all complaints referred to the CRE by CRCs in the respective years. With the exception of Birmingham, whose CRCs referred 35 per cent of all CRC-originating complaints in 1987 and only 11 per cent in 1986, the regional variations outside London do not appear to be substantial although in Leicester CRCs referred 23 per cent of total CRC complaints in 1984 but only 10 per cent of that total in 1987. The number of CRC-referred complaints over the eleven-year period was exceptionally high in 1984 and 1987. The latter is explained largely by the very high number of referrals through Birmingham in that year, while in the former year of 1984 the increase in CRC-originating complaints was evenly spread throughout the regions. With the exception of 1984 there appears to be no correlation between regions in respect of increase or diminution per year of complaints originating through the CRCs. The very low percentage originating from London CRCs, may be explained, at least in part, by a greater knowledge in London of the whereabouts of the headquarters of the CRE resulting in increased direct access, in part by the relatively low staffing levels of CRE field officers in London making contacts with local CRCs more difficult and possibly also by both the greater willingness of London CRCs to assist individual complainers directly or to use other agencies such as law centres. The CRE annual reports also refer to complaints originating from CABs, from neighbourhood law centres and from solicitors, but at present such referrals do not constitute a significant proportion of all complaints.

FORMAL INVESTIGATIONS

Of the forty-four reports of formal investigations conducted by the CRE between 1977 and 1988 eleven relate to housing. By the end of 1988 the CRE had initiated a further five housing investigations (CRE, 1989(f)). The housing investigations are summarised in Table 12.6. In its 1978 Annual Report (CRE, 1979: 5) the CRE set out how it examined each of the major areas in which it might be conducting investigations and asked four basic questions:

1. How important is this area in terms of equal opportunity for ethnic minorities and good race relations?
2. What is the extent of ethnic minority disadvantage in the area, and how much is unlawful discrimination the cause of that disadvantage?
3. Are formal investigations an appropriate and effective method of eliminating discrimination and promoting equal opportunity in this area? (Broadly speaking the more organised an area is and the larger the institutions involved, the more susceptible it is to effective investigation. Where, however, there are innumerable small units, a wide-ranging investigation is much more difficult.)
4. What is being done by other agencies in this area, e.g. Government departments, local Community Relations Councils and the Runnymede Trust, bearing in mind that it is only the Commission, after conducting an investigation, that has the power to issue a Non-discrimination Notice?

In looking at these priorities the Commission made reference to its view of the main reason for being given power to conduct formal investigations as tackling unlawful discrimination. But it recognised that such investigations may be linked with an examination of other causes of disadvantage, and the recommendations for tackling these, although not in themselves legally binding, will be just as important, if not more so, as any Non-discrimination Notice that might be issued.

In commenting on the priorities adopted by the CRE, the Home Affairs Committee in its 1981 report on the CRE (Home Affairs Committee, 1981a: para. 5.8) observed that the criteria and sense of priorities in deciding where to investigate seemed sensible. The committee observed that from the list of current investigations there was an evident regional spread and also a spread among sectors in the employment field. With reference to the then current investigations the committee observed that it was surprising that they should be so heavily concentrated on employment, given that the Commission had only sixteen of its thirty-seven investigative staff in the employment section, compared with seven in housing and eleven in education and services. It concluded that the staff should be reallocated or that the housing and education services were not pulling

Table 12.6: Formal investigations: housing reports issued or awaited 1979–88

	Name	CRE area office	Year	Type	Result	Promotion
1	Brymbo Community Council (Wrexham)	Birmingham	1981	Inducement to discriminate	Finding of discrimination S. 21 Non-discrimination Notice	Publicity for report
2	Collingwood Housing Association	Manchester	1983	Allocations	Possible discriminatory practices: S. 28	Acceptance of recommendations; publicity for other housing associations
3	Cottrell and Rothan	London	1980	Accepting discriminatory instructions from vendors	Finding of discrimination Non-discrimination Notice issued	Publicity; Guide for estate agents
4	Hackney Borough Council	London	1984	Allocation	Finding of discrimination Issuance of Non-discrimination Notice S. 20	Acceptance of recommendations; continuing ethnic monitoring; publicity
5	Mortgage Lending in Rochdale	Leeds	1985	Mortgage lending	Indirect discrimination	Recommendation to building societies, local authorities and estate agents in area; publicity

	Location	Year	Subject	Finding	Recommendations
6 Council Housing and Work Permit Holders (Kensington and Chelsea, Barnet and GLC)	London	1982	Allocation	Finding of discriminatory practice (S. 28)	Recommendations; publicity
7 Walsall Metropolitan Borough Council	Birmingham	1985	Allocations and dispersal policy	Finding of discrimination	Acceptance of recommendations; publicity
8 G.D. Midda/ D.S. Services Ltd, accommodation agency	London	1980	Accommodation agency services	Discrimination SS. 20, 21 and 30. Non-discrimination Notice	Equal Opportunity policy; publicity
9 Allen's Accommodation Agency	London	1980	Accommodation agency services Racial preferences	Finding of discrimination. Issue of Non-discrimination Notice	Acceptance of recommendations; publicity
10 Tower Hamlets Borough Council	London	1988	Allocations treatment, homelessness	Finding of discrimination. Non-discrimination Notice	Acceptance of recommendations; publicity; ethnic monitoring for five years by CRE
11 Richard Barclay & Co., estate agents	London	1988	Discriminatory treatment	Finding of discrimination. Non-discrimination Notice	Action considered under Estate Agents Act 1979; publicity. Good practice guide issued 1989
12 Liverpool City Council	Manchester	1989	Nominations for housing association lets	Finding of discrimination. Non-discrimination Notice	Acceptance of recommendations; publicity

their weight. They also suggested that the decision to investigate might have been made too hastily or on insufficient, hearsay evidence but qualified such observations by the acceptance that such decisions were bound to appear by their selective nature to be arbitrary and in the then absence of completed reports finer judgments were impossible. The committee also observed that the Home Secretary had the power to direct the Commission to undertake an investigation which had not been used and recommend that the Home Secretary should bear in mind the possibility of directing the Commission to undertake an investigation. Alluding to the large team, including staff seconded from UKIAS, involved in the investigation into the Immigration Department of the Home Office, the committee, noting that there was a backlog of over thirty uncompleted cases, recommended that in future the Commission should take stock of its staff resources before plunging into a major new investigation. Moreover, while the committee acknowledged that there remained one or two 'responsive' investigations into comparatively small fry and that the procedures for investigations themselves caused delay, most of the causes of delay in completing investigations rested with the Commission, which had yet to develop that style of brisk and systematic investigation which might be expected of it. In the committee's view it was necessary for the Home Office and the Commission, together, to conduct a thorough review of the Commission's practices in the conduct of investigations with a view to ensuring that there was no repetition of the prolonged delays which had hitherto marred the Commission's investigative record. The committee heard evidence from major organisations under investigation, who gave confidential submissions which were unanimous in deploring the length of time taken by investigations. The committee remarked laconically that if the CRE were to investigate itself as the committee had done in the report, the results might well not have seen the light of day until 1984. Such delays, in the committee's view, not only infringed the natural rights of the body under investigation but seriously diminished the impact of a completed report.

In the Government's reply to the Home Affairs Committee report on the CRE (Home Office, 1982a), the various recommendations (pages 7 and 8, recommendations 14 to 19 inclusive) were either accepted or noted, including the suggested repeal of Section 49(4) of the 1976 Act, which was accepted to be a major delaying factor. This section, which is commented on later, provides that respondents should have the opportunity of making representations to the Commission before an investigation begins. In cases where the investigation proposed is into the acts of specifically named persons, the committee observing that this provision served little or no worthwhile protection to a respondent as judicial review represented a considerably more reliable safeguard.

In response to the various recommendations relating to investigations the CRE undertook to review training procedures for its equal opportunity division staff so as to ensure that they were aware of the need for discretion and courtesy, to pay particular attention to any publicity given to investigations where there have been no allegations of discrimination against the respondent, to show greater discretion in the publicity which it gave to uncompleted investigations, to take stock of staff resources before plunging into a major new investigation and, in consultation with the Home Office, to undertake a review of current practice to ensure that there was no repetition of prolonged delays. However, the CRE observed that in its view the committee had failed to understand the extent and complexity of large investigations. Even as research exercises alone they would be lengthy and demanding, for the issues involved were difficult and sensitive and the CRE was often hampered by a lack of records. The CRE had to make sure, as far as possible, not only that it got things right but that everything it did, including the issues of Non-discrimination Notices, could withstand challenge in the courts and tribunals. In the CRE's view, then, the committee had underestimated the delays caused by the machinery as well as by respondents.

With reference to Section 49(4) it is noteworthy that there was no equivalent sub-section in the Sex Discrimination Act 1975 as first passed. Nor was any such provision contemplated in the White Paper which preceded the 1976 Act: it was added as a House of Lords amendment and became also an amendment to the 1975 Act. Its effect was stated at the time by the Government when accepting the Lords' amendment as being 'to give a person against whom a complaint is made a right to information...' (HL Deb., Vol. 91, Col. 603). At the time of the Home Affairs Committee report and the Government's response, Section 49(4) was seen as a procedural safeguard applicable to a particular

391

type of formal investigation. However, the case of Hillingdon London Borough Council v CRE (1982) AC 775 seriously affected the power of the Commission to conduct investigations. In short, the effect of Hillingdon London Borough Council was to question whether it would any longer be possible for the CRE to carry out formal investigations into named persons or groups of a strategic nature in such areas as housing, education or employment, unless information was obtained prior to the investigation which led the Commission to believe that discrimination was occurring. The doubt expressed in the Hillingdon case was confirmed in the case concerning the Prestige Goup (1984) IRLR 335, by which time the Commission had on hand seven other investigations of a similar type and felt bound to accept their invalidity also, although several of the respondents had nonetheless agreed to the publication of reports of the Commission's findings.

In the Hackney investigation (CRE, 1984b) the existence of widespread discrimination was demonstrated by detailed statistical analysis. The report of the investigation was widely publicised and formed the basis of a promotional drive to bring about change in other local authorities. The Minister of State, Sir George Young, endorsed ethnic monitoring of service provision in this context as a result of the Hackney findings. However, since Prestige the Commission has been restricted in conducting such a broad investigation into a named authority.

The difficulty this raises is that it is only when the system is looked at as a whole that the existence of discrimination will become apparent and it is therefore most improbable that any individual would be able to bring to the Commission the necessary evidence of suspicion wide enough to justify a far-ranging accusative type of formal investigation. Consequently future investigations, where they involve named respondents, are likely to be confined to responsive situations, where there is existing evidence of discrimination sufficient to form a suspicion on the basis of which an investigation would be justified. Such investigations will be confined in scope to that area where the suspicion is backed by evidence before the investigation is embarked upon. Clearly this has significant implications for the strategic role of the CRE in conducting formal investigations, and while the Rochdale inquiry into mortgage lending (CRE, 1985c) illustrates the potential utility for general investigations the respondents are not named. Such investigations have inherent drawbacks including problems associated with the issuance of Non-discrimination Notices and the question of what sanctions it might be appropriate to apply, if any, in the event of patterns of discrimination emerging from the inquiry.

With reference to the balance of formal investigations undertaken, seven of the twelve listed in Table 12.6 were in London and five related to local authorities either directly (Liverpool, Hackney, Tower Hamlets, Walsall), or indirectly (Kensington and Chelsea, Barnet and the GLC in respect of work permit holders). Two related to accommodation agencies, two to estate agents, one related to mortgage lending and one to inducement to discriminate by Brymbo Community Council. Clearly there is a bias towards London and the public sector and both raise questions regarding whether or not the CRE is consistently following its own set of priorities as set in 1978 and as modified in 1986, where it stressed a movement to an emphasis on outcomes and on areas where there was a real prospect of effecting change quickly.

With reference to the London bias, in many London boroughs such as Camden, Lambeth, Newham and Brent amongst others, there is evidence that significant initiatives have been taken. Although questions may well remain as to how satisfactory and speedy progress has been and there may be some justification for correlating such progress to the outcome of complaints and investigations in the London area, there is certainly an argument that future investigations in the public sphere should be pursued outwith London to provide balance and a broader regional impact on the promotional work following investigation.

Similarly, given the increased importance of housing association tenancies by reason of new build and the assumption of ownership of local authority housing provided for by the Housing Acts of 1988, there is a strong argument that further strategic investigations should focus on that area. It is recognised, however, that such a strategy may well be inhibited by the restrictions imposed on formal investigations resulting from the court cases referred to above.

The AMA survey Housing and Race (AMA, 1985: 7) of sixty-nine housing authorities, of which sixty-one responded, found as a 'fundamental and recurring' conclusion that although significant progress had been made by a few local authorities towards ensuring racial equality in housing the majority had a long way to go in terms of policy and practice and many had not yet got started. There was a cluster of fifteen authorities, ten of them in London, which were actively seeking to eradicate racial discrimination in housing. The London borough of Lewisham's anti-racism plan for housing sought to act on and implement the recommendations of the CRE report of its formal investigation into Hackney and the AMA has endorsed the Hackney report as well as the Race and Council Housing in Liverpool publication (CRE, 1984a). This report illustrates that the AMA amongst other associations has taken and continues to take action in order to improve the promotion and adoption of equal opportunity policies and their monitoring in the housing area.

This being the case it would appear prudent for the CRE, in determining its strategy for formal investigations in housing, to consult national and local bodies if it is committed, as the 1978 report suggested, to using the mechanism of formal investigations as a complement to activities undertaken by other institutions in the housing sphere. There is no evidence that this has been done on a systematic basis.

PROMOTIONAL ACTIVITIES IN HOUSING

The promotional work of the CRE has already been examined in Chapter Eleven. Table 12.6 identifies some key initiatives conducted by the CRE during the period 1979 to 1989. Clearly these summaries are not comprehensive but should be sufficient to indicate the breadth and width of CRE activity in its promotional work in housing and, moreover, given that it reflects the CRE's own view of its priorities, the table should not significantly misrepresent the focus of the CRE thrust in its promotional work.

The summary of promotional activity provided by Table 12.6 confirms that the CRE does attempt to use formal investigations as a key to promoting private and public initiatives in housing.

THE 1976 ACT: CRE PROPOSALS FOR CHANGE

Introduction

Section 43 of the 1976 Race Relations Act in establishing the Commission for Racial Equality charged it, inter alia, by way of Sub-section 1(c) with the function of keeping 'under review the working of this Act and, when they are so required by the Secretary of State or otherwise think it necessary', drawing up and submitting to him proposals for amending it. While the Secretary of State has not required the CRE to draw up proposals for change, in July 1983 it issued a consultation document entitled The Race Relations Act 1976: Time for a Change? (CRE, 1983c). The CRE, having gone through a wide process of consultation, then issued its final proposals entitled Review of the Race Relations Act 1976: Proposals for Change in July 1985 (CRE, 1985b). The consultation process itself had, evidently, been fairly lengthy and involved the distribution of some 18,000 copies of the discussion document, a series of conferences and the submission of over 200 comments on the proposals. Although the purpose of this section is not to review all the recommendations for change but merely to focus on those with a direct or indirect impact on housing it is important to note that the CRE viewed the Race Relations Act 1976, in its final submission to the Secretary of State, as 'generous in its intentions' and it did not therefore seek a radical departure from the main themes of the legislation (CRE, 1985b: 1).

In referring to the extent of continuing discrimination in Britain the CRE referred to the survey conducted between February 1984 and March 1985 by the Policy Studies Institute in collaboration with the Commission, involving London, Birmingham and Manchester (PSI, 1985). Here controlled testing had revealed that over one-third of employers discriminated against black applicants. This research provided a measure of the extent of direct discrimination only and consequently provided no indication of the extent to which black people were further disadvantaged in the job market by indirect discrimination. The report agreed with the Commission's own experience that, in spite of some valuable initiatives by individual institutions and employment agencies, the extent of racial discrimination in recruitment to jobs had remained fairly constant over the ten-year period since the previous PSI survey.

At that time, no such similar investigation or research had been conducted in order to discover the extent of discrimination in housing, whether in the public or private sector, but the Commission concluded that its formal investigations in Hackney and its research report on Liverpool, in revealing widespread discrimination in public housing, was indicative of discriminatory practices on a large scale on which the Act had had very little impact. Although the Commission acknowledged that both promotion and persuasion had a large part to play in eliminating discrimination it placed emphasis on the formal investigations and individual complaints as providing an essential basis for the promotional work, referring to the Select Committee's conclusions to which reference has been made above (Home Affairs Committee, 1981a: para. 14).

In the attempt to eliminate discrimination, persuasion and promotion alone will not suffice and the incentive to change has often in practice to be related, at least initially, to the existence of the Commission's law enforcement powers: underpinning the Commission's promotional effort there has to be the prospect that, failing voluntary change, the law will be used effectively to cause that change to occur and consequently the state of the law relating to discrimination is of fundamental importance, not as an end in itself but as a mechanism for effecting change and bringing about good practice.

We have noted how the Commission's enforcement powers fall into three categories:

1. The power to assist individuals alleging discrimination.
2. The power to bring legal proceedings in respect of allegations of pressure or instructions to discriminate; unlawful advertisements; and persistent discrimination.
3. The power to conduct formal investigations and issue Non-discrimination Notices.

In the Commission's view the most pressing cause for concern, although by no means the only one, was the way in which the present law was working in the area of formal investigations. However, it referred to Lord Denning MR's opinion in the Court of Appeal in CRE v Amari Plastics (1982) 1 QB 1194 that 'the machinery of the Act is so elaborate and cumbersome that it is in danger of grinding to a halt'. The proposals for change, made by the CRE, addressed seven areas:

1. Improving the definition of discrimination.
2. Reducing the number of exemptions from the provisions of the Act.
3. Providing for specialist tribunals to hear all discrimination cases.
4. Redefining the Commission's formal investigation powers.
5. Providing for the Commission to have access to an independent tribunal of fact across the range of its law enforcement activities.
6. Improving the remedies for dealing with proven discrimination.
7. Improving a number of the mechanisms for bringing about change.

The definition of discrimination

With reference to direct discrimination the Commission pointed out the distinction between 'motive' and 'grounds' for a direct discriminatory act. Thus, for example, a person discriminates unlawfully where he or she declines to allocate a tenancy to a black person in order to meet supposed tenants' dislike of black neighbours even though he or she does not share that dislike. A refusal to allocate a tenancy would demonstrate grounds which are racial even though the motives might not be. In the CRE's view the basic text of the Act, which in fact refers only to racial grounds, should be made clear without further reference to existing case law (see Sections 1.1. and 1.2).

With reference to indirect discrimination the CRE referred to three definitional problems of particular importance: first, the 'condition or requirement' referred to in Section 1(1)(b) as interpreted by the Court of Appeal in Perera v Civil Service Commission (1983) IRLR 166 is too narrow to be effective in the sense that a condition or requirement would only be said to exist where it amounted to a complete bar if it was not met (see also Meer v Tower Hamlets London Borough Council (1988) Times, 3 June). Currently where a discriminatory criterion, if only one of a number of criteria, is expressed as a preference it would not constitute a condition or requirement in terms of the Act despite its adverse affect on ethnic minorities, although evidently if the preference was exercised with the intention of adversely affecting any particular ethnic group it would amount to direct discrimination. In the CRE's view the adoption of a test relating to policies, practices or preferences would more nearly meet the intentions of Parliament when the Act was first passed.

With reference to the phrase 'considerably smaller' in Section 1(b)(1) the CRE referred to the decision in Wong v GLC (EAT 524/1979, unreported), where it was decided that if a requirement or condition ruled out all of one racial group then, as none of that group 'could comply' and as nil is not a proportion at all, the considerably smaller test could not be applied. In Kidd v DRG (UK) (1985) ICR 405, Waite J expressed the view that the question of how large a proportion must be before it can be called 'considerable' was very much a matter of personal opinion on which views were likely to vary over a wide field. The CRE, referring to US experience, suggested the adoption of a four-fifths rule to the effect that the Act should be satisfied where a 20 per cent difference between the racial groups involved was demonstrated.

However, more important than the above two considerations, in the CRE's view, was the meaning of 'justifiable' in Section 1(b)(2). Having described the watering down of the test applied to 'justifiable' from 'reasonable commercial necessity' to a reason accepted as 'tolerable' by a 'right thinking person' (Ojutiku and Oburoni v MSC (1982) IRLR 418), to which reference has been made previously, the CRE argued for the substitution of a test based on what is 'necessary'. A conclusion that the test 'justifiable' is one of fact and will depend on the subjective judgment of the court or tribunal (see Mandla v Dowell-Lee (1983) 2 AC 548) will leave the decision unchallengeable on appeal and will lead to widely differing decisions being recorded.

With regard to victimisation by reference to Kirby v MSC (1980) 3 All ER 334, the CRE concluded that the protection against victimisation for invoking the Act is incomplete and consequently advocated that the remedy for victimisation should be redefined so that there is protection against a person's suffering any detriment whatever as a result of his or her doing any of the acts listed in Section 2.1(a)-(d) (involvement in allegations, proceedings, etc., under the Act).

Despite the close connection between religious discrimination and ethnicity demonstrated by the House of Lords decision in Mandla v Dowell-Lee (above) the CRE felt that it may not be appropriate for the same piece of legislation as that covering racial discrimination to tackle the issue of religious discrimination (note, however, that Title VIII of the 1968 Civil Rights Act in the US in addressing fair housing provision also includes discrimination on the basis of religion).

Exemption from the Act

The CRE referred to the fact that at present a wide range of actions, Governmental in nature, are outside the ambit of the Act. The definition relating to the provision of 'goods, facilities and services' (Section 20) should make it clear that it extends to all areas of Governmental and regulatory activity, whether central or local, such as acts in the course of immigration control, in the prison and police services and in planning control. With reference to the latter we have already noted the amendment to the Town and Country Planning Acts of 1971 and 1972, extending to England and Wales and Scotland respectively, to the extent that discrimination on racial grounds in planning will be covered by the 1976 Act.

By reference to Section 41, which protects discrimination carried out in pursuance of any enactment regardless of its antiquity and extends beyond this to circulars and ministerial pronouncements, and similarly by reference to Section 75(5), which provides exemption from the Act for certain discriminatory employment roles in the Crown service and prescribed public bodies by reference to birth, nationality, descent or residence, the CRE alluded to the incongruity of a piece of civil rights legislation such as the Race Relations Act being made subordinate in its own terms to a wide range of rules, existing or future, with which it conflicts. The CRE, therefore, advocated that the 1976 Act should be made superior to earlier Acts and all subordinate legislation and other forms of rule-making: where Government required as a matter of policy that discrimination should be permitted on grounds of birth, nationality, descent or residence, this should be provided for expressly by statute.

In the context of exemptions the CRE also made reference to the fact that work experience trainees are not regarded as 'employed' and are not therefore covered by the main provisions of the Act. Although this exemption is clearly related to employment matters in so far as it may affect those involved with trainees in housing associations or housing authorities the CRE recommendation to bring such trainees directly within the protection of the Act as though they were employees rather than relying on the more limited and little known protection given by designation under Section 13 is clearly relevant to housing: following the decision in Daley v Allied Suppliers (1983) IRLR 14 where the EAT held that a person on a work experience training programme was not protected by the employment provisions of the Act from discrimination, the Government made a Designation Order under Section 13 of the Act so that trainees on YOPs or YTS were, to a limited extent, brought within the terms of the legislation. Clearly, however, this piecemeal and partial response to a lacuna in the legislation, while welcome, did not address the main weakness of the Act in this area.

The only direct reference to housing made by the CRE in its proposals for change affecting the substantive provisions of the Act were in relation to Section 21, where it argued that the Act should be altered to make clear that it encompasses the quality of housing, including the desirability of the location as well as the physical qualities of the particular property (CRE, 1985b: 42). In making the above recommendations in respect of the definition of discrimination and the areas covered by the 1976 Act in respect of unlawful discrimination the CRE was clearly guided predominantly by the experience gained from employment cases, whether before tribunals or before the Appeal Courts. Nonetheless the proposals would have a direct impact on housing in extending the scope of the Act in this area.

Proving discrimination and adjudication

First, the CRE argued that a person against whom discrimination is alleged, in circumstances consistent with less favourable treatment on racial grounds, should be required to establish non-racial grounds for that treatment since he or she is the person best able to show the grounds for his or her own actions. The emphasis here, clearly, is on establishing non-racial grounds which would go beyond the mere assertion that another, non-racial interpretation of the events which took place was possible or even plausible: this recommendation goes to the heart of the matter in both employment and housing cases where institutional practices, demonstrated by statistical evidence, may show a pattern of racial disadvantage over a period of time but in the past such evidence has often been insufficient to persuade a court or tribunal in a particular case that a discriminatory act has occurred when faced with an unsubstantiated innocent explanation.

With reference to Section 65 of the Act, which provides that where a questionnaire has been issued to a respondent but he has either omitted to reply or given an evasive or equivocal answer the tribunal of fact may draw any inference that it considers just, the CRE argues that it should be a duty placed on the tribunal of fact to draw an inference that it considers just rather than merely a discretion to do so. In the past, tribunals have been very reluctant to adduce any inferences whatsoever.

Most importantly, in the area of proving discrimination and adjudication, the CRE recommends that a discrimination division within the industrial tribunal system should be established to hear not only employment cases but also non-employment cases and also sex discrimination cases. It is advocated that the discrimination division would be able to call upon the services of High Court judges - and presumably judges of the Court of Session, although this is not stated in the proposals - for more complex cases and the discrimination division should have full remedial powers. The CRE observed that the present designated County Court jurisdiction had fallen almost into disuse, probably because of the expense and the long delays inherent in the procedures. The judicial statistics show fifteen complaints in 1982 and eighteen in 1983 in relation to the whole range of subject matters covered by Part 3 of the Act, including housing. The purpose of a specialist tribunal would be to develop rapidly an understanding of the way in which discrimination occurs and of ways of dealing with it. In the Commission's view, familiarity with the ways in which discrimination occurs is a more important attribute in the tribunal of fact, in cases which are now subject to the County Court jurisdiction, than is general familiarity with the subjects covered by Part 3 (education, housing and the provision of goods, services and facilities). It is therefore a logical step to transfer such cases to that jurisdiction where cases alleging discrimination are more frequently heard even though this means an extension to the traditional remit of the tribunals. This move would bring with it the advantage of the less daunting procedure of the tribunals. The reason for arguing that a discrimination division should not deal with race cases alone but also with sex discrimination cases was that otherwise it might be regarded as 'black peoples' tribunals set up to protect a particular section of society'. In tandem with this proposal was a recommendation by the CRE (CRE, 1985b: 18) that legal aid be made available for all racial discrimination cases. The CRE's own survey of industrial tribunal cases showed that twice as many employers (70 per cent) had legal representation as applicants in race cases between July 1978 and June 1980 (CRE, 1986b). The Commission, as noted earlier, represented 29 per cent of cases, increasing the then chances of success from 5 per cent to 25 per cent.

Formal investigations and law enforcement

The CRE's view that Section 49(4) of the Act should be repealed has already been noted: this would mean that the effect of <u>Prestige</u> would be reversed and the Commission's powers to conduct a formal investigation for any purpose connected with the carrying out of its duties (Section 48) would thereby be clearly established. With reference to defects in the machinery relating to formal investigations the CRE referred to the following:

1. They are ill-designed to resolve disputes of fact according to traditional notions of fairness appropriate to law enforcement.
2. They place the decision on whether the remedy of a Non-discrimination Notice is appropriate in the hands of the investigating commission.
3. Remedies are aimed only at general practices and not also at reparation or redress for individuals involved.

While the Commission recognised the need to keep an investigative fact-finding function and also an involvement in law enforcement it argued for change in respect of the latter. The system devised should permit the Commission, wherever it has unearthed it, to put evidence of discrimination before an independent tribunal of fact for a decision after full opportunities for cross-examination as to whether discrimination has occurred and consideration of what remedies are appropriate. The tribunal of fact should not be concerned with the way in which evidence has been obtained (i.e. with the quality of, and formalities surrounding, any investigation, but only with the substantive issues (a) was there discrimination? and (b) what should be done about it? In the Commission's view this would have the following three advantages: first, the whole process would be shortened; second, the respondent would be able to tackle the evidence fairly and by cross-examination if desired, without the feeling that he was trying to dislodge the findings of a public body which he believed to be partisan; and, third, it would give the Commission access to an independent tribunal of fact across the whole field of its law enforcement activities and this would remove any suggestion that the Commission was both prosecutor and judge in the same cause. The Non-discrimination Notice, at present a remedy in the hands of the Commission but one which does not enable the Commission to prescribe particular changes in practice, should be replaced by an equivalent order emanating from an independent tribunal which, in turn, should be able to order particular changes in practice. Although the advantages of having an independent tribunal adjudge the legitimacy of a Non-discrimination Order are manifest the concrete advantages in relation to the scope of a Non-discrimination Order over the current Non-discrimination Notice in housing would appear to be less critical, particularly in the light of the Non-discrimination Notice issued following the investigation into Tower Hamlets in 1987. Here the Commission found discrimination and specified requirements in the following areas:

1. A review of procedures and practices to check that they were non-discriminatory in respect of homeless and emergency applicants as well as in relation to applicants and tenants who were rehoused by the council.
2. A change in the system operated by the council in relation to separated families.
3. Arrangements for the guidance and training of staff in respect of equal opportunity issues.
4. Administrative arrangements to be made by the council to ensure that the Notice was complied with.
5. Monitoring of the Notice by the CRE for five years.

Self-evidently such a Non-discrimination Notice is wide-ranging and the effect of this notice, as well as the notice following on the Hackney investigation which preceded it, demonstrates what may be achieved by negotiation although it is accepted that where negotiations break down a

Discrimination Order backed by fuller remedial powers would be preferable.

Remedies

In addition to the powers of an independent tribunal of fact to deal with past discrimination and potential future discrimination by way of a Non-discrimination Order, the CRE recommended that the Commission should have power to join in any proceedings in which discrimination was alleged, to draw the attention of the tribunal of fact to the potential for future discrimination in the situation under consideration. The Commission also recommended that, in any case brought by it, a tribunal of fact should have the power to award compensation to any person it found to have suffered unlawful discrimination, either named or otherwise sufficiently identified, provided that any such person joined the proceedings within a specified time and sought the compensation. The Commission has not recommended that this be extended to any proceedings where it had exercised its power to join in as 'amicus curiae' but no reason for this omission was stated.

As far as monetary remedies are concerned the provision of compensation should be improved, in the CRE's view, as follows: first, there should be a prescribed norm figure by way of compensation for injury to feelings; second, compensation should be payable where indirect discrimination is proved and the present exemption in Section 57(3) should be removed; third, the tribunal of fact should be able to award continuing payments of compensation until a stipulated event such as promotion or engagement occurred; and, lastly, the statutory limit to compensation in employment cases should be removed.

Mechanisms for bringing about change

The Commission recommends that its code-making power provided under Section 47 of the Act should not be restricted to the field of employment but should be extended to include other areas such as housing and education. In respect of the former this has already been effected by the Housing Act 1988 and a draft code has already been drawn up (CRE, 1989b) but otherwise amendments will have to await, it is anticipated, a fuller review of the Act. The CRE also recommends that the

Secretary of State should be given powers to prescribe ethnic record-keeping concerning, amongst other things, the recipients of housing or other service provision by a public body. The orders prescribing the keeping of records should be capable of limitation by (1) area of the country, (2) types of activity, (3) the duration of the record-keeping. There should be a power vested in the Commission to require returns to be made where record-keeping has been prescribed. Safeguards against abuse of the information should be enacted. Where there is agreement between the Commission and a body on specific practices to be adopted, the Commission should have the power to accept legally binding and enforceable undertakings by that body to adopt those practices. Undertakings should be recorded in a public register.

The general statutory duty imposed on local authorities by Section 71 of the Act, in the CRE's view, should be amended to conform to those imposed on the Commission by Section 43 with regard to each of the various functions of the authority. Those duties are 'to work towards the elimination of discrimination and to promote equality of opportunity and good relations between persons of different racial groups generally'. Moreover the duty should be extended to all bodies carrying on a service or undertaking of a public nature (for definition see Section 75(5) of the Act). The CRE also recommends that public bodies should be required by law to publish, in their annual reports or separately, annual programmes and reports to enable the public to evaluate their work in the field of race. The Housing Act 1988, as has been noted already, extends Section 71 of the Act to include the Housing Corporation, Housing Action Trusts and Scottish Homes as a consequence of the process of privatisation of local authority housing. However, one of the difficulties in respect of Section 71 relates not only to its somewhat nebulous drafting in placing a duty on local authorities but also to powers of enforcement, which are nominally in the hands of the relevant Secretary of State, but in practice no enforcement action has ever been taken. It would appear appropriate, therefore, although omitted from the CRE's recommendations, for the CRE to be named as a responsible body for enforcing the observance of Section 71. In terms of housing, in addition to the obligations placed on the bodies mentioned above, the Department of Environment and the Scottish Development Department would, on amendment, have duties relating to the promotion of good race relations and this might secure a greater commitment to equal opportunity than has been evidenced in the past. Broadening the scope of direct and indirect discrimination together with improved enforcement powers and the issuance of a code of practice in housing would all secure more effective coverage in the private sphere.

CONCLUSIONS

With the exception of the piecemeal amendments referred to in the previous section, as at June 1990 the Government had made no response to these proposals intimated in July 1985 and it would seem therefore that a strengthening of the legislation related to unlawful discrimination in housing, as elsewhere, cannot be expected under this current Government. Although the Labour Opposition promised a commitment to fair housing in its Manifesto prior to the 1987 election and is currently reviewing all strategies, the question remains as to whether or not the recommendations as adopted would have a significant impact on the strategies for preventing the continuance of discrimination in housing and elsewhere. In respect of employment, commentators such as Lustgarten (1978, 1987) have questioned not so much the need for a strengthening of the legal provisions against discrimination but the dangers of seeing them in a limited context outwith the need for wider administrative and political strategies against discrimination and in an environment which has been generally negative if not hostile. Bindman (1980: 258) has argued that the most pressing need is not for changes in the law but for a substantial strengthening of the legal and economic powers and inducements to apply effective equal opportunity policies.

> This will only come about if there is greater readiness by the courts and tribunals to enforce the law if more resources are provided for law enforcement, and if the Government demonstrates its commitment to racial equality by using its executive powers.

However, as Solomos (Solomos and Jenkins 1987: 49) has argued, there are broader structural limits on the operation of the 1976 Act, namely the impact of economic recession, and it is clear that the pursuit of 'equality of opportunity' as a goal of Government policy cannot be made sense of outside the pressures placed on the stated objective by the economic and political forces of society as a whole (Offe 1984). In race relations there is an obvious contradiction between promise and experience and this contradiction, Offe argues, can lead to a seemingly endless series of short-term and at times self-defeating strategies aimed at securing immediate requirements for legitimacy from one interest or the other. This is characterised by crisis management rather than producing planned social reform. The Scarman Report led to a flurry of reports, pronounce-ments and research studies, and the advocacy of a series of measures which were seen as the best insurance against a recurrence of the violent confrontations that took place in areas such as Brixton and Liverpool in 1980 and 1981. Although the various recommendations of the Scarman Report and the analyses which followed it were far from fully implemented, the experience of continuing racial unrest as demonstrated by the urban protests in 1985 in Handsworth, Brixton, Toxteth and Tottenham has pointed to the limits of what was achieved and what might have been achieved by such mechanisms.

CHAPTER THIRTEEN

ANTI-DISCRIMINATION POLICY:
TOWARDS RECONSTRUCTION

POSITIVE DISCRIMINATION OR POSITIVE ACTION?

There is much confusion about what the concept of positive, or reverse, discrimination means and what it might achieve. There is also strong opposition to its introduction not only from those opposed to all legislation concerning equality of opportunity but also from those who recognise and sympathise with the need for legislative protection for rights denied on racial grounds.

It is in the US where the concept has evolved and where the opposition has marshalled the arguments. The decision in Bakke was clearly a turning point not only in constitutional terms but also in terms of public support. Essential to that decision - and the concept of positive or reverse discrimination - was that preferential access was determined solely on the grounds of race irrespective of the comparative need or merit of those in competition (Edwards, 1987). This may be contrasted with the following:

Preferential treatment based on area
The Urban Programme, for example, in identifying areas of relative deprivation may, in conferring benefits on all residents of a specified area, disproportionately advantage those belonging to a particular ethnic/social or economic group. Race, however, is not the sole criterion although it may be a reference point where a specific racial group has failed to benefit from the programme in proportion to its size.

Preferential treatment based on need
Just as the area-based approach assumes need by reference to physical location, so other criteria subsuming need may be utilised. Thus Section 11 funding in providing 75 per cent Central Government grants to local authorities to meet

certain staffing costs is based on an assumption of need in respect of New Commonwealth immigrants. The present guidelines extend support to those local authorities which have a high proportion of second-generation immigrants. While it may be argued that the 'New Commonwealth and Pakistan' tag is an incontrovertible racial criterion, such a reference point is qualified. First, the reference to 'Immigrants and their descendants' links the difficulties of local authority provision in mainstream services to assumed needs of immigrants. Such needs, for example in English language, social service support and housing provision, are unequivocal and the beneficiaries are to be qualifying groups rather than individuals identified by racial origin. Second, the immediate beneficiaries are local authorities, whose responsibility it is to assess the needs of such groups before securing the necessary staffing to meet them. Third, the programme is designed to be transitional: consequently to the extent that Section 11 does provide positive discrimination in favour of a group identified by ethnic origin, such provision is relatively short-term. Fourth, there is no assumption that such financial support to local authorities will actually result in such groups securing a better service than any other group. It merely facilitates equal provision, and access, for such groups. The additional resourcing provided by Section 11, therefore, is designed to alleviate disadvantage rather than confer additional benefits.

Provision of targets
A number of local authorities, notably some of the inner London boroughs, and a few major employers provide targets whether in respect of recruitment, promotion or services - such as housing allocation. Such targets should be distinguished from quotas, which are fixed numbers. Although targets are similarly based on projections of what a fair and reasonable proportion of particular racial groups might be, they are not fixed or immutable. Accordingly they permit individual assessment of merit or need. Ethnic origin and other racial grounds are, consequently, not the reference point for determining eligibility but are used as a general yardstick for assessing whether the particular system employed is delivering equitable benefits. A target therefore provides an indicator as to whether or not racial discrimination takes place. Clearly where targets become the sole or critical reference for the conferment of benefits they have crossed the bridge dividing them from quotas and

then fall within the category of positive discrimination.

The law only permits selection on racial grounds in a very limited number of circumstances - most frequently in respect of training and special language provision (Lambeth London Borough Council v CRE (1989) IRLR 379). Edwards (1987) distinguishes positive action from positive or reverse discrimination. Where positive action is used as a general description of steps designed to promote equality of opportunity then it is distinguishable from positive discrimination. The former is designed to attract members of ethnic minority groups to, or qualify them for, 'benefits' because they have historically been under-represented. The actual 'benefits' themselves retain objective criteria, not based on racial grounds, concerning eligibility. The latter concept, as noted above, while similarly designed to compensate for past under-representation uses racial grounds, directly, as a reference for eligibility.

While there are sound arguments for the latter approach there are also more profound objections and some practical difficulties, not only in deciding what constitutes a fair quota but also in respect of unintended consequences. For example, individuals who have secured benefits through preferential selection in employment may lack the training and skills of other employees selected by 'open' competition. While this is not a necessary consequence of positive discrimination, where it does occur it may lead to a lack of confidence by others in the abilities of those preferentially selected as well as self-doubt by the beneficiaries themselves. Moreover, while it may be straightforward to demonstrate the need for general progress to compensate for past racial disadvantage, at the individual level - as the Bakke case illustrates - compensatory unequal treatment which discounts individual merit and performance is much harder to sustain and justify, especially to those adversely affected.

While Edwards (1987: 185) attempts to list the 'utilities' of positive discrimination these apply equally to positive action, i.e.:

1. Production of more ethnic minority professionals in order to serve better the needs of such communities.
2. Improvement of the status of minority groups in society and sensitive services.
3. Improvement of the social, physical and environmental conditions of ethnic minorities.

4. Counteracting racial disadvantage.
5. Providing role models, raising the self-esteem, self-respect and expectations of minority group members.
6. Promotion of social and racial harmony.
7. Prevention of civil unrest.

Edwards referred to ethnic minority enrolment in the universities and increasing recruitment to welfare bureaucracies but such concerns may be subsumed within the first two groups. Significant omissions include:

8. Increasing efficiency by facilitating and promoting access according to merit.
9. Improving professionalism generally: the demands of objectivity in one area are likely to provide a springboard for more general demands for equity.

Edwards (1987: 186) then listed disutilities relating to positive discrimination.

1. Loss in efficiency or a drop in standards.
2. Loss of self-esteem in respect of beneficiaries.
3. Loss of status of the minority groups in the eyes of the majority.
4. Resentment and the danger of a backlash from the majority.

The first three disutilities would not apply to positive action, as herein defined. The fourth cannot be discounted but would be groundless: it should not weigh much in any balance of worth.

It is not my intention to provide a critique of the discussion of positive discrimination given by Edwards, nor indeed to suggest that I agree with his analysis or conclusions. What he rightly questions are the generalised assumptions regarding the benefits or disbenefits of particular approaches. In housing, therefore, it is highly speculative to make presumptions about tenure preference either from the basis that historical patterns are predictive of real choice or conversely that they primarily reflect racial disadvantage even where evidence of racial discrimination is overwhelming. It would be prudent, therefore, to focus attention, at least in the short term, on ameliorating access to provision in all types of tenure rather than insisting on positive discrimination to secure absolute proportional representation in each. Such an approach is not to deny the value of targets but merely to suggest that they should be based on actual demand while positive action should secure that such demand reflects real choice and not past patterns of tenure occupancy.

Similarly it must be recognised that fixed quotas even if designed to compensate for inherited racial disadvantage may equally impose constraints on access. Thus higher demand by ethnic groups for council housing or housing association tenure than is reflected in quota or target figures may reflect restricted access to other forms of housing. While the targets are sufficiently flexible to reflect demand in tandem with need, quotas, being fixed reference points, may effectively result in unlawful discrimination not only against the white majority but against those very groups they were designed to serve.

The great attraction of fixed quotas is that they will usually effect significant gains and obviate areas of discretion, which have been shown to disadvantage such groups. However, in addition to the disadvantages listed by Edwards, in their provision of a simplistic remedy they also distract attention from any analysis of the causes of disadvantage which, in the longer term, must be addressed. This may be illustrated by reference to the findings of the CRE in respect of the Rochdale investigation concerning mortgage lending (CRE, 1985c). Building societies may have determined that ethnic minorities were historically under-represented in the group of clients who obtained mortgages. Quotas reflecting proportional numbers would be fixed and, within the existing criteria for determining the suitability of properties as security, ethnic minority applicants would be given a larger share of the loans - if necessary by applying less rigorous criteria in respect of income gearing to the size of loan. But this would not have solved the underlying problem. Some ethnic minority applicants would be given loans in circumstances in which loans to white applicants would have been refused. More importantly the discrimination against those owning houses without gardens - a specific issue identified by the CRE - would, in all probability, have remained.

Conversely if targets had been set the shortfall in ethnic minority loans would not have been made up in this fashion (because it clearly involves positive discrimination) and the post-mortem would ask why. Hopefully the search for an explanation by the building societes would identify the cause and appropriate changes in the rules would be effected. If not the continuing failure to meet target figures at periodic reviews would ensure a continuing search for rational explanations (including direct and indirect discrimination) and remedial action. Because quotas demand no such inquiry, in any complex selection process their rigid application may result, simultaneously, in positive discrimination as a necessary counterbalance to negative discrimination occurring elsewhere in the system, with a merely random chance that any individual within the quota group would, in experiencing both, end up with equal access to the facility. It is more likely that some of the quota group would become unfairly advantaged and some would remain unfairly disadvantaged.

Consequently while quotas may result in positive discrimination they may also obscure the need for positive action to tackle racial discrimination where it is actually occurring. For those reasons the CRE was right, in the author's view, not to advocate change in the 1976 Act to allow for positive discrimination, at least in the area of housing provision.

THE TOPOGRAPHY OF DISPUTE RESOLUTION

In the opening chapter reference was made to the limits of law in providing an effective forum for the resolution of disputes. Despite the exposition of the weaknesses of the Race Relations Acts in subsequent chapters, the impact of the 'best' law and the most effective policing is evidently merely one aspect of promoting equality of opportunity. Even if the law was able to provide optimum resolution of disputes in discrimination cases there is evidence that factors such as time, patience and resources result in extra-judicial resolution of disputes acquiring a momentum and pattern of its own (Galanter, 1983). The law, however, has been seen as a strong bargaining endowment (Mnookin and Kornhauser, 1979) affecting not just the manner in which people bargain, negotiate and resolve but also the substance of any resolution which may be 'cast in the shadow' of the legal norms applicable to judicial resolution.

However, as Dhavan (1988) has argued, discrimination cases are frequently loaded in respect of moral principal. In particular the 'shame and blame' attached to the respondent - whether real or perceived - in respect of successful claims has been shown to influence the extent to which he or she will defend high moral ground. Inevitably entrenched positions are frequently adopted in the early stages and then opportunities for settling outside the judicial arena are sacrificed.

This has two immediate consequences. First, the respondent may be willing to resource a defence action beyond its economic worth - but commensurate with its perceived moral value - thereby diminishing the claimant's prospects of success. Consequently the shadow of the law in such disputes is short and thin, providing a poor negotiating base for the claimant. Second, irrespective of such historical shadows, the respondent is ill disposed to the process of negotiation because even where a 'fair and square' deal is in the offing, the risk of the 'shame and blame' label adversely affecting his or her 'goodwill' or that of the housing authority as employer or service provider may be seen as unacceptable.

In the instance of direct discrimination claims it may well be argued that the extent to which public opinion demands a high moral (and non-discriminatory) stance is beneficial in reinforcing the law with a moral code of normative behaviour. Conversely, however, in indirect discrimination cases where racially discriminatory practices may be 'justifiable' (1(1)(b)ii) by reference to non-racial grounds, where the intention of discrimination on racial grounds may be absent (or irrelevant in law), the rationale of 'shame and blame' is much less obvious. Given that such cases are notoriously difficult to prove and that a judicial resolution may depend on nice value judgments, such labelling structures a significant impediment to both judicial and extra-judicial resolutions.

Frequently claimants, in whatever civil dispute, will 'lump it' or 'clump it', that is commence and abandon a claim: Dhavan (1988) has argued that, for a variety of reasons, including those relating to the bargaining endowment of the Race Relations Act 1976, ethnic minority claimants faced with difficulties of language, access to advice, lack of familiarity with the judicial system and weak economic bargaining power, are even less likely to pursue complaints than others. Whether or not non-judicial settlements are either numerically significant or normatively 'just' in the sense of reflecting judicial decisions, that information which is available - and here the study by Dhavan is timely but incomplete - indicates that in race cases informal dispute resolution fails to match the wholly unconvincing record of judicial settlements.

It may be myopic to stress the impact of judicial resolution - the shadow of the law - on extra-judicial negotiation but it seems improbable that the latter will provide a meaningful alternative without radical changes in the context in which such negotiations take place. A significant improvement in the processes and performance of judicial resolution may, while not dictating extra-judicial negotiations, provide an improved 'bargaining endowment'. Simultaneously should the courts and tribunals understand more fully and emphasise in their pronouncements the moral, as well as legal, distinction between direct and indirect discrimination this would undoubtedly affect the climate in which negotiations are possible. Evidently the views expressed at the time of the passage of the 1976 Act and still current that 'Britain is not a racist society' off-loaded not only the 'shame and blame' label but also the issue of public responsibility and accountability which a recognition of indirect or institutional discrimination entails. Equally if institutional practices could not be held responsible - because of the linkage with guilt - the burden was placed firmly on the victim. As a result the unravelling of this process of labelling is not just an issue of semantics but is necessary in changing the whole climate of dispute resolution.

Many of the difficulties in using the Race Relations Act 1976 as an effective backdrop to implementing public policy and securing non-judicial dispute resolution were anticipated in the debates during its passage through Parliament. The ambiguous role of the CRE, however, as the enforcement agency in respect of supporting claimants and as the arbiter in respect of Non-discrimination Notices was glossed over. In addition it was apparent that the right of claimants to gain direct access to courts and tribunals rather than being absolutely dependent on the Race Relations Board was a double-edged sword in that the CRE had no duty to support a claim nor any formal locus to conciliate a resolution of a dispute before it entered the courts. Moreover, Government had failed, whether consciously or otherwise, to establish any support mechanism for complainants beyond the limited accessibility of the CRE. For example the potential role of the CRCs in these areas was ignored. As we have seen (Chapter Eleven) the evolving relationship between the CRE and CRCs has, as yet, failed to achieve a coherent support network for law enforcement. But would non-statutory agencies provide a solution?

LEGAL DEFENCE FUND

A number of commentators (Runnymede Trust, 1982; Dhavan, 1988) have argued for the establishment of some central organisation, separate from the CRE, for the purpose of promoting, supporting and facilitating the pursuit of legal remedies in race cases. Indeed the Women's Legal Defence Fund (WLDF) launched on 18 April 1989 may prove to be a model worthy of emulation (Guardian, 19 April 1989: 4). Clearly this fund is reflective of American experience.

While the headquarters of WLDF are in London, it has or proposes a network of regional offices - Birmingham, Liverpool, Cardiff, Belfast, Glasgow, Edinburgh, Newcastle, Sheffield and Norwich. It has a support team of over 200 barristers, solicitors, law students, law teachers and trade unionists and has a principal objective of doubling the number of successful discrimination cases by 1992. Like the CRE, the Equal Opportunities Commission (EOC) is financed by the Home Office and, because of funding difficulties, it is reviewing its support for individual cases, choosing to focus this on the more complex Equal Pay provisions. To fill the gap the Fund will give advice and information, prepare cases and conduct them in court or tribunals and provide training in law and advocacy. As with race cases, claimants in sex discrimination cases experience considerable moral and psychological pressure in initiating and pursuing claims. A distinct benefit of this network may well be its lack of a bureaucratic and impersonal public interface, such as both Commissions have been criticised for. The local offices of the Fund provide a better spread of access points (excepting, in particular, Scotland, where offices in Dundee, Aberdeen and Inverness would have been desirable) than the EOC and the resultant potential for providing more immediate and personal support than is currently available through the EOC.

Given the limitations of the CRE in respect of manpower, resourcing and access and the under-utilisation of the Race Relations Act 1976, the case for the CRE supporting an equivalent body to WLDF is strong. Obviously, it may be argued, the existence of an established network of CRCs with strong local contacts and the experience, albeit uneven, of pursuing race cases suggests that this alternative, not available to the EOC, has the advantage of building on existing practice. Nonetheless, as the section on the CRCs indicates, their role in providing support for the pursuit of legal claims is less than convincing, because they are heavily committed to other case work - such as immigration and nationality - have a role in promotional activities and frequently lack not only in-house· legal expertise but also access to sympathetic and cheap lawyers. There is no reason, therefore, why an equivalent defence fund (or Anti-discrimination Centre serviced by local offices as advocated by Dhavan, 1988) should not be established in the race field to work in tandem with the CRCs.

The question remains, however, whether there are a sufficient number of committed lawyers who would provide the essential backbone to such networking. The experience of the Equal Opportunities Law Group in Scotland (for background see MacEwen, 1981), which was established with similar objectives in 1982, is not encouraging. Apart from legal academics the response from practising solicitors was woefully thin and largely confined to those whose interest was technical but who lacked an evangelical commitment to securing equality of opportunity. Such experience was replicated in the short-lived Civil Rights experiment, based in London, supported by the Gulbenkian Foundation. As Dhavan's study of race claims demonstrates such a commitment is generally a necessary prerequisite for the successful pursuit of such cases because of the substantial legal, moral and psychological obstacles faced by claimants and their representatives. Accordingly it would be necessary for the CRE in association with local CRCs to explore the potential of a fund very thoroughly before making any financial allocation. It remains a truism that the nature of the legal remedies and the availability of and limits to legal aid and assistance remain a significant structural constraint on the establishment of a Legal Defence Fund. Without the financial backing and explicit support of the Law Societies, the Bar and the Faculty of Advocates the prospects of success are less than for the Women's Defence Fund. Despite sex discrimination in the legal profession and its male dominance, there is a sufficiently large pool of women lawyers together with a much larger and more evenly spread client group enabling a commercial return for such specialism, to promise, alongside voluntarism as a key element in its establishment, the ultimate economic viability of the fund. In contrast the existing voluntarism of the CRCs may remain the favoured option with the CRE. Certainly the radical restructuring proposed in the CRE/CRC relationship (CRE, 1989a) should facilitate the housing of a sharper cutting edge to CRC legal support but without the involvement of the legal establishment this alternative will be difficult to translate into reality. Despite the difficulties alluded to, a Legal Centre, with local networks manned by lawyers, even on a voluntary basis, and linking with CRCs, must not be abandoned by default. The existing Neighbourhood Law Centres, like the CRCs and CABs fail to keep adequate records of their involvement with race discrimination cases (Dhavan, 1988; MacEwen, 1988) but the information available suggests that they are ill-geared to make a substantial contribution in this area.

As noted earlier, the CRE and Law Centres Federation have made attempts to investigate mechanisms for improving such involvement from Law Centres but a Race Legal Defence Fund would not steal much thunder from a fairly cloudless sky of inactivity.

Obviously the extent to which Lord MacKay's proposals for legal reform result in non-lawyers being granted advocacy certificates and the proposed farming out of areas of legal advice by competitive tender in England and Wales may effect a radical shake up in lawyering (Lord Chancellor's Department, 1989). Few think, however, that such changes will either benefit the less well-off or inject resourcing into areas of 'low' demand expertise. It may be, however, that such changes will facilitate the specialisms of some CRCs in legal work and, in allowing them access to the County and Sheriff Courts, encourage the pursuit of housing claims in these forums. What is more difficult to gauge is how quickly such proposals will be effected, the extent to which the strength of institutional opposition will be able to safeguard the status quo and what ripple effects the changes will have on areas of law - such as anti-discrimination legislation - which were far from the forefront of Governmental concern in clearing the Augean stables of restrictive practices. Such factors make predictions hazardous and render a restructuring of CRE support problematic, being subject to a seemingly ever increasing number of dependent and independent variables. It would seem wise, then, in a period of flux, for the CRE to hedge its bets and to explore a range of alternatives in constructing a wide network of supportive agencies. Such an approach does not predicate a wilful dissipation of CRE resources but rather a broad selection of starter-pack initiatives to be taken up and developed by others, to minimise losses and optimise gains. Such eclecticism in pursuing legal claims, moreover, is complementary to the promotional networking advocated, which in turn will serve a number of discrete political objectives. Clearly such initiatives must be set within some corporate strategy.

DEVISING A CORPORATE STRATEGY

It is one thing to conclude that the present legislation and its enforcement is inadequate, that piecemeal changes are marginalised and that the Governmental approach is contradictory and evasive and another to construct, from the disaggregated evidence of racial discrimination and disadvantage to which we have referred, a corporate strategy which is sufficiently robust to promise substantial and sustained improvement in practice. Nonetheless this critique in respect of housing suggests that any strategy must address a number of key objectives, namely:

1. A fundamental change in the attitudes, ethos, structure, authority, policy formulation and policy review of Central Government, principally through the Home Office.
2. A substantial injection of ethnic minority employees, at all levels, in those institutions responsible for devising and implementing housing policy, from the DoE, the Welsh Office and the Scottish Office, through to the local authorities, the housing association movement, major developers, building societies and private landlords.
3. A major revision to CRE strategies for promotion and enforcement, based on extensive and supported networking through Central and Local Government, the CRCs, CABs, Law Centres and advice agencies as well as voluntary organisations and associations at national, regional and local levels.
4. The integration of annual equal opportunity planning at the local authority level with that, to be devised, at the national level, with the CRE as principal adviser, coordinator and watchdog.
5. The legal requirement of all housing agencies with a stock of more than say 200 houses to keep and publish

419

statistical returns on ethnic minority applicants and tenants by access route (homeless, waiting list, referral etc.) and, in respect of the latter, by area and type of housing.

6.   A review of access to the courts and tribunals, a review of the training, attitudes and approach of the professions (including the Bench), the accessibility of legal aid and advice, the potential for class actions and the award of damages.

7.   Substantial implementation of the CRE's proposals for change in respect of the Race Relations Act 1976: the economic risk of unlawful discrimination must become unacceptable.

8.   Adequate resourcing of the CRE and a strengthening of its consultative role in respect of all Central and Local Government policy together with improved access to information.

9.   The adoption of a mandatory code of practice in respect of housing by all public-funded agencies.

10. A total revision of the requirements relating to contract compliance in respect of all publicly funded agencies.

Reference has been made to Peter Mason's observation that Government had not been cynical in its approach to race relations in the sense that however negligent it had proved to be, and despite the obvious ambivalence regarding the distinction between formal equality (the colour-blind approach) and normative equality (the pursuit of social justice), it did not wish to see an alienated and disillusioned section of the community identified purely by colour or ethnic origin. The subsequent chapters have neither confirmed nor refuted the accuracy of that belief but have demonstrated that an emphasis on social control and tension management has dominated policy. Despite the failure to stem racial harassment and the obvious shortcomings in policy implementation, a significant Central and Local Government effort has attempted to address this issue. There is no evidence, however, of a like commitment by Central Government to securing equal access to public and private housing. Douglas Hurd, as Home Secretary, announced on Friday, 28 October 1988, the introduction of a Code of Practice applicable to the Immigration and Nationality Department to prevent discriminatory practices. In doing so he alluded to the 'good record' of the Civil

Service in respect of implementing equal opportunity policies.

The CRE (1988a: 7) in its Annual Report for 1987, addressed to Douglas Hurd, stated the following:

> Contract compliance was also a major weapon [in Government strategy to eliminate racial discrimination] of which Central Government, in spite of our repeated representations, has made no use, in striking contrast to its policy in Northern Ireland. The Government's own equal opportunity policies were also meant to set an example to the rest of the country - a lead that has only recently been at all evident.

All the evidence, moreover, supports the CRE interpretation of events and not that of Douglas Hurd. Still, he may not be cynical: it may be that he believed another, unpublished, version of history written by the Civil Service with its practised economy of reporting. 'What is Truth?' asked Pontius Pilot and did not wait for an answer: in Government truth, on occasion, does not appear to be far removed from political expediency.

The strategy for achieving change depends crucially on Central Government's perception of issues. If Government does not know what the truth is in respect of racial discrimination in housing, the solution is to inform it repeatedly and convincingly, assembling all available evidence to that end. Conversely if the political reality is that there is no electoral mileage in the truth such assemblage will continue to be ignored. In that event a 'rational' approach may appear irrational. An emotive and emotional call for action may appear necessary and justifiable. To those concerned with race and ethnicity the psychology of the Nuremburg Rally, even where the evangelical creed is premised on reality, has little attraction. It may be, however, that a more thorough, persistent and pervasive approach to Parliamentary lobbying through all available networks and contacts is a necessary prerequisite for Government to place racism, racial discrimination and racial disadvantage at the forefront of its concerns.

For such an approach to be effective it will be insufficient to rely on the CRE, the CRCs or on the activities of ethnic minority groups themselves. The NCCL, the CABs and the plethora of public and voluntary agencies involved in the field of welfare rights and housing have shown too little concern for equal opportunity and have proved an inefficient and ineffective vehicle for creating sustained pressure for change. Certainly the present ethos of individualism promoted by the Government has changed social attitudes, diminishing the coherence of community participation and ostracising challenge to the New Right orthodoxy while the processes of structural change in community identity, referred to in the opening chapter, have desensitised our concern for equality and created the conditions which make residential apartheid and racial harassment an everyday reality. Consequently if a national anti-discrimination forum were to be established for the purpose of co-ordinating national and local networks, the extent of racism in the UK would ensure virulent and active opposition in an attempt to denigrate and marginalise its impact. Here the involvement of broad sectional interests, from the chambers of commerce to the TUC, from the CRE to community and parish councils, from the mosques to the churches, from nursery schools to the universities, from tenants' associations to the local authority housing department, would provide a key not only to its strength but also to its social legitimacy. If the facts of racial disadvantage have, in Harvey's terminology, a structural ontology in the sense of permeating all the structural components of reality from birth to the grave, the sustaining strength of a national forum would be the constancy and pervasiveness of its ideology and message.

It must be admitted, however, that such a quest for equal opportunity based on redistributive social justice may prove an elusive goal, as it runs counter to the growing support for the ideologies of the New Right.

IDEOLOGIES AND SOCIAL CHANGE

While the opening chapter has attempted a brief description of the ideology of the Race Relations Act 1976 at its inception under a Labour Government, any conclusion regarding a critique of the Act in practice is bound to be located in the wider context of ideology and its interplay with contemporary social policy and practice. There is clear evidence that the Thatcher Government has not only dismantled the Butskellism which dominated post-war social policy and legislation - as exemplified by changes in social security and trade union legislation - but has also influenced social values in respect of the legitimacy of State intervention and State managerialism. Thus while the sale of local authority houses was strongly opposed when introduced in 1980, the extent to which the public has embraced the transfer of council housing to private ownership and has identified self-interest in the promotion of home ownership has challenged the validity of local authority control to such a degree that wholesale transfers of estates to the private sector are now enshrined in the Housing Act 1988. It is suggested that the latter legislation has become possible only because of a significant ideological change in values effected under this administration. The extent to which that change may have been based on false premises and assumptions is beyond the boundaries of this book to explore. But a recognition of this change, even by the Labour Party, is surely evidenced in its adoption of share ownership as an acceptable mechanism for spreading wealth.

The consensus of the New Right, in accepting the reality of class exploitation and disadvantage in nineteenth-century capitalism - although not in terms of the Marxist explanation of its cause in dialectical materialism - sees an amelioration of conditions, in all sections of the community, both in practice and in opportunity, in the form of an expansion of the property and share-owning democracy. These two facets of wealth demonstrate, it is argued, that as the economy expands there is a filter effect so that both the rich and the poor get richer. Moreover within this ideological structure poverty is no longer to be seen as the stigma of class. The principle of the uninhibited market ensures flexibility and mobility, accommodating aspirant yuppies of whatever social origin. It is only where pockets of multiple disadvantage in which a combination of social and economic factors such as poor housing, single-parent families and lack of employment opportunities due to structural change are beyond the beneficial outreach of the market that the State must intervene as a last and temporary support in a period of transition. Within such an approach the residualisation of council housing is not the regrettable by-product of the expansion of owner-occupation and, at least potentially, the private rented sector but an essential element of the process. Indeed if local authorities continued to offer a vast choice in housing stock at reasonable or subsidised rents, this would not merely compete with private-sector housing but make investment unattractive through an unfair (subsidised) market.

Even the ideologues of the Tory right accept that this process will leave casualties and that the extremes between the richest and poorest will increase: but in absolute, if not comparative, terms generally the less well-off, it is argued, will benefit. The inventive, the industrious, the imaginative, the intelligent and the well organised (at an individual level) will prosper. Inevitably those who do not prosper will be seen not merely as victims of structural change but frequently as the feckless and undeserving. Although this may be a simplistic and reductionist view it may be sufficiently accurate to identify the difficulty of locating equal opportunity policy within the structure. Again, with crude simplicity, a successful equal opportunity policy which eliminates race or colour as a factor in respect of social and economic mobility would not be expected to result in an even distribution of jobs and incomes reflecting that of the broader community. It is in the nature of free market forces that the process, while giving Brownie points to the characteristics of the enterprising, previously described, awards substantial benefits to those with existing income and wealth. The market does not espouse Rawls's <u>Theory of Justice</u> (1972), which would start all off at 'even stevens' and demand that the market be designed to reflect an acceptable risk of failure: we start from where we are on the economic ladder, ensuring not only that the rich get richer quicker but also that they can insure against relative failure and cushion themselves against loss. On the gambling table the extent of resources has a direct relationship with the odds on failure or success. Consequently the extent to which ethnic minority groups are located in worse housing, have lower incomes, are more frequently unemployed and have less disposal capital (Brown, 1984) will, in the absence of any discrimination, guarantee comparative failure, as it does in respect of other economically disadvantaged groups. But it is in the nature of racial discrimination that those who are enterprising and black will not prosper as well as those with like enterprise from other groups (assuming like advantage and disadvantage). As a result the experience of past racial discrimination will be crystallised in the new 'equal opportunity market'. At this juncture it must be emphasised that this ideal market assumes that there is a total absence of direct discrimination in terms of Sections 1 and 2 of the Act and also an absence of indirect discrimination in so far as it is not justifiable.

In ideological terms the differences in the meaning of 'justifiable' are the mirror image of the differences in political value judgments. We have already seen how the courts, in their interpretation of 'justifiable', have moved from 'necessity' to 'commercial convenience' and this shift is in keeping with the shift in more general social values in attributing some objective equality to the market system. Charles Wilson's observation 'What's good for General Motors is good for the country' has been transmogrified into 'What, in the sole opinion of General Motors is commercially convenient is absolutely justified irrespective of the consequential disadvantage suffered by black people'. Accordingly embedded in the definition of racial discrimination in the 1976 Race Relations Act, and crucial to its application and extent, is a reference to contemporary social values and that reference currently enshrines the view that the imposition of any requirement or condition may reflect the market imperative. In housing the test of justifiability in allocations will have commercial considerations as an increasingly important reference point with the privatisation of local authority housing. In respect of houses retained by local authorities for society's casualties, commercial viability may not be an obvious criterion, other than in respect of ability to pay the rent. However the ability to pay the rent, taking into account the innovations regarding eligibility, amount and cut-off in respect of housing benefit and expected local authority rent regulation, will become increasingly problematic. Of more critical importance to ethnic minorities, however, may be the increased competition for allocation between the avenues first of the general waiting list and second of the homeless persons route, now, as was noted especially in respect of the London boroughs, disproportionally from black households. In such competition local authorities may argue that preference for the former is 'justifiable' in respect of length of time on the waiting list, in which case the homeless will be housed increasingly and for longer periods in temporary accommodation. But even in this confined area if the consequences of applying the market or economic test of justifiability were recognised as unfair and inequitable there is little prospect that social justice in the sense of determining the equity of process by the results would be substituted. Discrimination on the basis of class, on the basis of income, on the basis of schooling, on the basis of appearance or accent, or any other labelling to determine social worth, is not unlawful and it consequently constitutes an unchallengeable part of the fabric of everyday decision-making, frequently explicitly and invariably implicitly. Although such discrimination may be as arbitrary as discrimination based on colour or ethnic origin, the potential victim may, and frequently does, camouflage the characteristics, none of which is immutable, which he or she perceives will be seen as detrimental. Without wishing to justify such arbitrary discrimination or even the evasion tactics adopted as a response, it may be observed that the process does permit an escape route for those with mental agility. The same is not true in respect of discrimination based on race directly. Accordingly if such arbitrariness is to be avoided it must be confronted rather than evaded.

The difficulty with the concept of indirect discrimination in respect of race, as in respect of gender, is that it questions the relevance of secondary criteria such as class, language, accent, school and dress. While ultimately, in the House of Lords, the determination of the courts in <u>Mandla</u> demonstrates that such arbitrariness - in that instance dress - may be open to challenge, the nature of social discrimination in Britain is so pervasive that any challenge is bound to be piecemeal and incremental. If there are those who doubt the pervasive nature of such discrimination, the dominance of Oxbridge graduates amongst the echelons of the upper reaches of the Civil Service in England, Wales and Scotland should provide convincing evidence of the extensive outreach of this particular masons' club. Of course some would argue that the Oxbridge success story is based on objective merit - Oxbridge is the best and attracts the best. But a more considered assessment, while accepting that Oxbridge continues to attract and produce graduates of distinction, must conclude that the label, irrespective of ability, is a significant social, and economic, asset and a passport to opportunity denied to others of equal merit.

Indirect discrimination, consequently, constitutes a challenge to the fundamental freedom of choice in decision-making by incorporating non-racial criteria as a legitimate reference point. Thus the Britishness of the black citizen cannot, according to the formal requirements of the law in respect of indirect discrimination, be confined to colour or race but must encompass accent, education, association, dress and religion - all critical reference points for discrimination based on class. Ineluctably, while direct discrimination may be severable from considerations of class, indirect discrimination refuses to recognise such boundaries. To that extent the Race Relations Act 1976 has, perhaps unwittingly, entered an arena which, with the limited exception of the Sex Discrimination Act 1975, has been left otherwise untouched by the legislature. The ambivalent segregation of race and of the characteristics of ethnicity - language, religion, culture, dress, accent, etc. - is enshrined in the distinction between direct and indirect discrimination. The inevitable result has been that the latter concept - encapsulating prejudices beyond race and otherwise untouched - has proved unpalatable to the courts. They, in an indirect search for meaningful interpretation, have in turn applied the same ambivalence and lack of objectivity reflected in the broader community. Let us examine how this process may operate.

Case A

Family A, of Asian origin, live in a shorthold tenancy and have been given notice to quit. They cannot afford alternative rented accommodation near work and have been refused a mortgage (on the grounds that the house under offer lacks a front garden). They apply to the local authority as homeless. They are given emergency accommodation by the local authority and will, eventually, be accommodated in poorer housing, having been treated in the same way as any other applicant in their circumstances. The family would have been entitled, through waiting time points, to better accommodation from the general waiting list but a previous application by the husband was refused because the wife and two children had two years to wait in Bangladesh before being given entry clearance and the local authority has a policy of only acknowledging housing need in respect of those actually resident.

Patently this policy, so far as it has a disproportionally adverse impact on ethnic minority families, constitutes indirect discrimination. The local authority will contend that it is 'justifiable' by reference to non-racial criteria, i.e. actual residence and the shortage of suitable housing. The local authority will also contend that if the housing needs of the family are to be assessed by including residents abroad - whose time of arrival is unknown - the result would be uncertainty and unfairness: it is only reasonable to assess the needs of actual residents in preference to potential residents. In this situation, should the matter be challenged by application for judicial review (which is unlikely), the courts, first, would have to decide whether the policy has a proportionally adverse impact on ethnic minorities. Following the decision in Orphanos v Queen Mary College (1985) IRLR 349, the courts might choose to compare the treatment of Bangladeshis with all non-Bangladeshis (including Pakistanis and Indians who may have suffered like disadvantage). If that hurdle is surmounted what criterion will be applied to 'justifiable'? Following Mandla v Dowell-Lee (1983) IRLR 17, we know that the question of justifiability is one of fact, not of law. The question may be rephrased. Could a reasonable authority have come to the conclusion it did on the facts?

Case B

Family B, from Pakistan, apply to the local authority for a council house. Owing to Mr B's work commitments, Mrs B visits the housing department for interview. Her English is poor and no interpreters are available because of cut-backs in local authority expenditure. With guidance from the local authority staff, she is nominated for an unpopular estate, her preference being overlooked owing to a misunder-standing. A year later Mr B visits the local authority, corrects the error, but loses waiting time points because they are not transferable from one area of choice to another. Effectively the family wait a year longer for a house than anyone without language difficulties. Unless the local authority exercises discretion in favour of family B, they have little prospect of a remedy for the following reasons: there is no obligation to provide an interpreting service and, even if such were available, mistakes might still be made. The rule relating to non-transferability of waiting time points from one area to another may have only a marginally disproportionate impact, if any, on ethnic minorities and this would be difficult to prove. The policy regarding interpreters and transferability may be justifiable by reference to necessity (regarding the former) and convenience (regarding the latter).

Case C

Mr C, from India, suffers constant racial abuse at work. He gets involved in a fight and is dismissed. Although he has a right of appeal his union refuses to represent him on the grounds that, where two members are in dispute, it will represent neither, and he fails to lodge his appeal timeously. Following his dismissal he falls into rent arrears and is evicted. He and his family apply to the Homeless Unit of his local authority. Family C are given temporary accommodation but are found to be intentionally homeless: the local authority policy is to require all applicants threatened with dismissal to pursue all avenues of appeal as a causal link in the homelessness chain. The rule has been applied fairly. The fact that Mr C would not have got involved in a fight had he not been subject to racial abuse is not a factor the local authority considers relevant in determining 'intentionality'. Although the rule may appear harsh and arbitrary, the local authority refuses to exercise any discretion because of the vast numbers of families on the general waiting list and in the homeless categories. The rule may not result in indirect discrimination (there is no record of the ethnic origin of applicants who have been refused because of it) and, even if it does, it may appear justifiable. Consequently an application for judicial review is not a realistic prospect.

The above examples are mundane and unremarkable in themselves and the detriment suffered, at least in respect of families A and B, may be considered marginal. Moreover the same or equivalent experiences, particularly in respect of cases B and C, will have happened to many non-ethnic minority families. But the fact is that the experience of (1) members of the family being resident abroad, (2) language difficulties, and (3) racial abuse and harassment is a permanent facet of the life of many ethnic minorities and cumulatively the experiences result in a perpetuation of disadvantage. Viewing it in isolation we might write off the experience of families A, B and C as 'tough' - they are not the only social category within the broader community to suffer. We might observe that for family A, if there was fault, it lay with the entry clearance system and the building society rules and for family C with the poor practices of C's employer and trade union. It might be argued that the family were at least partially to blame for their own apparent fecklessness. The question might arise

'Why should we apply more onerous standards in the housing departments to secure equality of treatment than has been apparent in the treatment of white applicants?' If the housing authority had secured better treatment for the families would that not have constituted preferential treatment in a fashion not available to white families suffering like disadvantage through the application of the same or other arbitrary rules? Because the answer to these questions is unsatisfactory, they have assumed importance beyond their merit in the scales of justice. Surely the question to be asked is, if you were one of the families would you consider the outcome fair? If not, and the result led to indirect discrimination on the grounds of race or ethnic origin, should we not expect the courts to find a remedy in the Race Relations Act 1976? In arguing that others are also subject to arbitrary or unfair treatment two responses are overlooked. First a remedy should not be denied by justifying arbitrary practice in one area because of its existence in another – the remedy should be to tackle all such practices. Second it ignores the fact that ethnic minority families will also suffer from the other arbitrary practices without remedy where they have no correlation, directly or indirectly with racial grounds. These cases illustrate, however, that we have yet to come to terms with indirect discrimination and, with too great facility, justify our tolerance of the status quo by reference to other aspects of inequality.

The inability or unwillingness to change policy, moreover, may reflect purely economic considerations. Where resources are static or diminishing a local authority may consider a change impossible without additional resourcing. Of course 'impossibility' may, in turn, reflect value judgments concerning the relative merit and importance of existing programmes and resultant resource allocation even to the extent that statutory obligations may not be met. In McLaughlin v Inverclyde District Council (1986) GWD 2.34, it was held that the availability of local authority resources was a relevant factor when considering whether a long delay in implementing obligations to repair was legally justifiable. Although this case, determined by the Sheriff Court in Scotland, did not create any binding precedent, it does demonstrate the potential for the courts to attenuate legal obligations when resourcing makes performance problematic: the alternative, and one preferred in logic and law by the author, would be to require performance and allow the responsibility to be seen as that of Central Government where, as in this case, it clearly lay. By the same token, Government does appreciate that local authorities may, in the situation of increasing Central Government control of expenditure, be unable to meet the statutory requirements imposed; hence the proposal to redefine homelessness as rooflessness.

In its Annual Report for 1987 the CRE made reference to lobbying in respect of six separate Bills - the Criminal Justice Bill, the Education Reform Bill, the Housing Bill, the Immigration Bill, the Local Government Bill and the Local Government Finance Bill. In each Bill (all subsequently enacted with insubstantial change) the CRE expressed concern about the adverse impact of the proposals on ethnic minority groups: this ranged from diminishing the opportunity of black defendants to have a jury representative of a multi-racial society, parental choice in schools exacerbating the 'Dewsbury' syndrome and allowing racial prejudice to determine schooling opportunities, the removal of the right to family unity of immigrants settled before 1973, the limits placed on local authority contract compliance by a requirement, in respect of equal opportunity, to seek answers from potential contractors only in respect of Government-approved questions and the introduction of the Community Charge in respect of its disproportionate impact on ethnic minorities. Certain concessions, none of which fully met the concern expressed, were made but, more significantly, there was not one proposal, not one part, not one section, not one clause in any of the Bills considered by the legislature in that year, attributable to Government initiative, which furthered the promotion of equality of opportunity. There must surely be a distinction in kind between Government actively promoting the interests of disadvantaged groups and the concessions, frequently paltry and ungracious, made in respect of legislation which will have an adverse impact on such groups.

Some would contend that, cumulatively, the evidence of Government commitment is not merely absent but supports the contention that, implicitly or explicitly, Government is active in dismantling the impact, limited as it has proved to be, of the Race Relations Act 1976: because, if nothing else, the composite consciousness of Government's intentions is difficult to locate, such is not the focus of my concern. Rather it is that what is practised, irrespective of motive and what may or may not be preached, is the material reality on which the Government should be judged. My assemblage of facts and opinion may do injustice to the defence but it is my contention that the evidence constitutes, at least, a case to answer. That the case is largely unanswered may depend not merely on the availability of evidence but on the persistence and strength of the prosecution. Do we care sufficiently to translate our misgivings about the adequacy of Government policy in achieving equality of opportunity into action? Clearly the answer is 'No'. When Government, of any political persuasion, has ignored or denied the interests of minorities, of whatever nature, it is, perhaps a forgivable response to state a case and expect a remedy. But the strength of the case relies, in such instances, not so much on the immutable logic of the evidence but more frequently on the weight of public opinion. Governments can afford to ignore justified criticism and the evidence which supports it when the electorate are largely either indifferent to the outcome or antagonistic to the unauthorised version of events. Consequently, in essence, it is not the Government which is on trial but the weight of public opinion. Equality, an integral aspect of freedom, requires eternal vigilance for which we are each severally responsible and have, in my book, been found wanting.

A conclusion that race is currently of marginal political importance because, generally, the electorate does not identify with the cause of equality of opportunity demands an explanation. I have suggested that an explanation may be located, in part, in shifting social values which reflect structural social change. There can be little doubt that racism was a cultural facet of class identity. The question remains as to whether the dynamics of social change are capable of accommodating ethnic diversity and whether they will strengthen or diminish the prospects of racial equality.

THE DESCHOOLING OF CLASS

Raymond Plant in debating the case for redefining social values based on class argues that the Marxist tradition in characterising the market economy as exploitative and dehumanising towards the working class has failed to reflect significant historical changes in the composition and homogeneity of 'class' and, in tandem, to recognise the emergence of a diversity of competing interest groups whose legitimacy fits uneasily within the rigidity of class divisions (Plant, 1988). The difficulties of a class-based approach are seen as principally threefold. First, the present industrial working class is too small a base from which to gain power. Second, the solidarity of a working class based on community regarding work, residence and neighbourhood is now largely no more than nostalgic. Third, given the failure of Marxist predictions about the shape of class in capitalist society, the European socialist movement has sought common interest with other classes or groups: any abandonment of a more pluralist approach in favour of a narrowly based and largely illustory working-class identity is unreal and has little electoral legitimacy.

Consequently the contemporary socialist movement must be seen to represent a coalition of interest groups - whether or not defined as class - which may be identified by religion, language and ethnicity as well as by reference to production. Moreover it must appeal to individuals because many no longer identify either with class or with a specific group interest. The difficulty with this shift of appeal is its inherent lack of coherence. There cannot be an assumption that all such interests eschew contradiction and conflict. Consequently unless resolution is to be found in the political expediency of meeting the demands of those with the greatest muscle at the expense of those with the least, there remains a political and philosophical vacuum in determining the criteria for the adjudication of interest-group and individual claims. The Conservative philosophy turns to market forces not because they guarantee a fair distribution of resources but because, first, as we have already noted, the efficiency of the market, it is argued, will secure optimum growth, realising a larger cake with the potential of a larger absolute if not proportional slice for the worse-off and, second, as the whole process of rational adjudication of competing needs and demands cannot be satisfactorily resolved, the random unpredictability of the market constitutes a fair gamble in which the State does not fix the odds in favour of one group at the expense of another.

Evidently, in respect of ethnic minorities, this interest group has gained even less in private housing than in public housing between 1979 and 1988 when market forces have been central to Government policy. Since 1978 personal disposable income has risen by 14 per cent in real terms while Supplementary Benefit (now Income Support) levels have fallen from 61 per cent of disposable income in 1978 to 53 per cent in 1987 (Plant, 1988: 10). Consequently the incomes of the poorer have not kept pace, let alone increased, during that period. Of course if the proportion of the population in that category had fallen substantially during this period, their loss could be written off as an unfortunate but insignificant by-product of more general gains. In fact the number of people suffering such loss, by reason of dependence on Supplementary Benefit, increased from 5.7 million in 1979 to 7.7 million in 1984.

As a result market processes would have ensured absolute as well as comparative losses, to which we have already made reference, in income for the majority of the ethnic minority population in Britain in this period even in the absence of racial discrimination through effective enforcement of the Race Relations Act 1976.

Similarly by reference to homelessness, another clear symptom of poverty, the increases referred to in Chapter Seven provide corroborative evidence of absolute disadvantage to lower socio-economic groups. Moreover although being better-off is no passport to good health, as the Black Report demonstrates there is a clear correlation between poverty and health and a further correlation between the incidence of particular diseases, such as rickets, sickle cell anaemia and schizophrenia and ethnic origin. Such incidence may have genetic and cultural causes, such as diet, but living conditions and discrimination may also be important factors. Thus in respect of schizophrenia a study by Nottingham psychiatrists (Guardian, 31 October 1988) found that while higher rates of mental illness had been found in first-generation immigrants - Hispanics in the US, early Polish immigrants in Britain and first-generation French Canadians - these rates had levelled off around those of the adopted country in the second generation. In contrast second-generation Caribbean immigrants in the UK are up to ten times more likely than white children to develop schizophrenia in adulthood. Dr Bebbington observed: 'These rates suggest something has gone dreadfully wrong with the Caribbean second generation'. Such findings have not gone unchallenged (Francis et al., 1989). But few disagree with the conclusion of Dr Harrison that 'Many black people in Britain, faced with discrimination and limited opportunities, suffer crises of identity, which could precipitate the illness in those already vulnerable'. The crisis of funding in the Health Service, consequently, is of particular significance to the poor who have no choice of opting for private medicine and for ethnic minority groups who, for whatever cause, suffer from the twin connections of poverty and ethnic origin with illness. While there may be evidence to suggest that more money has gone to the NHS (and that its appetite for resources is virtually insatiable) and that the Government continues to see it as an important social safety net, current policy raises issues regarding even the restrictive Conservative view of social justice and equity. For example if the tax cuts have failed to secure an improved base level for the poor in absolute and comparative terms the deployment of such lost revenue for welfare support becomes a case against which a Conservative view has difficulty marshalling facts and arguments. Therefore within the Conservative hypothesis that welfare rights must be confined to a narrow band of need in order to permit the maximum freedom of choice - or in the context of economics the minimum restrictions on market forces - decisions must be made regarding the level of State intervention or, in other words, how much revenue should be deployed for welfare as opposed to defence, law enforcement, environmental protection, infrastructure provision, etc., and within the welfare budget what areas should be prioritised. Indeed, once the concept of a welfare safety net is accepted, value judgments based on social justice and equity must be made within subject area headings. Any decision to freeze family allowance, to charge for eye testing or health checks, to withdraw student subsistence grants and substitute loans cannot be based on any market principles of equity, but must look beyond for some theoretical framework.

However, before outlining such change it is important to emphasise that while we may choose to classify ethnic minorities, lower-income groups, the unemployed, single-parent families, those on Income Support and other groups dependent on various aspects of the Welfare State by reference to class, it can no longer be assumed that such groups will perceive a common identity and will cohere in respect of political allegiance. With such political fragmentation, there will be increased competition from political parties for support defined by group rather than class interest. While this may be seen as a healthy challenge to the assumptions of class support for the Conservative or Labour parties almost irrespective of the policies pursued, it has also resulted in electoral success in recent years being dependent on comparatively less support, in absolute terms, from the electorate. Effectively the most obvious beneficiary of this process has been the Tory Party and its continuing success will depend on its consolidation of existing support, despite the fact that it remains a minority. Consequently so far as it continues to appeal to or appease this support it need not seek broader support, including that of ethnic minority groups.

The choice for disenfranchised groups is not obvious. While they may be wooed by different parties and, for that reason, play a more prominent part in framing policies affecting their interests, their prospects of achieving electoral success seem remote. For minority interests in such circumstances, proportional representation may seem the only viable mechanism for transferring political support on to the bargaining counter of executive power, even where such opportunities are acknowledged to be limited. Within such an analysis - and stressing that political perceptions of interest determine the reality - the attraction of the SLD and the Greens becomes more apparent despite the obvious contradiction that they must gain success within the existing structure before they can change the rules to benefit their future prospects. Alternatively if those currently and persistently disenfranchised perceive greater prospects of power in constitutional rather than electoral reform, Labour-voting Scotland being the most obvious example, a shift to nationalism and the SNP in particular may prove a more attractive option.

The Labour Party's policy review published on 18 May 1989 offers some prospect of an effective race rallying point (Labour Party, 1989). Indeed its very prominence on the agenda of reform together with the proposal to relax the arbitrary rigour of immigration controls should more than counteract the adverse media coverage over a black section of the party and the central veto exercised over constituency selection of Parliamentary candidates. The review proposes the transfer of the CRE's promotional role to a Ministry in Central Government while strengthening the CRE's responsibilities in enforcement. There can be little doubt that Central and Local Government should assume responsibility for promotional work. What remains in doubt, however, given the nature of such bureaucracies as are illustrated in Chapter Eleven, is the potential for radical change to be effected from within. Other proposals include: (1) mandatory use of contract compliance by National and

Local Government under the Department of Employment, drawing on the experience of the Northern Ireland Fair Employment Act, (2) the creation of a national equality forum, (3) the creation of a Department of Legal Administration responsible for the working of all courts and tribunals and for legal appointments and training, (4) the extension of legal aid to an array of tribunals, (5) the establishment of Equality Tribunals with power to increase compensatory awards (Wintour, 1989).

Cumulatively these proposals, along with related measures such as reduction of tax levels for the low-paid and the introduction of a limited array of class action, represent a package capable of redressing the withered concern for racial disadvantage. What they fail to do, however, is to build on the existing network of agencies such as the Community Relations Councils, ethnic minority organisations, Law Centres and CABs and their links with the CRE. Consequently, particularly in respect of housing, the loopholes relating to identifying indirect discrimination and providing effective support for victims, together with the difficulties which would remain concerning the standard of proof and the defence of justifiability, suggest that the proposals are not capable of translating good intentions into practical reality: their ability to attract electoral support also remains untested.

But it is apparent that ethnic minority groups reflect a broad spectrum of interests and allegiances going well beyond the stance of political parties on race and immigration. The fragmentation of political ideology into discrete political groupings has not only reflected and accelerated the break-up of class identity so far as they continue to reflect interests severable from those of other sections of the community. The potential for a regrouping of sectional interests at the political level, however, depends on the theoretical framework for the Welfare State within which such interests are located and the changing climate of opinion effected by the Government towards welfarism.

THE DEINSTITUTIONALISATION OF SOCIAL WELFARE

There are grounds for believing that the apparent absence of such a framework has been a political concern only recently but will become of increasing importance to both the right and left as an emerging battleground for electoral support.

Such a proposition requires some evidence: it may not be self-evident that this political divide was any less apparent in 1979 when the Conservative Government assumed office.

In that year the electoral appeal was the promise of a resurgent economy based on the market, and a stripping away of unnecessary, inefficient and financially debilitating bureaucracy in Central and Local Government to create a leaner and meaner industrial and service base which could compete within the EC and in world markets. The appeal was to the wage-earner from a leader with a 'gut populism, a strong will and North Sea Oil to lubricate her casualties' (Corrigan et al., 1988). Despite increased unemployment, the administration has achieved, whether by design or otherwise, an increase in real wages, a degree of control in respect of inflation and the prospect of an expanding economy. While the casualties may have been substantial in electoral terms, despite the increasing North/South divide, they had not dented the heartland of popular Tory support. However, in addition to North Sea oil, the sale of the profitable sectors of nationalised industries, the revenue from which along with taxation is diminishing, the Government has had to look increasingly to cuts in public expenditure. The strategy has not been, hitherto, a dismantling of the framework of the Welfare State such as the NHS, State education, public housing or the provision of State benefits, but rather a process of guerilla warfare in which the formal institutional power bases have survived but suffered such casualties and losses of territory that their room for manoeuvre and counter-offensive is severely restricted. Government tactics have to a large extent, not been those of directing change but cutting off the supply of resources, particularly in respect of local authorities, whilst simultaneously resourcing alternatives to the existing institutional arrangements.

This approach has had two immediate consequences. First, it camouflages cuts in real terms by the process of making them progressive and allowing some flexibility as to what aspects of the service will be under-resourced. Initially, it may be suggested, this has the benign effect of demanding the excision of mismanagement and waste, but ultimately departments, quangos and local authorities must cut services. Second, the process deflects responsibility from Central Government to the institutions concerned. At the local authority level, quite apart from dismantling the GLC and the metropolitan counties in the name of efficiency, the 60 per cent cut in resource allocation for public housing has resulted, at least initially, in criticism addressed at local authorities in their failure to maintain programmes of cyclical maintenance, refurbishment and repairs, the cessation of new build, the increased rigidity in the application of rules in respect of the homeless and their restrictions on improvement grant provision. Inevitably local authorities have attempted, generally, to attenuate the impact of cuts by dispersing the disbenefits within the discretion permitted by Central Government. The result, equally predictable, has been that the housing authority itself has been criticised for its own determination of priorities in so far as this has been detrimental to the particular interests of an identifiable group. The lack of homogeneity in respect of groups affected has secured a dissipation of focus in respect of the causes of disadvantage. While many may have applauded those cuts affecting others and have been convinced by Government arguments concerning their spendthrift policies of such authorities as the GLC and Brent, with the cuts biting deeper and wider few interest groups have now survived unscathed. These factors have, to a degree, bought time in respect of public demand for a Government view of social justice but as the IoH report on homelessness illustrates (IoH, 1988), while a shift from the perception of dependence on the 'Nanny State' may have been effected, there remains a large residue of responsibility in resource allocation in Government hands which is now bound to respond to demands for equity. Of course, as we have already noted, such demands are multifarious and contradictory but as they now stem in increasing number from the Conservative heartland - from the two-parent family to the old age pensioner - they cannot be ignored.

It would be misleading, however, to suggest that the Government does not have any set of social priorities (beyond balancing budgets and effecting public expenditure cuts to fuel the economy) or that these have been wholly implicit, witness the call to Victorian values and the appeal to the family as the primary social unit in respect of rights and obligations. What has remained implicit is the extent to which continuing State and local authority provision in welfare services is to be maintained and what level of resourcing, and service may be expected. But even here, while that question is not directly answered, some indications of resourcing and priorities are provided by reference to five-year public expenditure plans and the new framework of resourcing and management being applied to local authority responsibilities in education and housing. Effectively this process is one of selectivity and of creaming off both the best schools and the best local authority estates. If the former do not receive preferential treatment from the local authority they may choose to become either independent schools with direct aid from Central Government and the latter may choose alternative private-sector landlords; in the instance of local authority housing such choice may be determined by a failure to vote, as the Housing Act 1988 now provides that such failure is recorded as support for privatisation.

This process of residualisation does not merely affect estates and schools but also tenants and pupils because it is the worst-off who have no bargaining power: in the inner London boroughs, for example, the primary concern of parents is not the absence of a uniform national curriculum or the quality of teaching provided, it is the inadequacy of resourcing in respect of buildings, infrastructure, equipment and books which, together with the difficulties of attracting teachers to fill vacant posts, has prevented other than piecemeal improvement. Similarly the least popular estates are likely to reflect the physical condition of the buildings, the lack of maintenance, the poorer community infrastructure and the lack of employment opportunities in the neighbourhood: in neither instance does opting out to the private sector represent a viable alternative. It could be argued, nonetheless, that this process has only short-term disadvantages for the poorest. Once local authorities have lost 80 per cent of their housing stock (and possibly their schools) and have got rid of surplus employees, their ability to upgrade the residue will be greatly enhanced. Clearly, however, the nature of local authority expenditure plans determined by Central Government will ensure that this does not happen, the residualisation of provision going in tandem with the residualisation of resourcing. Again the concept of welfare vouchers - credits which the recipient may apply to whatever needs he or she determines are essential - may have an egalitarian appeal but if, cumulatively, they diminish real choice and opportunity they merely reinforce relative disadvantage.

As Jacobs (1988) has observed, the issues of unemployment, housing, health and education as well as those relating to poverty and social security all tend to produce competitive attitudes in the absence of effective collective campaigning. In, for example, queues for the social security office together with the general lack of accountability of the DSS and other Government departments, the scapegoating of ethnic minorities both by competing claimants and by the bureaucracies concerned results. Moreover this emphasis on competition, on individualism and on private enterprise has manifested inequality not merely in relation to relative income but also in respect of the distribution of resources between inner cities and the provinces and between the North and the South.

RECONSTRUCTION, RACIAL EQUALITY AND HOUSING

In effect, then, the social divisions manifested in the past ten years are not confined to class or income group but extend to housing tenure and geographical location. Where community interest and political identity have converged in opposition, such as in the inner cities and Scotland and the North East, this opposition has been effectively dismantled at Local Government level through Central Government financial controls and has had insufficient political support outside such areas to dent Tory confidence in the tactics of divide and rule having continuing success at national elections. Such confidence has led to the dismantling not merely of the GLC and the metropolitan authorities but also of the instruments of Central Government whose voice contradicted or challenged the tenets of the reborn Tory faith in the economic and social equity of the market: thus the Select Committee on Scottish Affairs and the Sub-committee on Race Relations and Immigration have been disbanded along with the University Grants Committee, the Supplementary Benefits Commission, the Scottish Housing Advisory Council and the power and influence of the Wages Councils. The Conservative Party has wrested from the Opposition the support of those who have benefited from the new wave of the property and share-owning democracy and effectively transferred, in the short term, to local authorities the perception of responsibility for public expenditure increasingly required from diminishing revenues from taxation, North Sea oil and denationalisation. This has had two major consequences in the light of the political fragmentation and economic growth which have taken place during the life of the government: there has been no urgency in placating the 'wets' within the Tory Party or in seeking further support beyond and there has been no need to define or quantify the nature and the extent of Central Government support for the Welfare State generally and public housing specifically, even while the processes of privatisation and residualisation have made significant inroads into the existing framework. This process has clearly disbenefited ethnic minorities at a number of discrete levels, whether by reference to socio-economic group, by reference to location in the inner cities, in respect of employment opportunities and schooling, or by reference to child and housing benefit. The area-based inner city initiatives such as the Urban Programme have proved a marginal palliative, the monies directed to the inner cities not even matching losses sustained by local authorities. Their purpose has focused on the containment of urban unrest and social control. Equal opportunities so far as they equate with redistributive justice clash with the primacy of the market and, with the exception of proposals for Northern Ireland in respect of contract compliance, the legislative programme demonstrates a total lack of commitment to anti-discriminatory objectives, almost irrespective of how they are defined. The association between class and political allegiance, tenuous as it has been in the past so far as it safeguarded ethnic minority interests, has been substantially eroded. The strain of long-term Opposition has encouraged self-doubt and equivocation in the Labour Party and until the major policy review has borne fruit and has demonstrated sufficient coherence to act as an effective electoral rallying point both group and individual defections are likely to accelerate the processes of fragmentation previously described.

Such a scenario is fairly desperate for those currently disenfranchised who seek to achieve some influence in changing the political priorities of Government through the electoral process directly. Charter 88, with its call for constitutional reforms, including a redefinition of citizens' rights, whatever its specific merits or demerits, reflects a growing unease with the policies of the Government which is shared by some of its beneficiaries in economic terms. Whether this and like non-partisan loose groupings of interest, especially those which more clearly represent the needs of the unemployed, the homeless and the economically powerless, will emerge into a coherent network of opinion capable of negotiating with the Government and, more promisingly, with the various political parties in Opposition may be given long odds. But it would seem inevitable in a period when the Government of the day has other priorities and the prospects of the parties in Opposition achieving electoral success seem wholly unpredictable, that extra-party political power bases will be seen as one of the few lifelines remaining. If these developments take place they would afford a further access channel for a restructured race relations lobby.

Because the Government has launched a multi-faceted attack on the Welfare State a multi-faceted response is needed. Clearly anti-discrimination law as it affects housing is merely one victim in this process. As Cranston has observed (1985: 335) attempts at reform through the legal system have a better chance of success if they are part of a broader political strategy. Building effective community groups through campaigns for legislative change provides a base to support reform activity. The courts themselves, for all the need for reform as illustrated by the Civil Justice Review (1988, Cm 394), have a limited role in bringing about social change even if access and judicial training and attitudes were to be significantly improved. Moreover changing the law itself, particularly in respect of the Race Relations Act 1976, may effect improvements but the Act as it stands appears to provide a foundation for the pursuit of equal opportunity in housing as in other areas. Its weaknesses are not, principally, in drafting or in inter-pretation but in the dominant social attitudes and approaches which marginalise its effective enforcement. Such attitudes and approaches cannot be changed by Act of Parliament but they are susceptible to political and professional ideology, conduct and leadership. It is, perhaps, in these areas where community networking will prove most influential in determining what kind of multi-racial Britain will emerge in the year 2000.

REFERENCES

Abel-Smith, B. and Townsend, P. (1965) The Poor and the
    Poorest, London: Bell Publications.
Abrams, P. and Brown, R. (eds.) (1984) UK Society: Work,
    Urbanism and Inequality, London: Weidenfeld &
    Nicholson.
Alder, J. and Handy, C. (1987) Housing Association Law,
    London: Sweet & Maxwell.
Association of Metropolitan Authorities (1985) Housing and
    Race: Policy and Practice in Local Authorities, London:
    AMA.
Association of Metropolitan Authorities (1986) Programme
    for Partnership: an urban policy statement, London:
    AMA.
Association of Metropolitan Authorities (1987) Racial
    Harassment: Report No. 1 from the Local Authority
    Housing and Race Equality Working Party, London:
    AMA.
Association of Metropolitan Authorities (1988a) Homeless-
    ness: Report No. 2 from the Local Authority Housing
    and Racial Equality Working Party, London: AMA.
Association of Metropolitan Authorities (1988b) Allocations:
    Report No. 3 from the Local Authority Housing and
    Racial Equality Working Party, London: AMA.
Austerbury, H. and Watson, S. (1983), Women on the
    Margins: a Study of Single Women's Housing Problems,
    City University, London: Housing Research Group.
Bains (1972) The New Local Authorities: Management and
    Structure, London: DoE.
Ball, M. (1983) Housing Policy and Economic Power, London:
    Methuen.
Ball, M. (1988) 'The Limits of Influence: ethnic minorities
    and the Partnership Programme', New Community, Vol.
    15, No. 1 (October), London: CRE Publications.
Banham, J. (1988) 'Urban Renewal and Ethnic Minorities:

444

the challenge to the private sector', New Community, Vol. 15, No. 1 (October), London: CRE Publications.

Banton, M. (1985) Promoting Racial Harmony, Cambridge: Cambridge University Press.

Bate, R. and Burton, T. (1989) 'Department making little progress over land supply figures', Planning, No. 814, 14 April, page 12, London: Ambit Publications Ltd.

Batley, R. and Edwards, J. (1978) The Politics of Positive Discrimination, London: Tavistock Press.

Benson (1979) The Report of the Royal Commission on Legal Services (The Benson Report) Cmnd 7648, London: HMSO.

Ben-Tovim, G. (ed.) (1980) Racial Disadvantage in Liverpool - an Area Profile, Liverpool: Liverpool University Press.

Benyon, J. and Solomos, J. (eds.) (1987) The Roots of Urban Unrest, Oxford: Pergamon Press.

Bersano, G. (1970) Blacks and Physical Planning, Detroit: Wayne State University Press.

Biddiss, M.D. (ed.) (1979) Images of Race, Leicester: Leicester University Press.

Bindman, G. and Lester, A. (1972) Race and Law, Harmondsworth: Penguin Books.

Bindman, G. (1980) 'The law, equal opportunity and affirmative action', New Community, Vol. VIII, No. 3: 248-60, London: CRE Publications.

Bindman, G. (1985) 'Reforming the Race Relations Act', New Law Journal, 15 November, London: Butterworth.

Blackaby, S. and Paris, C. (1979) Not Much Improvement: Urban Renewal Policy in Birmingham, London: Heinemann.

Boddy, M. (1976) 'The structure of mortgage finance: building societies and the British social formation', Transactions of the Institute of British Geographers, New Series, Vol. 1, No. 1.

Boddy, M. (1981) 'The Property Sector in Late Capitalism: the Case of Britain', in Urbanisation and Urban Planning in Capitalist Society, M. Dear and A.J. Scott (eds.), 267-86, London: Methuen.

Bonnerjea, L. and Lawton, J. (1987) Homelessness in Brent, London: Policy Studies Institute.

Boparai, N. (1987) 'Intentionally Homeless Bangladeshis', Foundation, No. 3 (December), London: London Race and Housing Research Unit.

Boseley, S. (1987) 'Foreign Bodies', Guardian, 18 November,

References

Manchester and London: Guardian Publications.

Brown, A. (1981) 'Planning for Blacks in Britain', Town and Country Planning, February.

Brown, C. (1984) Black and White Britain: the Third PSI Survey, London: Heinemann.

Brown, C. and Gay, P. (1985) Racial Discrimination: 17 years after the Act, London: Policy Studies Institute.

Bullock, A. and Stallybrass, O. (1977) Dictionary of Modern Thought, London: Fontana Books.

Bulpitt, J. (1983) Territory and Power in the UK, Manchester: Manchester University Press.

Burney, E. (1967) Housing on Trial: a Study of Immigrants and Local Government, London: Oxford University Press.

Cabinet Office (1988) Action for Cities, London: HMSO.

Castles, S. (1984) Here for Good: Western Europe's New Ethnic Minorities, London: Pluto Press.

Central Housing Advisory Committee (CHAC) (1971) Housing Associations, London: HMSO.

Central Statistical Office (1988) Social Trends 18, London: HMSO.

Chambliss, W.J. (1977) 'On lawmaking', British Journal of Law and Society, Vol. 6, No. 2: 149, Cardiff.

Cohen, P. and Bains, H. (1988) Multi-racist Britain, London: Macmillan.

Cole, I. and Wheeler, R. (1987) New Directions in Housing Policy: the challenge for building societies, Hertfordshire: Chartered Building Societies Institute.

Commission for Racial Equality (CRE) (1974) Unemployment and Homelessness: a Report, London: Community Relations Commission (CRE 1983 reprint).

Commission for Racial Equality (1979) Annual Report for 1978, London: CRE Publications.

Commission for Racial Equality (1980) Report of Formal Investigation into Allan's Accommodation Bureaux, London: CRE Publications.

Commission for Racial Equality (1983a) Report of Formal Investigation into the Immigration and Nationality Department of the Home Office, London: CRE Publications.

Commission for Racial Equality (1983b) Collingwood Housing Association Ltd, Report of a Foraml Investigation, London: CRE Publications.

Commission for Racial Equality (1983c) The Race Relations Act 1976: Time for a Change?, London: CRE

Publications.

Commission for Racial Equality/Royal Town Planning Institute (1983) Planning for a Multi-racial Britain, London: CRE Publications.

Commission for Racial Equality (1984a) Race and Council Housing in Liverpool: a Research Report, London: CRE Publications.

Commission for Racial Equality (1984b) Race and Council Housing in Hackney, Report of Formal Investigation, London: CRE Publications.

Commission for Racial Equality (1984c) Race and Environmental Health, Institute of Environmental Health Officers/CRE, London: CRE Publications.

Commission for Racial Equality (1985a) Ethnic Minorities in Britain: Statistical information on the pattern of settlement, London: CRE Publications.

Commission for Racial Equality (1985b) Review of the Race Relations Act 1976: Proposals for Change, London: CRE Publications.

Commission for Racial Equality (1985c) Race and Mortgage Lending: Formal Investigation of Mortgage Lending in Rochdale, London: CRE Publications.

Commission for Racial Equality (1985d) Walsall Metropolitan Council: practices and policies of housing allocation, Report of a Formal Investigation, London: CRE Publications.

Commission for Racial Equality (1986a) Annual Report for 1985, June, London: CRE Publications.

Commission for Racial Equality (1986b) Industrial Tribunal Applications under the Race Relations Act 1976: a Research Report by Vinod Kumar, London: CRE Publications.

Commission for Racial Equality (1987a) Living in Terror, London: CRE Publications.

Commission for Racial Equality (1987b) Educating in Terror, London: CRE Publications.

Commission for Racial Equality (1987c) Racial Attacks: a Survey of Eight Areas of Britain, London: CRE Publications.

Commission for Racial Equality (1987d) Annual Report 1986, London: CRE Publications.

Commission for Racial Equality (1988a) Annual Report 1987, London: CRE Publications.

Commission for Racial Equality (1988b) Racial Discrimination in a London Estate Agency, Report of a Formal

References

Investigation into Richard Barclay & Co., London: CRE Publications.

Commission for Racial Equality (1988c) Report of Formal Investigation into Tower Hamlets Borough Council, London: CRE Publications.

Commission for Racial Equality (1988d) Report of Formal Investigation into St George's Hospital Medical School, London: CRE Publications.

Commission for Racial Equality (1988e) Housing and Ethnic Minorities: Statistical Information, London: CRE Publications.

Commission for Racial Equality (1989a) A New Partnership for Racial Equality, London: CRE Publications.

Commission for Racial Equality (1989b) Code of Practice: Rented Housing, Consultative Draft, London: CRE Publications.

Commission for Racial Equality (1989c) A Guide for Estate Agents and Vendors, London: CRE Publications.

Commission for Racial Equality (1989d) A Guide for Accommodation Bureaux, Landladies and Landlords, London: CRE Publications.

Commission for Racial Equality (1989e) Racial Discrimination in Liverpool City Council: report of formal investigation into the Housing Department, London: CRE Publications.

Commission for Racial Equality (1989f) Annual Report 1988, London: CRE Publications.

Commission for Racial Equality (1989g) Local Authority Contracts and Racial Equality, implications of the Local Government Act 1988, London: CRE Publications.

Commission for Urban Priority Areas (1985) Faith in the City, Report of the Archbishop of Canterbury's Commission, London: Church House Publishing.

Consumers' Association (1987) 'House buying and selling', Which, April: 171, London: Consumers' Association Ltd.

Coote, A. and Phillips, M. (1979) 'The Quango as referee', New Statesman, 13 July: 50-3, London: New Statesman Publications.

Corrigan, P., Jones, T., Lloyd, J. and Young, J. (1988) Socialism, merit and efficiency, Fabian Tract 530, London: Fabian Society.

Cowen, H. (1983) 'Homelessness among black youth: policies and planning', Goucestershire Papers in Local and Rural Planning 19, Gloucester: Department of Town and Country Planning, Gloucestershire College of Arts and

Technology.

Cranston, R. (1985) Legal Foundations of the Welfare State, London: Weidenfeld & Nicolson.

Crenshaw, K. (1988) 'Race, Reform and Retrenchment: the transformation and legitimisation in anti-discrimination law', Harvard Law Journal, Vol. 101, No. 7, Cambridge, Mass.: Harvard University Press.

Cross, M. and Johnson, M. (1982) 'Migration, settlement and inner city policy' in J. Solomon (ed.) Migrant Workers in Metropolitan Cities, Strasbourg: European Science Foundation.

Cullingworth, B. (1969) Council Housing Purposes, Procedures and Priorities, London: HMSO.

Dalton, M. (1984) 'The New Commonwealth and Pakistan Population in Scotland, 1981' in Scottish Council for Racial Equality Annual Report 1983-1984, Glasgow: SCRE.

Dalton, M. and Daghlian, S. (1989) Race and Housing in Glasgow: the Role of Housing Associations, London: CRE Publications.

Davies (1974) 'West Indians and Council Housing in Bristol', Unpublished Paper, Brunel University.

Davies, J.G. (1985) Asian Housing in Britain: a Success or a Grim Story?, Research Report 6, London: Social Affairs Unit.

Davis, D.B. (1966) The Problem of Slavery in Western Culture, Cornell University Press and (1970) Harmondsworth: Penguin Books.

Deakin, N., Higgins J., Edwards, J. and Wicks, M. (1983) Government and Urban Poverty, Oxford: Blackwell.

Department of Employment (1979) Department of Employment Gazette, Vol. 87, No. 9 (September), London: Department of Employment.

Department of Employment (1987) Department of Employment Gazette, Vol. 95, No. 9 (September), London: Department of Employment.

Department of Employment (1988) Employment Gazette, Vol. 96, No. 3 (March), London: Department of Employment.

Department of the Environment (DoE) (1968) Old Houses into New Homes, London: DoE.

Department of the Environment (1975) Race Relations and Housing: Observations on the Report on Housing of the Select Committee on Race Relations and Immigration, Cmnd 6232, London: HMSO.

References

Department of the Environment (1977) Policy for the Inner Cities (White Paper), Cmnd 6845, London: HMSO.

Department of the Environment (1978) Building Societies Support Scheme for Local Authorities, Note on the operation of the scheme, London: DoE.

Department of the Environment (1981) Single and Homeless, London: DoE.

Department of the Environment (1983) English House Conditions Survey 1981, Part 2, Report of the Interview and Local Authority Survey, London: HMSO.

Department of the Environment (1983a) Housing and Construction Statistics 1972-82, London: HMSO.

Department of the Environment (1986a) Homelessness Statistics, London: HMSO.

Department of the Environment (1986b) Evaluation of Environmental Projects, funded under the Department of the Environment, Inner Cities Directorate: the Urban Programme, London: HMSO.

Department of the Environment (1987a) Statistical Service Housing Survey, London: HMSO.

Department of the Environment (1987b) The Urban Programme 1980/87: A report on its operation and achievements in England, London: DoE.

Dhavan, R. (1988) 'Why So Few Cases? A Study of the Race Relations Act 1976', unpublished research sponsored by CRE, London.

Doling, J. and Davies, M. (1982) 'Ethnic minorities and the protection of the Rent Acts', New Community, Vol. 3: 487-9.

Duffield, P. (1986) 'Session 3', in Y. Ahmed (ed.) Planning for Ethnic Minorities Seminar, Sheffield: Royal Town Planning Institute (Yorkshire Branch).

Duncan, S. (1977) 'The housing question and the structure of the local market', Journal of Social Policy, Vol. 6, No. 4.

Edwards, J. and Batley, R. (1978) The Politics of Positive Discrimination: an Evaluation of the Urban Programme 1967-1977, London: Tavistock.

Edwards, J. (1987) Positive Discrimination, Social Justice and Social Policy, London: Tavistock Publications.

Emery, C.T. and Smythe, B. (1986) Judicial Review, London: Sweet & Maxwell.

English, J. et al. (1976) Slum Clearance, London: Croom Helm.

Eversley, D. (1975) 'The Landlords "Slow Goodbye"', New

Society, 12 January 1975, London.

Farnsworth, R. (1989) 'Urban planning for ethnic minority groups: a summary of initiatives taken by Leicester City', paper presented to Bristol Polytechnic conference, October 1988, published in Planning Practice and Research, Vol. 4, No. 1, spring: 16-22, London: Pion Ltd.

Federation of Black Housing Organisations (FBHO) (1987) Black Housing, Vol. III, No. 5 (December) 4, London: FBHO.

Field, J. and Hedges, B. (1979) Ethnic Groups and Housing Stress, Social and Community Planning Research, London: SCPR.

Flett, H. (1979) 'Dispersal policies in council housing: arguments and evidence', New Community, Vol. 7: 184-94, London: CRE Publications.

Foot, P. (1965) Immigration and Race in British Politics, Harmondsworth: Penguin Books.

Francis (1971) Report of the Committee on the Rent Acts (The Francis Committee), Cmnd 4609, London: HMSO.

Francis, E., Pilgrim, D., Rogers, A. and Sashidharan, S. (1989) 'Race and "schizophrenia": a reply to Ineichen', New Community, Vol. 16, No. 1 (October), London: CRE Publications.

Gaffney, J. (1987) Interpretations of Violence: the Handsworth Riots of 1985, Coventry: Centre for Research in Ethnic Relations, University of Warwick.

Galanter, M. (1983) 'Reading the Landscape of Disputes', Los Angeles Law Review, Vol. 31, No. 4, Los Angeles: University of California.

Gardner, J. (1987) 'Section 20 of the Race Relations Act 1976: "Facilities and Services"', Modern Law Review, Vol. 50: 345-53.

Gay, P. and Young, K. (1988) Community Relations Councils: Roles and Objectives, Policy Studies Institute, London: CRE Publications.

Giddens, A. (1979) Central Problems in Social Theory, London: Macmillan.

Gifford (1986) The Broadwater Farm Inquiry, London: Hackney Borough Council.

Gillon, S. (1981) 'The racial dimension', Town and Country Planning, June.

Glazer, N. and Young, K. (1986) Ethnic Pluralism and Public Policy, Aldershot: Gower Publishing Co.

Gooday, I. (1988) Letter to author from Department of the

References

Environment (H11/834/S) dated 29 June 1988.

Gordon, P. (1986) Racial Violence and Harassment, London: Runnymede Trust.

Goulbourne, S. (1985) 'Minority entry to the legal professions: a discussion paper', Policy Papers in Ethnic Relations No. 2, Coventry: Centre for Research in Ethnic Relations, University of Warwick.

Greater London Council (GLC) (1985a) Greater London Housing Statistics 1984, London: GLC.

Greater London Council (1985b) Homelessness in London, GLC Housing Research and Policy Report No. 1, London: GLC.

Greater London Council (1986a) Temporary Accommodation - Counting the Cost, GLC Housing Research and Policy Report No. 4, London: GLC.

Greater London Council (1986b) Private Tenants in London: GLC Survey 1983-84, GLC Housing Research and Policy Report No. 5, London: GLC.

Greater London Council (1986c) Planning for the Future of London: Race and Planning Guidelines, London Against Racism, London: GLC.

Greater London Council (1986d) Tackling Racism in Homelessness: Implementation of Conference Proposals, report by IoH and GLC Housing and Ethnic Minorities Committee, London: GLC.

Greenberg, J. (1959) Race Relations and American Law, New York: Columbia University Press.

Greenwood, J. (1988) 'Homelessness and the Scottish New Towns', unpublished dissertation for Diploma in Housing, Heriot-Watt University/Edinburgh College of Art, Edinburgh.

Gregory, J. (1987) Sex, Race and Law, London: Sage Publications.

Greve, J., Greve, S. and Page, D. (1971) Homelessness in London, Edinburgh: Scottish Academic Press.

Greve, J. (1985) Investigation into Homelessness in London: Interim Report, Leeds: University of Leeds.

Gribbin, J.J. (1978) 'The implications of the 1976 Race Relations Act for housing', New Community, Vol. VI, Nos. 1 and 2, winter 1977-78, London: CRE Publications.

Griffith, R. and Amooquaye, E. (1989) 'The Place of Race on the Town Planning Agenda', Planning Practice and Research, Vol. 4, No. 1, spring (5-8), London: Pion Ltd.

Gujral (1986) 'Credit due for Black Housing Association's

plan' (letter) in Inside Housing, Vol. 3, No. 40 (17 October) p. 4, London: IoH Services.

Hamnet, C. and Randolf, B. (1986) 'The British Condo Experience', in Smith and Williams (eds.) The Gentrification of the City, London: Allen & Unwin.

Hancock, D. and MacEwen, M. (1989) Ethnic Minorities and Public Housing in Edinburgh, London/Edinburgh: CRE/SEMRU, joint publication.

Handy, C. and Alder, J. (1987) Housing Association Law, London: Sweet & Maxwell.

Harnett, C. (1988) Letter to author from Home Office (RDI/88) dated 10 October 1988.

Harvey, D. (1973) Social Justice and the City, London: Edward Arnold.

Harvey, D. and Chatterjee, L. (1974) 'Absolute rent and the structuring of space by government and financial institutions', Antipode, Vol. 6, No. 1: 22-36.

Harvey, D. (1978) 'The urban process under capitalism: a framework for analysis', International Journal of Urban and Regional Research, Vol. 2, No. 1, 100-31.

Harvey, D. (1982) The Limits to Capital, Oxford: Basil Blackwell.

Hawes, D. (1985) 'Survey of Essex housing authorities', unpublished, referred to in Foundations, No. 3, December 1987: 14, London: London Race and Housing Research Unit.

Hawes, D. (1987) Building Societies - the Way Forward, Bristol: School for Advanced Urban Studies.

Henderson, J. and Karn, V. (1987) Race, Class and State Housing, Aldershot: Gower.

Hepple, B. (1987) 'The Race Relations Acts and the process of change', New Community, Vol. XIV, No. 1: 32-44, London: CRE Publications.

Higgins, J. (ed.) with Deakin, N., Edwards, J., Wicks, M. (1983) Government and Urban Poverty, Oxford: Basil Blackwell.

Hill, C.S. (1965) How Colour Prejudiced is Britain? Panther, London: Victor Gollancz Ltd.

Hill, M. and Issacharoff, R. (1971) Community Action and Race Relations, London: Oxford University Press.

Himsworth, C. (1986) Public Sector Housing Law in Scotland, 2nd edn, Glasgow: Planning Exchange.

HMSO (1961) Housing in England and Wales, Cmnd 290, London: HMSO.

HMSO (1971) Housing Associations, Report of the Central

References

Housing Advisory Committee, London: HMSO.

HMSO (1974) Land, Cmnd 5730, London: HMSO.

HMSO (1975a) Race Relations and Housing, Government Observations on the Report on Housing of the Select Committee on Race Relations and Immigration, Cmnd 6232, London: HMSO.

HMSO (1975b) Equality for Women, Cmnd 5725, London: HMSO.

HMSO (1980) Government Observations on the Report of the Select Committee on Race Relations and Immigration on the Organisation of Race Relations Administration, Cmnd 6603, London: HMSO.

HMSO (1983) Streamlining the Cities (White Paper), London: HMSO.

HMSO (1985) Lifting the Burden (White Paper), London: HMSO.

HMSO (1987a) General Household Survey 1985, London: HMSO.

HMSO (1987b) Social Trends, London: HMSO.

HMSO (1988) Social Trends, London: HMSO.

HMSO (1989) Annual Report of the Nationality and Immigration Department of the Home Office 1988, London: HMSO.

Home Affairs Committee (HAC) (1981a) First Report, Session 1981-82, Commission for Racial Equality, Vol. 1, HC46-1, London: HMSO.

Home Affairs Committee (1981b) Fifth Report from the Home Affairs Committee, Session 1981-1982 HC424: Racial Disadvantage, London: HMSO.

Home Affairs Committee (1986) Racial Attacks and Harassment, The Third Report from the Home Affairs Committee, Session 1985-86, HC409, London: HMSO.

Home Office (1965) Immigration from the Commonwealth, Cmnd 2739, London: HMSO.

Home Office (1975) Racial Discrimination, Cmnd 6234, London: HMSO.

Home Office (1976) Urban Deprivation, Racial Inequality and Social Policy, London: HMSO.

Home Office (1981) Racial Attacks: Report of a Home Office Study, London: Home Office.

Home Office (1982a) The Government Reply to the First Report from the Home Affairs Committee, Session 1981-1982 HC46-I: Commission for Racial Equality, Cmnd 8547, April, London: HMSO.

Home Office (1982b) The Government Reply to the Fifth

Report from the Home Affairs Committee, Session 1980-1981 HC424: Racial Disadvantage, Cmnd 8476, January, London: HMSO.

Home Office (1986) Government Response to the Home Affairs Committee Report on Racial Attacks and Harassment, Cmnd 45, December, London: HMSO.

Home Office (1989) The Response to Racial Attacks and Harassment; a guidance for the statutory agencies, Report of the Inter-departmental Racial Attacks Group, London: Home Office.

House of Commons (1971) Race Relations and Housing, Report of the Select Committee on Race Relations and Immigration, London: HMSO.

House of Commons (1978) Organisation of Race Relations Administration, Report of the Select Committee on Race Relations and Immigration, London: HMSO.

House of Commons (1982) First Report of the Environmental Committee 1981-2, The Private Rented Sector, III, Appendices, London: HMSO.

Housing Advice Switchboard (HAS) (1983) 'The Flat Merchants', referred to in LARH (1988), London.

Housing Corporation (HC) (1975) 'The selection of tenants by Housing Associations for subsidised schemes', Circular 1/75, Housing Corporation, London.

Housing Corporation (1989) Performance Expectations, HA Guide to Self-monitoring (June), and HC47/89 (circular), 'New Arrangements for Monitoring Registered Housing Associations' (July), London: Housing Corporation.

Hughes, D. (1987) Public Sector Housing Law, 2nd edn, London: Butterworth.

Immigration Law Practitioners Association (ILPA) (1987) Immigration Bill: Parliamentary Briefing No. 2, London: ILPA.

Inside Housing (1988) Weekly Journal of the Institute of Housing (dates various: identified in text), London: IoH Services Ltd.

Inside Housing (1989) 'Race Code will be extended', Vol. 6, No. 6, 10 February, London: IoH Services Ltd.

Institute of Housing (IoH) (1983) Working Together, Professional Practice Guide No. 1, London: IoH.

Institute of Housing (1986) Race and Allocations, Guide for Local Authorities, London: IoH Services.

Institute of Housing (1987a) The Key to Equality: the 1986 Women and Housing Survey, Report by IoH Women in

References

Housing Working Party, D. Levison and J. Atkins, London: IoH.

Institute of Housing (1987b) Institute of Housing Yearbook 1987, London: IoH.

Institute of Housing (1988) Who will House the Homeless? London: IoH Services Ltd.

Institute of Housing (1989) 'Homeless Law Unchanged and Cash Injected', Inside Housing, Vol. 6, No. 4-5, 17 November; IoH Services.

Institute of Race Relations (IRR) (1977) Policing against Black People, London: IRR.

Jacobs, B.D. (1986) Black Politics and the Urban Crisis in Britain, Cambridge: Cambridge University Press.

Jacobs, B.D. (1988) Racism in Britain, London: Christopher Helm.

Jenkinson, J. (1985) 'The Glasgow Race Disturbances of 1919', Immigrants and Minorities, Vol. 4, No. 2.

Johnson, M. (1987) 'Housing as a process of racial discrimination', in S. Smith and J. Mercer (eds.) New Perspectives on Race and Housing in Britain, Glasgow: Centre for Housing Research, University of Glasgow.

Johnson, M. and Cross, M. (1984) Surveying Service Users in Multi-racial Areas, Research Paper No. 2, CRER, Coventry: University of Warwick.

Jones, M.M. (1972) Prejudice and Racism, New York: Addison Wesley.

Karn, V. (1976) Priorities for Local Authority Mortgage Lending: a case study of Birmingham, Birmingham: University of Birmingham Centre for Urban and Regional Studies.

Karn, V. (1978) 'The financing of owner-occupation and its impact on ethnic minorities', New Community, Vol. VI, Nos. 1 and 2, winter 1977/78, London: CRE.

Karn, V. et al. (Kemeny and Williams) (1985) Home Ownership in the Inner City, Aldershot: Gower Publishing.

Karn, V. and Henderson, J. (1987) Race, Class and State Housing, Aldershot: Gower.

Katznelson, I. (1973) Black Men, White Cities, London: Oxford University Press.

Kemeny, J. (1981) The Myth of Home Ownership: Private versus Public Choices in Housing Tenure, London: Routledge & Kegan Paul.

King, M. and May, C. (1985) Black Magistrates, London: Cobden Trust.

Knewstub, N. (1989) 'Renton urges more integration in rebuke to Muslin protesters', Guardian, 15 April, p. 4, London: Guardian Publications.

Kushnick, L. (1981) 'Parameters of British and North American racism', Race and Class, Vol. 13: 187-206.

Labour Party (1989) Meet the Challenge, Make the Change: a new agenda for Britain, Final report of Labour's Policy Review for the 1990s, London: Labour Party.

Lambert, C. (1976) Building Societies, Surveyors and the Older Areas of Birmingham, Birmingham: University of Birmingham Centre for Urban and Regional Studies.

Law Society of Scotland (LSS) (1989) Code of Conduct for Scottish Solicitors, Edinburgh: LSS.

Lawless, P. (1988) 'British inner urban policy post-1979: a critique', Policy and Politics, Vol. 16, No. 4 (October), 261-75: Bristol, School of Advanced Studies.

Layfield, F. (1976) Local Government Finance: Report of the Committee of Inquiry (The Layfield Report), Cmnd 6453, London: HSMO.

Layton-Henry, Z. (1984) The Politics of Race in Britain, London: Allen & Unwin.

Layton-Henry, Z. and Rich, P.B. (1986) Race, Government and Politics in Britain, London: Macmillan.

Legates, R.T. and Hartman, C. (1986) 'The anatomy of displacement in the United States', in Smith and Williams (eds.) Gentrification and the City, London: Allen & Unwin.

Lester, A. and Bindman, G. (1972), Race and Law, Harmondsworth: Penguin Books.

Lomas, G. (with Monck) (1975) The Coloured Population of Great Britain, London: Runnymede Trust.

London Against Racism in Housing (LARH) (1988) Anti-racism for the Private Rented Sector, London: LARH.

London Housing Unit (LHU) (1988) Just Homes, London: LHU.

London Housing Unit (1989) One in every Hundred, London: LHU.

London Race and Housing Research Unit (LRHRU) (1987) Progress Report of the London Race and Housing Research Unit, October 1985 to April 1987, London: LRHRU.

London Research Centre (LRC) (1987) London Housing Statistics, 1986, London: LRC.

Loney, M. and Allen, M. (1979) The Crisis of the Inner City, London: Macmillan.

References

Lord Chancellor's Department (1989a) The Work and Organisation of the Legal Profession, Cm 570, London: HMSO.

Lord Chancellor's Department (1989b) Conveyancing by Authorised Practitioners, Cm 571, London: HMSO.

Lord Chancellor's Department (1989c) Contingency Fees, Cm 572, London: HMSO.

Lustgarten, L. (1978) 'The new meaning of discrimination', Public Law, summer: 178-205.

Lustgarten, L. (1980) Legal Control of Discrimination, London: Macmillan.

Lustgarten, L. (1987) in Solomos and Jenkins (eds.) Racism and Equal Opportunity Policies in the 1980s, Cambridge: Cambridge University Press.

McCrudden, C. (1987) 'The Commission for Racial Equality: Formal Investigations in the Shadow of Judicial Review' Regulation and Public Law, London: Weidenfeld & Nicolson.

McDonald, E. (1988) 'Weak links in the house chain', Observer, 24 July 1988: 9, London: Observer Newspapers.

MacEwen, M. (1976) Ethnic Minority Housing Survey in Edinburgh, Edinburgh: Lothian Community Relations Council.

MacEwen, M. (1980a) 'Ethnic Minorities and Public Housing Services in Scotland' in Scottish Council for Racial Equality First Annual Report, 1979-1980, Edinburgh: SCRE.

MacEwen, M. (1980b) 'Race relations in Scotland: ignorance or apathy', New Community, Vol. VII, No. 3, winter, London: CRE.

MacEwen, M. (1981) 'Equal Opportunity and the Law: New Bottles for Sour Grapes', Journal of the Law Society of Scotland, Vol. 26, No. 8 (August), Edinburgh: Law Society of Scotland.

MacEwen, M. (1985) 'Local Authority Duty to Promote Equality of Opportunity', SCOLAG No. 105 (June), Dundee: SLAG.

MacEwen, M. (1986) Racial Harassment, Council Housing and the Law, SEMRU/Edinburgh College of Art/Heriot-Watt University Research Paper No. 11, Edinburgh: SEMRU.

MacEwen, M. (1987a) Housing Allocations, Race and Law, Scottish Ethnic Minorities Research Unit (SEMRU)/ Edinburgh College of Art/Heriot-Watt University

Research Paper No. 14, Edinburgh: SEMRU.

MacEwen, M. (1987b) 'Public Sector Housing Allocations', Journal of the Law Society of Scotland Vol. 32, No. 88 (August), Edinburgh: Law Society of Scotland.

MacEwen, M. (1988) Community Relations: Scotch Mist, Domestic Colonialism or What?, SEMRU/Edinburgh College of Art/Heriot-Watt University Research Paper No. 23, Edinburgh: SEMRU.

MacEwen, M. and Hancock, D. (1989) Ethnic Minorities and Public Housing in Edinburgh, London/Edinburgh: joint SEMRU/CRE publication.

McKay, D.H. (1977) Housing and Race in Industrial Society, London: Croom Helm.

Malpass, P. and Murie, A. (1987) Housing Policy and Practice, 2nd edn, London: Macmillan.

Marx, K. (1962) quoted in Rex (1986).

Mathieson (1980) Law, Society and Political Action: towards a Strategy under late Capitalism, London: Academic Press.

Merrett, S. (1979) State Housing in Britain, London: Routledge & Kegan Paul.

Miles, R. (1982) Racism and Migrant Labour, London: Routledge & Kegan Paul.

Miles, R. and Phizacklea, A. (1984) White Man's Country: Racism in British Politics, London: Pluto Press.

Miles, R. and Muirhead, L. (1986) 'Racism in Scotland' in Scottish Government Year Book 1985, Edinburgh: Paul Harris Publishing.

Miles, T. (1989) 'Rule Change on Homelessness', Observer, 22 October: 3, London: Observer Publications.

Milner-Holland (1965) Report of the Committee on Housing in Greater London (The Milner-Holland Committee), Cmnd 2582, London: HMSO.

Mnookin and Kornhauser (1979): 'Bargaining in the Shadow of the Law: the Case of divorce', Yale Law Journal, Vol. 88: 950, New Haven: Yale University.

Modood, T. (1989) 'Alabama Britain', The Guardian, 22 May, p. 16, London: Guardian Publications.

Montagu, A. (1964) (ed.) The Concept of Race, New York: Free Press.

Moore, R. and Rex, J. (1967) Race, Community and Conflict: a Study of Sparkbrook, London: Oxford University Press for IRR.

Morris, P. (1975) Housing and Social Work: a Joint Approach, Scottish Development Department, Edinburgh: HMSO.

459

References

Mullins, D. (1988) 'Housing and Urban Policy' New Community, Vol. 15, No. 1 (October), London: CRE Publications.

Mullins, D. (1989) 'Housing and Urban Policy', New Community, Vol. 16, No. 1 (October), London: CRE Publications.

National Association for the Care and Resettlement of Offenders (NACRO) (1986) Black People and the Criminal Justice System, London: NACRO.

National Audit Office (1985) Report by the Comptroller and Auditor General, Department of Environment: the Urban Programme, National Audit Office, London: HMSO.

National Federation of Housing Associations (NFHA) (1982) Race and Housing: a Guide for Housing Associations, London: NFHA.

National Federation of Housing Associations (1983) Race and Housing: still a cause for concern, London: NFHA.

National Federation of Housing Associations (1985) Race and Housing: Ethnic Record Keeping and Monitoring, London: NFHA.

Niner, P. with Karn, V. (1985) Housing Associations Allocations: Achieving Racial Equality, London: Runnymede Trust.

Niner, P. (1987) 'Housing Associations and Ethnic Minorities' in Smith and Mercer (eds.) New Perspectives on Race and Housing in Britain, Glasgow: Centre for Housing Research, University of Glasgow.

Niner, P. (1989) Homelessness in Nine Local Authorities: Case Studies of Policy, DoE, London: HMSO.

Nixon, J. (1986) 'The House of Commons Home Affairs Sub-committee and Government Policy on Race Relations' in Layton-Henry, S. and Rich, P. (eds.) Race, Government and Politics in Britain, London: Macmillan.

North Hyde Park Residents Association (NHPRA) and South Headingley Community Association (1976), unpublished memorandum to the Department of the Environment review of housing finance - quoted in CRE (1985c): 51.

Nuffield Foundation (1986) Town and Country Planning: the Report of a Committee of Inquiry Supported by the Nuffield Foundation, London: Nuffield Foundation.

Offe, C. (1984) Contradictions of the Welfare State, London: Hutchinson.

Office of the Minister for the Civil Service (OMCS) (1987) Ethnic Origin Survey: Non-industrial Staff in the Civil

Service in London and the S.E. of England, London: Cabinet Office.

Ollerearnshaw, S. (1988) 'Action on equal opportunities in inner cities: the need for a policy commitment', New Community, Vol. 15, No. 1 (October), London: CRE Publications.

Ouseley, H. (1981) The System, London: Runnymede Trust/South London Equal Rights Consultancy.

Owen, J. (1988) 'Identifying good practice on race and planning; the National Development Control Forum's Initiative 1988: a view from the Districts', paper presented to a conference at Bristol Polytechnic, October.

Padfield, J. (1988) Letter to author from Welsh Office dated 27 October 1988.

PAG (1965) The Future of Development Plans: a Report by the Planning Advisory Group, London: HMSO.

Pareto, V. (1963) The Mind and Society - a Treatise on Sociology, London: Dover Books.

Paris, C. and Blackaby, B. (1979) 'Not much Improvement: Urban Renewal Policy in Birmingham', London: Heinemann.

Parker, J. and Dugmore, K. (1976) 'Colour and Allocation of GLC Housing', GLC Research Report No. 21, London.

Parmar, P. (1988) 'Gender, Race and Power: the Challenge to Youth Work Practice' in Cohen and Bains (eds.) Multi-Racist Britain, London: Macmillan.

PAS (1981) 'Public Attitude Surveys Research Ltd: Rochdale Survey', CRE, (1985c) London: CRE.

Paterson, I.V. (1973) The New Scottish Local Authorities: Organisation and Management Structures, Edinburgh: HMSO.

Peach, C., Winchester, S. and Woods, R. (1975) 'The Distributions of Colonial Immigrants in Britain' in G. Gappert and H. Rose (eds.) The Social Economy of Cities.

Phillips, D.A. (1986) 'What Price Equality? A Report on the allocation of GLC housing in Tower Hamlets', GLC Housing Research and Policy Report No. 9, London.

Planning (1989) 'Lingering Death for Structure Plans' in Planning, No. 804, 3 February 1989, London: Ambit Publications Ltd.

Plant, P. (1988) Citizenship, Rights and Socialism, Fabian Trust 531, London: Fabian Society.

Plowden (1967) Children and their Primary Schools (The

References

Plowden Committee Report), London: HMSO.
Policy Studies Institute (PSI) (1984) Black and White Britain: the Third PSI Survey, report by Colin Brown, London: Heinemann.
Policy Studies Institute (1985) Racial Discrimination: 17 Years After the Act (C. Brown and P. Gay), No. 646, London: PSI.
Policy Studies Institute (1987) Homelessness in Brent, report by L. Bonnerjea and J. Lawton, London: PSI.
Policy Studies Institute (1988) Community Relations Councils: Roles and Objectives, report by Gay and Young, London: CRE Publications.
Political and Economic Planning (PEP) (1967) Racial Discrimination, London: PEP and Research Services.
Political and Economic Planning (1968) Racial Discrimination in England, report of first PEP Survey in 1966 by W.W. Daniel, Harmondsworth: Penguin Books.
Political and Economic Planning (1977) Racial Disadvantage in Britain, report of second PEP Survey in 1974 by David J. Smith, Harmondsworth: Penguin Books.
Pound, R. (1943) 'A survey of social interest', Harvard Law Review, Vol. 57: 39, Cambridge, Mass.: Harvard University Press.
Provan, J.A. (1982) 'Allocation Practices of Housing Associations', unpublished M.Phil. thesis, Birmingham: University of Birmingham.
Quinney, R. (1974) Critique of Legal Order: Crime Control in a Capitalist Society, Boston: Little Brown & Co.
Race Relations Board (RRB) (1967) Report of the Race Relations Board for 1966-1967, London: HMSO.
Race Relations Board (1969) Report of the Race Relations Board for 1968-1969, London: HMSO.
Race Relations Board (1970) Report of the Race Relations Board for 1969-1970, London: HMSO.
Race Relations Board (1972) Report of the Race Relations Board for 1971-1972, London: HMSO.
Race Relations Board (1975) Report of the Race Relations Board for 1974, London: HMSO.
Radcliffe, Lord (1969) 'Immigration and Settlement: some general considerations', Race, Vol. II, No. 1: 33-51, London: IRR.
Ramsaran, Y. (1988) 'Private Black Tenants and the Bill', Foundation, No. 4, June: 10, London: LRHRU.
Rao, J. (1982) Hounslow: Survey of Housing Applicants, London: Hounslow Borough Council.

Rawls, J. (1972) A Theory of Justice, London: Oxford University Press.

Rex, J. and Moore, R. (1967) Race, Community and Conflict, London: Oxford University Press.

Rex, J. and Tomlinson (1979) Colonial Immigrants in a British City, London: Routledge & Kegan Paul.

Rex, J. (1973) Race, Colonialism and the City, London: Routledge & Kegan Paul.

Rex, J. (1986) Race and Ethnicity, Milton Keynes: Open University Press.

Richards, C. (1981) 'A Survey of Ethnic Minority Applications for Planning Permission in Glasgow', unpublished B.Sc. dissertation, Edinburgh College of Art/Heriot-Watt University, Edinburgh: Edinburgh College of Art.

Ridoutt (1980) 'Black People and Housing in Lewisham', unpublished Housing Department Report for the London Borough of Lewisham.

Robson, P. (1986) 'Bed and Breakfast in London: a Guide to the law and practice', GLC Housing Research and Policy Report No. 6, London: GLC.

Roof (1988) 'Free Marketeer as Social Engineer' (Steve Matt) Roof, Vol. 13, No. 3, May/June, London: Shelter Publications.

Rose, E.J.B. and Associates (1969) Colour and Citizenship, London: Oxford University Press (for Institute of Race Relations).

Royal Town Planning Institute (RTPI) (1981) Renewal of Older Housing Areas in the 1980s, Policy Paper by Housing Working Party, London: RTPI.

RSRG (1980) Britain's Black Population, Runnymede Trust and the Radical Statistics Race Group (RSRG), London: Heinemann.

Runnymede Trust (1982) Racial Discrimination: Developing a legal strategy through individual complaints, London: Runnymede Trust.

Runnymede Trust (1989a) 'House Lettings Policy Racist', Race and Immigration, No. 225, London: Runnymede Trust.

Runnymede Trust (1989b) 'Race and the Legal Profession' Race and Immigration, No. 226, June, London: Runnymede Trust.

Sandford, J. (1971) Down and Out in Britain, London: Peter Owen.

Scarman, Lord (1981) The Brixton Disorders, 10-12 April

1981, Cmnd 8427, London: HMSO.

Schaffer, R. and Smith, N. (1984) The Gentrification of Harlem, paper presented to the annual conference of the American Association for the Advancement of Science, 27 May.

SCOLAG (1988) 'This Month's Cases' (F.P. Davidson), No. 141 (June): 93, Dundee: Scottish Legal Action Group.

Scottish Council for Racial Equality/Royal Town Planning Institute (Scottish Branch) (1983) Planning in a Multi-racial Scotland, Edinburgh: SCRE.

Scottish Development Department (SDD) (1977a) Scottish Housing, Cmnd 6852, Edinburgh: HMSO.

Scottish Development Department (1977b) Assessing Housing Need: a Manual of Guidance, Edinburgh: HMSO.

Scottish Development Department (1980) Allocation and Transfer of Council Housing, Report of a Sub-committee of the Scottish Housing Advisory Committee, Edinburgh: HMSO.

Scottish Ethnic Minorities Research Unit (SEMRU) (1987) Ethnic Minorities Profile: a Study of Needs and Services in Lothian Region and Edinburgh District (3 volumes), Edinburgh: Department of Town and Country Planning, Heriot-Watt University/Edinburgh College of Art.

Scottish Ethnic Minorities Research Unit (1989) Housing Needs, Experiences and Expectations of Glasgow's Ethnic Minority Population: The Role of Housing Associations, M. Dalton and S. Daghlian (Draft Report), Glasgow: SEMRU.

Seebohm, (1968) Local Authority and Allied Personal Social Services, Report by the Seebohm Committee, Cmnd 3703, London: HMSO.

Select Committee on Race Relations and Immigration (1971) Race and Housing, Report of the Select Committee on Race Relations and Immigration, Session 1970-71, London: HMSO.

Select Committee on Race Relations and Immigration (1978) The Organisation of Race Relations Administration, Report of the Select Committee on Race Relations and Immigration, London: HMSO.

Shelter (1972) The Grief Report, London: Shelter.

Sherman, J. (1988) Letter from DoE (ICD) to author dated 3 October 1988.

Sim, D., Bowes, A. and McCluskey, J. (1989) 'Glasgow's Ethnic Minorities', Housing, Vol. 25, No. 5 (June), London: IoH.

Simpson, A. (1981) Stacking the Decks: a Study of Race, Inequality and Council Housing in Nottingham, Nottingham: Nottingham Community Relations Council.

Sivanandan, A. (1982) A Different Hunger: Writings on Black Resistance, London: Pluto Press.

Skellington (1980) Council House Allocations in a Multi-racial Town, Milton Keynes: Open University Press.

Smith, D.J. and Whalley, A. (1975) 'Racial Minorities and Public Housing', PEP Broadsheet No. 556, London: PEP.

Smith, D.J. (1976) 'The Facts of Racial Disadvantage', PEP Broadsheet No. 560, London: PEP.

Smith, D.J. (1977) Racial Disadvantage in Britain, Harmondsworth: Penguin Books.

Smith, N. and Williams, P. (1986) (eds.) Gentrification of the City, London: Allen & Unwin.

Smith, S.J. and Mercer, J. (1987) (eds.) New Perspectives on Race and Housing in Britain, Glasgow: Centre for Housing Research, University of Glasgow.

Smith, S.J. (1989a) 'Housing for Sons and Daughters', Runnymede Trust Bulletin, No. 225, May, p. 11, London: Runneymede Trust.

Smith, S.J. (1989b) The Politics of 'Race' and Residence, Cambridge: Polity Press.

Social and Community Planning and Research (SCPR) (1988) Queuing for Housing: a Study of Council House Waiting Lists, London: Department of the Environment.

Solomos, J. and Jenkins, R. (1987) Racism and Equal Opportunity Policies in the 1980s, Cambridge: Cambridge University Press.

Spicker, P. (1986) Local Authorities Allocations Criteria, London: Rowntree Studies Group.

Stearn, J. (1987) 'No Home? No Entry', Housing, Vol. 23, No. 10 (May/December), London: IoH Services.

Stearn, J. (1989) 'Housing the Homeless', Housing, Vol. 25, No. 9 (November), London: IoH Publications.

Stevens (1979) Law and Politics: the House of Lords as a Judicial body, 1800-1976, London: Weidenfeld & Nicolson.

Stevens, L., Karn, V., Davidson, E., and Stanley, A. (1981) Ethnic Minorities and Building Societies' Lending in Leeds, a report for Leeds Community Relations Council, Leeds: LCRC.

Stewart, M. and Whitting, G. (1983) Ethnic Minorities and the Urban Programme, SAUS, Bristol: University of

References

Bristol.

Steyn, Mr Justice (1988) 'Race Relations at the Bar', Counsel, November/December, London: Butterworth.

Street, H., Bindman, G., and Howe, G. (1967) Anti-discrimination Legislation: the Street Report, London: PEP.

Stuart, N. (1988) 'Portsmouth applies racist rule change', Roof, May/June 1988, London: Roof (Shelter).

Swann (1985) Education for All, Report of the Committee of Inquiry into the Education of Children from Ethnic Minority Groups, London: HMSO.

Taylor, L. (1987) 'Local Authority Housing and Ethnic Minorities - an investigation into the situation in Edinburgh', unpublished dissertation for Diploma in Housing, Heriot-Watt University/Edinburgh College of Art.

Thomas, A.D. (1986) Housing and Urban Renewal, London: Allen & Unwin.

Thwaites, F. (1986) 'Local authority housing and ethnic minorities - A preliminary study of demand and supply in Edinburgh 1981-1985', unpublished dissertation for Diploma in Housing, Heriot-Watt University/Edinburgh College of Art.

Town and Country Planning Association (1986) Whose Responsibility? Reclaiming the Inner Cities, London: TCPA.

Townsend, P. (1979) Poverty in the United Kingdom, Harmondsworth: Penguin Books.

Treasury, H.M. (1986) The Government's Expenditure Plans 1986-1987 to 1988-1989, Vol. 1, January, Cmnd 9702-I, London: HMSO.

Troyna, B. (1981) Public Awareness and the Media, London: CRE Publications.

UKIAS (1987) Immigration Bill: Background Information for MPs, London: UKIAS.

Uthwatt (1942) Final Report of the Expert Committee on Compensation and Betterment (The Uthwatt Committee), Cmnd 6386, London: HMSO.

Walker, D.M. (1975) The Principles of Scottish Private Law, Oxford: Clarendon Press.

Walker, M., Jesserson, T. and Senevirathe, M. (1989) 'Race and Criminal Justice in a Provincial City', paper presented to Bristol Criminology Conference 17-20 July, University of Sheffield.

Walsh, D. (1986) 'Racial Harassment in Glasgow: Research

Findings' in Scottish Council for Racial Equality (SCRE), Annual Report 1985/1986, Glasgow: SCRE.

Ward, R. (1975) 'Residential Succession and Race Relations in Moss Side, Manchester', Ph.D. Thesis, Manchester: University of Manchester.

Ward, R. (1982) 'Race, Housing and Wealth', New Community, Vol. 10: 3-15, London: CRE Publications.

Ward, R. (1984) (ed.) Race and Residence in Britain: Approaches to Differential Treatment in Housing, Monographs on Ethnic Relations, Birmingham: Aston University.

Watchman, P.Q. and Robson, P. (1983) Homelessness and the Law, Glasgow: The Planning Exchange.

Watchman, P.Q. and Robson, P. (1986) Developments in Homelessness and the Law 1983-1986, Glasgow, The Planning Exchange.

Watchman, P.Q. and Robson, P. (1989) Homelessness and the Law, 2nd edition, Glasgow: The Planning Exchange.

Watson, S. and Austerberry, H. (1983) Women on the Margins: a Study of Single Women's Housing Problems, City University, London: Housing Research Group.

Weir, S. (1976) 'Red Line Districts', Roof, July, London: Shelter Publications.

Wellman, D. (1977) Portraits of White Racism, Cambridge: Cambridge University Press.

Which? (1987) (April), London: CA Publications.

Williams, P. (1976) 'The Role of Institutions in the Inner London Housing Market: the case of Islington', Transactions of the Institute of British Geographers, n.s., Vol. 1, No. 1.

Williams, P. and Smith, N. (1986) Gentrification of the City, London: Allen & Unwin.

Wintour, P. (1989) 'Labour to Open up the Law', The Guardian, 3 May, p. 1, London: Guardian Publications.

Young, E. and Rowan-Robinson, J. (1985) Scottish Planning Law and Procedure, Glasgow: William Hodge & Co.

Young, K. and Connolly, N. (1981) Policy and Practice in the Multi-racial City, PSI Report No. 598, London: PSI.

Young, K. (1983) 'Ethnic Pluralism and the Policy Agenda' in Glazer and Young (1986).

NAME INDEX

Note: this is an index of individuals; Commissions of Enquiry, Reports, Organisations etc will be found in the General Index.

CASE INDEX

GENERAL INDEX

Note:
CRE = Commission for Racial Equality
CRC = Community Relations Council (now Racial Equality Council)
Acts referred to (eg '1976 Act') are Race Relations Acts

193, 206, 212-15 passim,
425; housing associations
267, 280; housing
professionals 360-1;
private rental sector
144, 285-97 passim, 300,
307-8; targets 409; urban
planning 146-7, 172; see
also individual Boroughs
London Accommodation
Bureau 95
Lothian Community
Relations Council 355-6,
357

maintenance see improve-
ment grants; repair
grants
Manchester 37, 182, 266,
273-8, 360, 384
media 22-3, 33, 51, 349
Middlesborough 360
Midlands see East Midlands;
West Midlands; and also
individual cities
Milner-Holland Committee
(1965) 98, 163
Mirpuris 30-1
morality, personal: and
discrimination 413-14,
416; and law 61-4
Moroccans 30-1
Morris Report (1975) 338
mortgages 17, 22, 86-7, 306-
16, 348, 393, 412-13;
defaulters 186; tax relief
144-5, 199, 302
Moss Side (Manchester) 182
movement, house 39-40, 50,
176
Musicians Union 58

National Committee for
Commonwealth
immigrants 75, 352

National Health Service
(NHS) 434-5
nationality see citizenship
Neighbourhood Law Centres
374-5, 378, 386, 418
New Zealand 34
Newham, London Borough of
36, 66, 393
Nigerians 31
Non-discrimination Notices
109, 130, 132-4, 140,
391, 393, 415; proposed
reform of 402, 403-4
North East 280
North West 36, 280
Northern Ireland 123, 436-7,
442
Nottingham 358, 360;
council housing 66, 237,
240, 241-2; private
rented housing 288;
urban planning 155, 182
Nuffield report (1986) 147,
149-50

older council housing 240-2
ownership, home 20, 40-4,
65, 143, 223, 285, 302-
18; see also estate
agents; improvement
grants; mortgages
Oxbridge graduates 426

Pakistanis 14, 34, 35, 37-9,
44; housing 216, 290
passport fraud 204
Pick a Landlord Scheme
229-30, 440
planning, urban 124-5, 143-
89, 399
Plowden Report (1967) 163
Poles 35, 434
police 32; community
policing 3, 340
politics and politicians: and

485

General index

www.ingramcontent.com/pod-product-compliance
Ingram Content Group UK Ltd.
Pitfield, Milton Keynes, MK11 3LW, UK
UKHW020859280225
455677UK00006B/111